PAVEMENT MATERIALS FOR HEAT ISLAND MITIGATION: DESIGN AND MANAGEMENT STRATEGIES

PAVEMENT MATERIALS FOR HEAT ISLAND MITIGATION: DESIGN AND MANAGEMENT STRATEGIES

HUI LI, Ph.D., P.E.
Research Scientist, Dept. of Civil and Environmental Engineering, University of California, Davis, US
Professor, School of Transportation, Tongji University, Shanghai, China

ELSEVIER

AMSTERDAM • BOSTON • HEIDELBERG • LONDON
NEW YORK • OXFORD • PARIS • SAN DIEGO
SAN FRANCISCO • SINGAPORE • SYDNEY • TOKYO
Butterworth-Heinemann is an imprint of Elsevier

B H

Butterworth-Heinemann is an imprint of Elsevier
The Boulevard, Langford Lane, Kidlington, Oxford, OX5 1GB, UK
225 Wyman Street, Waltham, MA 02451, USA

ISBN: 978-0-12-803476-7

Library of Congress Cataloging-in-Publication Data

A catalogue record for this book is available from the Library of Congress

British Library Cataloguing-in-Publication Data

A catalogue record for this book is available from the British Library

For information on all Butterworth-Heinemann publications visit our website at http://store.elsevier.com/

CONTENTS

LIST OF FIGURES

LIST OF TABLES

PREFACE

This dissertation book examines the effects of various cool pavement design and management strategies for improving the thermal environment and mitigating near-surface heat island effects through field measurements, modeling, and simulation. In this research, nine experimental test sections were designed, constructed, and instrumented, and the thermal performance of various types of pavements and management strategies (including high-reflectance pavement, high-thermal-resistance pavement, and permeable pavement with evaporative cooling) was empirically investigated. Different cooling effects were identified for each strategy along with their advantages and associated disadvantages. Relevant properties of pavement materials (e.g., albedo, permeability, thermal conductivity, heat capacity, and evaporation rate) were measured, in many cases using newly developed methods. With these fundamental material properties, a local microclimate model was developed, validated, and applied to conduct sensitivity analysis on some key parameters to evaluate the thermal impacts of various cool pavement strategies in different climate regions. In addition, the impacts of various strategies on outdoor human thermal comfort were evaluated for three different climate regions (Sacramento and Los Angeles in California and Phoenix in Arizona). One type of thermal load associated with building energy use was evaluated for Davis, California.

Findings indicate that using high-reflectance pavement will reduce pavement surface temperature and consequently might help improve the air quality through a reduction in the formation of ground-level ozone. However, increasing the pavement reflectance would affect the human thermal comfort during hot periods owing to an increase in the mean radiant temperature contributed by the increased reflected radiation striking human bodies. Enhancing the evaporation from the pavement through the use of permeable pavement and creating shading on pavement with trees or other devices (e.g., solar panels) are likely to be effective strategies to reduce pavement surface temperature and improve human thermal comfort in hot periods. However, to be effective in arid and semiarid climates such as California, the water level must be kept near the surface of the permeable pavement through infusions of wastewater such as waste landscape irrigation.

Some cool pavement strategies used to improve the summer thermal environment might make the cold winter slightly colder. Therefore strategies such as evaporation and shading only in summer, which can help reduce the summer hot temperatures but will not heavily reduce the winter cold temperatures, are desirable for some regions.

Based on the findings from the studies presented in this dissertation book, some preliminary recommendations on the application of cool pavement strategies for mitigating near-surface heat island effects are:

1. Pave less and plant more. For some areas such as parking lots and alleys, the sites could be partly paved, and more grass and/or trees could be planted on the sites to reduce negative impacts of pavement.
2. Pave smart if it has to be paved. Permeable pavements (integrated with irrigation systems during hot dry seasons), including pervious concrete pavement, porous asphalt pavement, and permeable interlocking concrete pavers and reinforced grass pavers, could be good alternatives for paving if applicable, to both manage storm-water runoff and potentially help mitigate near-surface heat island effects and improve the thermal environment.
3. Care should be taken with the application of high-reflectance pavements. High-reflectance pavements can be used in open areas to help mitigate the heat island effects. However, special attention should be given when applied in high-density areas or areas with frequent walking or cycling human occupancy.
4. Consider evaporation and shading. Evaporation and shading could be very effective strategies to help improve the thermal environment in hot climates.
5. The models developed in the study for local microclimate, thermal comfort, and building energy use can be used, if needed, and improved for evaluating the seasonal impacts of various pavement strategies in different contexts.
6. Life cycle cost analysis and/or cost–benefit analysis, as well as environmental life cycle assessment, should be performed to quantitatively evaluate the life cycle economic and environmental impacts of various cool pavement strategies in different climates.

This book is a compilation of many recent studies and it is recommended as a good read to both academics and policy makers. Read and enjoy.

ACKNOWLEDGMENTS

I have taken efforts in this dissertation project. However, it would not have been possible without the kind support and help of many individuals and organizations.

First, I would like to express my profound gratitude to my Ph.D. advisor, Professor John Harvey, for providing me an opportunity to do my study on cool pavements and for his continued assistance, support, guidance, and encouragement throughout the research, which led me to the right way.

I would also like to extend my appreciation to my other committee members, Professor Susan Handy, Professor Carl L. Monismith, and Dr Melvin Pomerantz, for their encouragement, guidance, and invaluable comments on this dissertation.

Grateful acknowledgment goes to Paulette Salisbury and Tom Tietz (California Nevada Cement Association), Craig Hennings (Southwest Concrete Pavement Association), David R. Smith (Interlocking Concrete Pavement Institute), Guy S. Collignon (Enviro-Crete, Inc.), Scott Erickson (Evolution Paving Resources), and Bruce Camper and Clark Bell (Pacific Coast Building Products, Inc.) for providing materials and construction assistance for the experimental test sections.

I would also like to thank Ting Wang, Lin Chai, Stephanus Louw, Doug McCarter, Mark Hannum, Irwin Guada, Dr Rongzong Wu, Dr James Signore, Dr David Jones, and many other people for help during the construction of the test sections and other activities of the research. I am also thankful to Chun Y. Kwan, Chongsuck Cho, Nevin R. Natividad, Siyuan Xian, and Dr Changmo Kim for their help with some measurements. Special thanks are given to David Rapkin for helping with the instrumentation and to Joe Holland for helping with the thermal images.

The research was supported by a doctoral dissertation fellowship granted from the Sustainable Transportation Center (the precursor of National Center for Sustainable Transportation) at the University of California at Davis, which receives funding from the U.S. Department of Transportation and Caltrans, the California Department of Transportation, through the University Transportation Centers program. The research activities described in this dissertation were also in part sponsored by the California Department of Transportation (Caltrans), Division of Research and

Innovation. Both sponsorships are gratefully acknowledged. The contents of this dissertation reflect the views of the author and do not reflect the official views or policies of the Sustainable Transportation Research Center, the U.S. Department of Transportation, the State of California, or the Federal Highway Administration.

Finally, yet most importantly, I would like to express my sincere gratitude and love to my family and friends for their constant understanding, encouragement, support, and help throughout my life.

CHAPTER 1

Introduction

Contents

The built environment has a profound impact on our natural environment, economy, health, and productivity. Designers, builders, operators, and owners are now seeking breakthroughs in building science, technology, and operations to create a sustainable built environment and maximize both economic and environmental performance [1].

Increasing attention is being paid to reducing the environmental impact associated with each phase of the life cycle of various civil infrastructures. Roadways and pavements, as one important sector of the transportation infrastructure system and the built environment, play a vital role in economic and social development. While promoting economic and social growth, researchers, agencies, industries, and other stakeholders in the field of the roadways and pavements are working together to reduce their environmental impact. One environmental impact that pavements can help reduce, where local climate and urban density make it important, is the heat island effect associated with the use phase of pavements [2].

Economic and social development of both urban and rural areas continues as the human population increases, especially in fast-developing countries such as China and India, which leads to more of the land surface being paved. In many urban areas a large percentage of the land area is covered with various types of pavement, including streets, parking areas, sidewalks, plazas, and playgrounds. For example, in 2000 it was estimated

Pavement Materials for Heat Island Mitigation
ISBN 978-0-12-803476-7
http://dx.doi.org/10.1016/B978-0-12-803476-7.00001-5

that for the city of Sacramento, California, approximately 39% of the urban land area was paved (streets, parking areas, sidewalks) when seen from above the vegetative canopy [3,4]. Conventional impervious pavements, especially new black impervious asphalt pavements, produce high-surface temperatures (up to 65–80 °C in hot summer depending on the climate region) and resulting high near-surface air temperatures. The high temperatures of pavements and near-surface air can produce severe negative impacts associated with the heat island effect during summertime in hot climates, including reduced human comfort and heath, increased energy use for cooling of buildings and vehicles, and impaired air and water quality [5], as well as accelerated pavement deterioration (e.g., rutting and aging of asphalt pavements and possibly thermal cracking of concrete pavements) [2].

Heat islands are attracting more attention from various organizations. For instance, LEED® 2009 (Leadership in Energy & Environmental Design, a green building certification program from the U.S. Green Building Council) awards points to new construction and major renovations for using various technologies and strategies for roofs and nonroofs (including roads, sidewalks, courtyards, parking lots, etc.) to reduce heat islands to minimize impacts on microclimates and human and wildlife habitats [1]. Some researchers from Lawrence Berkeley National Laboratory (LBNL) and some other institutes in the United States and other countries also have been focusing on cool pavement technologies to help address the problem of urban heat islands (UHIs), mainly by increasing pavement surface reflectivity (or albedo) [2,5].

In addition, mitigating heat islands and improving outdoor thermal environments are important for creating a walkable and livable sustainable community. Encouragement of walking and cycling for short-distance trips in local communities might be one of the promising strategies to reduce vehicle miles traveled and transportation-related energy consumption and associated environmental impacts, in addition to improving public transit systems. Moreover, as common physical activities, walking and cycling also provide an opportunity for improving human health and enhancing neighborhood communication [6,7]. There are various general factors that might affect an individual's willingness to walk and cycle, including safety, travel distance, street environment, and comfort [8–11]. With respect to the street environment and comfort, providing better infrastructure (e.g., smooth and durable pavement on the sidewalks and dedicated bike lanes/paths) and improving the built environment are of significant importance [10–12].

Improving the outdoor thermal environment and comfort levels during hot periods may also increase the probability of an individual choosing to walk or cycle rather than driving. However, very few studies have paid attention to the relationship between specific pavement design details and the thermal environment at the site level. Therefore, evaluating and then improving the street thermal environment through better pavement design strategies are of great interest, especially during the summertime in hot climates.

It should be noted, however, that the impact of higher pavement temperatures on urban heat islands is not always negative, and the significance of the impact varies for different locations and for different seasons. While producing a negative impact during hot seasons in hot climates, heat islands can also bring some benefits to building owners and occupants of the spaces near pavement surfaces by reducing building heating energy use and human thermal discomfort during cold weather and in cold regions [13]. Moreover, these effects are different at the global (meaning city-wide) and local (site-level) scales, which should be analyzed separately. When considering the overall effect on UHIs, large scales are appropriate, but when considering human thermal comfort and pavement life, the localized near-surface effects will be of more importance than the overall urban heat island effect.

Therefore, both negative and positive impacts of pavements should be taken into account when analyzing the influence of pavements on the heat island effect and other related environmental issues, especially localized near-surface effects (human comfort). To achieve this goal, a better understanding is needed concerning thermal interactions between pavement and its surroundings (including air, building, trees, vegetation), the factors influencing temperature, and the specific impacts of these factors on both pavement and near-surface air where most human activities occur, during both hot daytime and cold nighttime in both summer and winter and in regions of various climates.

With respect to the heat island effect, the pavement might be part of the problem, but it definitely can be part of the solution if well designed. Technologies exist, or can be developed, to better engineer the thermal performance and behavior of the pavement to more positively affect the environment, energy use, and human health and comfort. With the help of cool pavement strategies, the heat island effect would potentially be mitigated and also the pavement durability would potentially be improved to some extent.

To effectively and sustainably mitigate local heat islands through sustainable pavement technologies and strategies, it is of great importance to better understand the following fundamental issues:

1. What is the heat island effect?
2. What are the impacts and causes of the heat islands?
3. What are the potential mitigation measures?

1.1 HEAT ISLAND EFFECT

Generally, the term "heat island effect" describes the characteristic warmth of both the atmosphere and the surfaces in developed urban areas compared to their (nonurbanized) surroundings, usually the nearby underdeveloped or undeveloped suburban and rural areas. The annual mean air temperature of a city with 1 million people or more can be 1.8–5.4 °F (1–3 °C) warmer than its surroundings [5,14]. On a clear calm summer night, however, the temperature difference can be as much as 22 °F (12 °C) [5]. The heat island is an example of unintentional climate modification when urbanization changes the characteristics of the Earth's surface and atmosphere [2].

The heat island sketch pictured in Figure 1.1 shows how urban temperatures are typically lower at the urban–rural border than in dense

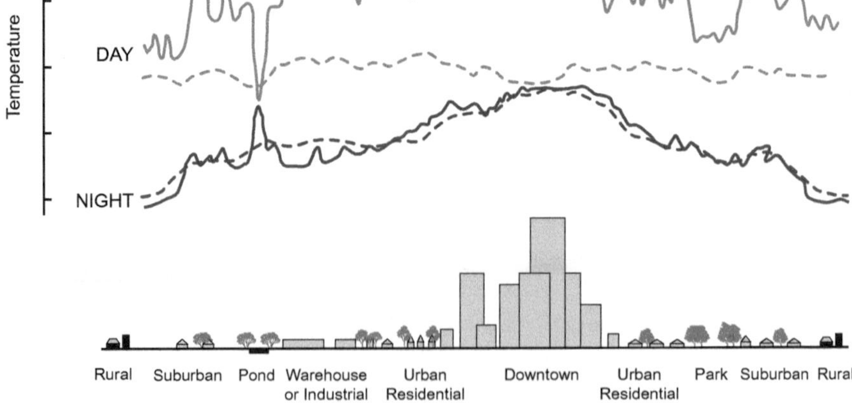

Figure 1.1 Heat island sketch [5].

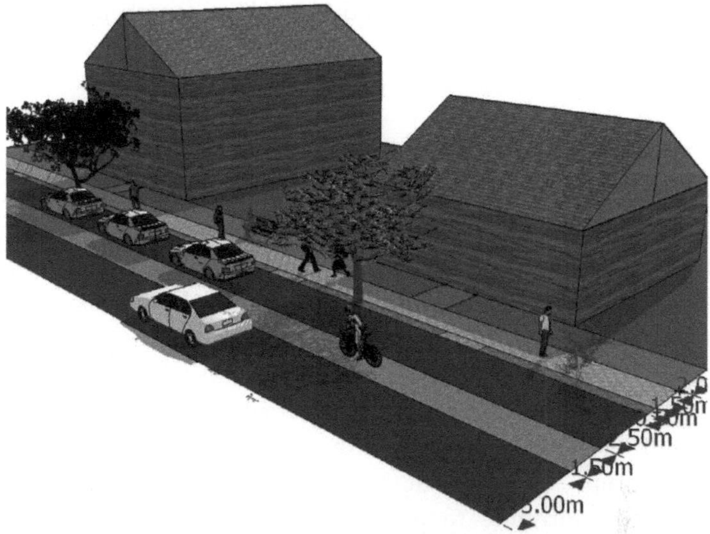

Figure 1.2 Localized pavement system with pedestrians, buildings, and vehicles.

downtown areas. The graphic also shows how parks, open land, and bodies of water can create cooler areas within a city.

In addition to the overall city-wide (urban-level) global effect, the term "heat island" also describes a site-level localized effect, that is, the built-up areas that are hotter than immediately nearby surrounding areas (e.g., a building or parking lot or an airport surrounded by open areas with bare natural soils or vegetation), creating hot spots. The local effect usually is much more severe than at the global scale owing to the higher local temperature and more direct impacts on pedestrians, buildings, and vehicles (as shown in Figure 1.2). The site-level localized effect of heat islands is the focus of this study.

1.2 POTENTIAL IMPACTS OF HEAT ISLANDS

In colder cities at higher latitudes and/or elevations, the wintertime warming effects of the heat island are seen as beneficial. In some urban areas during the summer, shade around high-rise buildings can create cooler areas for parts of the day. But in most cities, especially high-density cities, the effects of the summer heat island are seen as a problem [2]. The following discussion is based on regions with climates that are hot in the summertime.

Elevated temperature from urban heat islands, particularly during the summer, can affect a community's environment and quality of life. While some heat island impacts seem positive, such as lengthening the plant-growing season and reducing heating energy use in cold climates, most impacts are negative and listed as follows:

1.2.1 Compromised Human Health and Comfort

Increased daytime temperature of both pavement and near-surface air, reduced nighttime cooling, and associated higher air pollution levels can affect human health by contributing to general discomfort, respiratory difficulties, heat cramps and exhaustion, nonfatal heat stroke, and heat-related mortality. Heat islands can also exacerbate the impact of heat waves, which are periods of abnormally hot, and often humid, weather. Sensitive populations, such as children, older adults, and those with existing health conditions, are at particular risk from these events. Excessive heat events, or abrupt and dramatic temperature increases, are particularly dangerous and contribute to premature deaths, potentially resulting in above-average rates of mortality. The heat-related mortality may tend to increase (Figure 1.3) under the future potential climate change, under which more extreme surface weather conditions and heat events will probably be showing up [15].

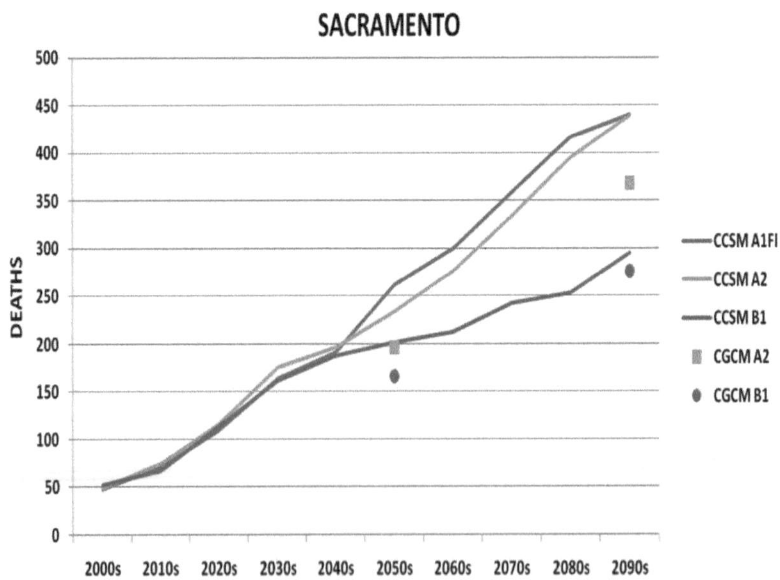

Figure 1.3 Annual heat-related mortality changing over time (predicted for Sacramento) [15]. *(Note: for different projected weather patterns, e.g., frequency and seasonality.)*

In addition to the impact on human health, the discomfort caused by the heat stress on the pavement will probably discourage people from walking or biking, which will reduce the likelihood of short-distance travel and physical activity of people by walking or biking over driving and would also have a health impact on people. This will be detrimental to creating a sustainable, livable, and walkable community.

1.2.2 Increased Energy Use

Elevated summertime temperatures in cities potentially increase energy demand for building cooling in hot regions (e.g. Figure 1.4). Research [5,16] by the LBNL, as an example, shows that electricity demand for building cooling increases 1.5–2.0% for every 1 °F (0.6 °C) increase in air temperature, starting from 68 to 77 °F (20–25 °C), suggesting that 5–10% of community-wide demand for electricity is used to compensate for the heat island effect. In addition to extra energy use for building cooling, the higher temperature also potentially increases the energy use for vehicle cooling if vehicles are parked or driving on hot pavements for some time.

Urban heat islands could increase peak demand as well as overall electricity demand. This generally occurs on hot summer weekday afternoons, when most of cooling systems, lights, and other appliances in offices and homes are running simultaneously.

As mentioned above, this effect is dependent on the local climate of the location where the pavements are applied and what the built environment surrounding the pavements is. However, only a few case studies have been conducted on the overall impact of whole urban heat island effects and very

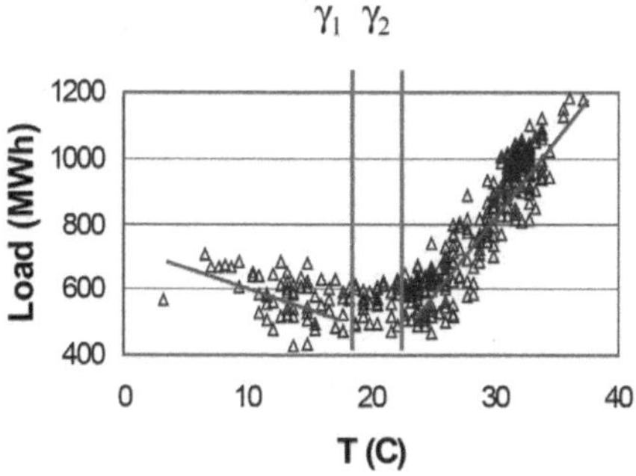

Figure 1.4 Example of electrical load versus air temperature for New Orleans, Louisiana [17]. *(γ_1 and γ_2 are the thresholds of low and high temperatures out of which the energy demand will rapidly increase.)*

few studies have been conducted on the specific effect of local heat islands caused by pavement heat.

1.2.3 Elevated Emissions of Air Pollutants and Greenhouse Gases

As described above, urban heat islands raise demand for electrical energy in summer. Companies that supply electricity typically rely primarily on fossil fuel power plants in the United States and even more so in China and India to meet much of this demand, which in turn leads to an increase in air pollutant and greenhouse gas emissions. The primary pollutants from fossil-fuel power plants include sulfur dioxide (SO_2), nitrogen oxides (NO_x), particulate matter, carbon monoxide (CO), and mercury (Hg) [17]. These pollutants are harmful to human health and also contribute to complex air-quality problems such as the formation of ground-level ozone (smog), fine particulate matter, and acid rain. Increased use of fossil-fuel power plants also increases emissions of greenhouse gases, such as carbon dioxide (CO_2), which contribute to global climate change.

In addition to their impact on energy related emissions, elevated temperatures can directly increase the rate of ground-level ozone formation. Ground-level ozone is formed when NO_x and volatile organic compounds react in the presence of sunlight and hot weather [18]. If all other variables are equal, such as the level of precursor emissions in the air and wind speed and direction, more ground-level ozone will form as the environment becomes hotter (e.g., Figure 1.5) [17,19].

1.2.4 Impaired Water Quality

High pavement surface temperatures can heat storm-water runoff. Tests have shown that pavements that are at 100 °F (38 °C) can elevate initial rainwater temperature from roughly 70 °F (21 °C) to over 95 °F (35 °C) [5]. This heated storm water generally becomes runoff, which raises water-body temperatures if it drains into streams, rivers, ponds, and lakes. Of course, this effect occurs primarily where there is rainfall at the same time as hot temperatures, which is common east of the Rocky Mountains, but uncommon in California.

Water temperature affects all aspects of aquatic life, especially the metabolism and reproduction of many aquatic species. Rapid temperature changes in aquatic ecosystems resulting from warm storm-water runoff can be particularly stressful, even fatal to aquatic life. However, if permeable pavements are used to reduce the storm-water runoff and infiltrate the

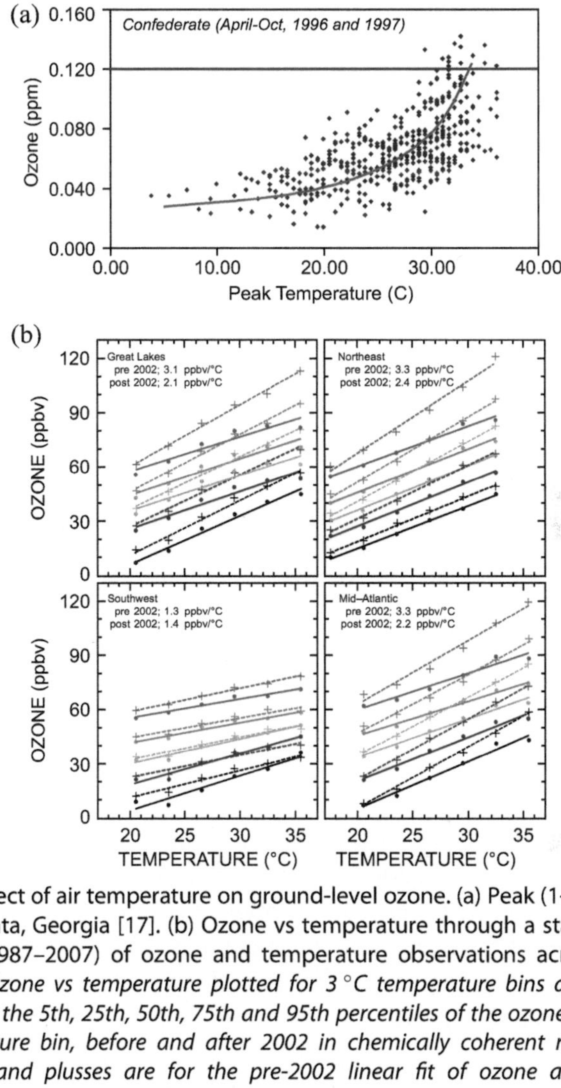

Figure 1.5 Effect of air temperature on ground-level ozone. (a) Peak (1-h) ground-level ozone in Atlanta, Georgia [17]. (b) Ozone vs temperature through a statistical analysis of 21 years (1987–2007) of ozone and temperature observations across the United States [19]. *(Ozone vs temperature plotted for 3 °C temperature bins across the range 19 to 37 °C for the 5th, 25th, 50th, 75th and 95th percentiles of the ozone distributions, in each temperature bin, before and after 2002 in chemically coherent receptor regions. Dashed lines and plusses are for the pre-2002 linear fit of ozone as a function of temperature; solid lines and filled circles are for after 2002. Color and position correspond to percentile (on top in red (dark gray in print versions) are 95th, next pair down in green (gray in print versions) is 75th, light-blue (light gray in print versions) is 50th, dark blue (darker gray in print versions) is 25th, and the bottom pair in black are the 5th percentile values.) Values are plotted at the mid-point temperature of the 3 °C temperature bin. The average slopes given on each panel indicate the climate penalty factors [19].)*

water down to the soil but not release it into water bodies such as rivers, ponds, etc., the effects on aquatic life will be mitigated and even eliminated.

1.2.5 Pavement Life

Pavement temperatures can have significant influence on pavement durability. For asphalt pavements in hot climates, high temperature in summer can significantly increase the risk of rutting (permanent deformation) and aging, and cracking, if not specifically designed well [20,21]. For concrete pavements, high temperatures and temperature gradients can significantly increase the probability of cracking caused by thermal stress [22,23]. However, for different pavement types, the effects of temperature on durability are different. Moreover, the exact effects on some pavements such as permeable pavements are still not very clear.

Cool pavements and the related cooling technologies would potentially reduce the pavement temperature and temperature gradient, and thus could potentially improve the pavement durability owing to mitigation of the thermal-related deteriorations such as rutting and/or cracking. In general, this could reduce the pavement maintenance costs and also bring other associated benefits such as reduced material use and user traffic delay.

1.2.6 Overview of an Open System for Evaluating the Pavement–Environment Interaction

The potential importance of heat island and pavement heat depends on various factors, such as the local climate, urban area scale, building density (affecting wind speed), pavement coverage, extent of air conditioning use, and nearby uses of the pavement. All these impacts are potential, but there are only some case studies on the overall effects of heat islands, and few of them investigated the specific effect of pavement heat. The importance of these impacts is different in different locations, and no systematic specific effects for different locations exist in the literature to date. A flowchart for an open system of evaluating the pavement–environment interaction is proposed and shown in Figure 1.6. The study presented in this dissertation will focus on evaluating the impacts of heat islands (specifically pavement heat) on human thermal comfort and building energy use. Other impacts will be out of the range of this study at present owing to the unavailability of relevant models and data. The open system is designed to consider possible unintended consequences and interactions between pavements and environment if identified later.

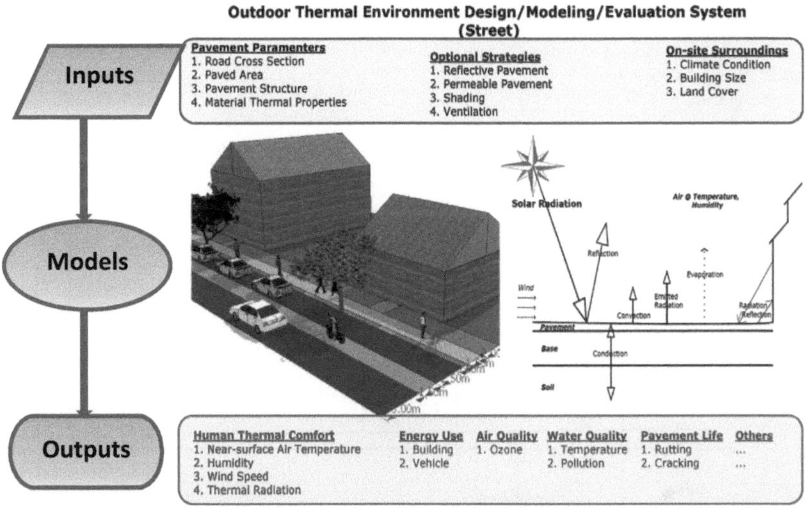

Figure 1.6 Flowchart of open system for evaluating pavement–environment interactions.

1.3 CAUSES OF HEAT ISLANDS

For identifying the causes of the heat island, it is useful to understand the concept of the "energy balance" at the Earth's surface and the main heat transfer processes, which will help identify and provide further understanding of the underlying causes of heat islands (Figure 1.7, also see Chapter 11.3 for more details on the surface heat balance).

As mentioned previously, there is no single cause of the heat island. Instead, many factors together combine to warm cities and suburbs, as listed in Table 1.1 [2].

In addition to the commonly identified and confirmed causes listed above, there are still some other possible factors that need more research work to identify and confirm the impacts, such as the thermal conductivity, heat capacity, and other properties of materials; level of air pollution (which influences the atmospheric radiation); etc. No consistent knowledge of the exact impacts of these factors on the heat island has been documented to date.

1.4 POTENTIAL MITIGATION MEASURES FOR HEAT ISLANDS

In the United States, the Environmental Protection Agency (EPA) has developed a three-pronged approach of (1) cool pavements, (2) urban forestry and vegetation, and (3) cool roofs and green roofs to mitigate the UHI [3].

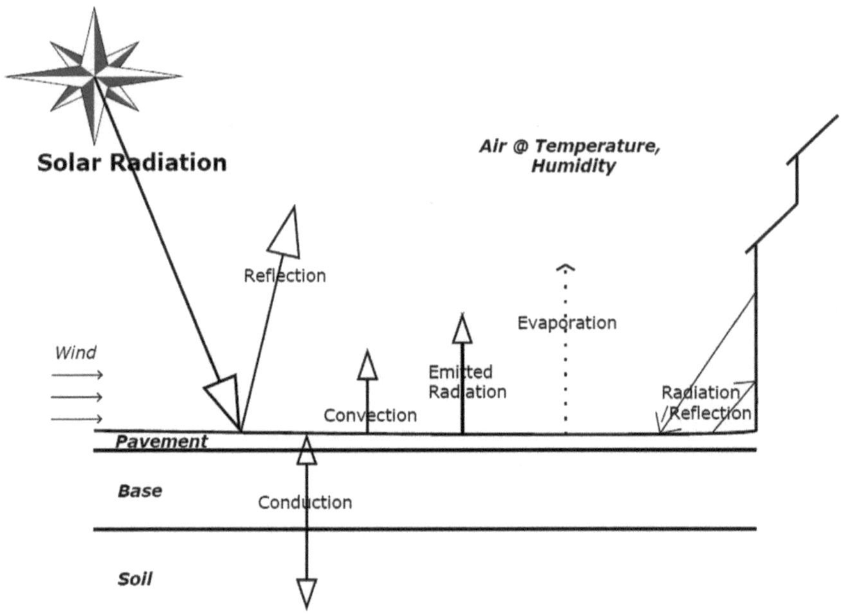

Figure 1.7 Energy balance on pavement surface.

Table 1.1 Characteristics related to heat island formation and their effects on the energy balance

Characteristic contributing to heat island	Effect on the energy balance
Lack of vegetation	Reduces evaporation
Widespread use of impermeable surfaces	Reduces evaporation
Low solar reflectance of urban materials	Increases net radiation
Urban geometries that trap heat	Increase net radiation
Urban geometries that slow wind speeds	Reduce convection
Increased energy use	Increases anthropogenic heat

Note: Revised based on Ref. [2].

Including the approaches developed by the U.S. EPA, communities can take a number of potential steps to reduce the local and/or atmospheric heat island effect, such as the strategies listed below based directly on the causes of the heat island listed in Table 1.1:

1. Increasing tree and vegetative cover;
2. Creating green roofs (aka "rooftop gardens" or "eco-roofs");
3. Installing cool—mainly reflective—roofs;
4. Using "cool" pavements;
5. Introducing water bodies into the urban area;

6. Reducing anthropogenic heat (released waste heat from heating/cooling, etc.); and
7. Improving urban geometry to improve air flow and enhance the natural ventilation.

As mentioned above, pavements are viewed as an important factor that contributes to heat islands. To effectively use cool pavements to mitigate heat islands, the following key issues need to be understood:

1. What are cool pavements by definition?
2. What are the potential cool pavement strategies and the cooling mechanisms?
3. What is the thermal performance of various strategies to create pavements?
4. What are the potential benefits and how can they be quantified when applied in various locations?
5. Which ones have the lowest life cycle environmental impacts (including materials production and construction)?
6. Which ones have the lowest life cycle costs?
7. Are there any potential unintended and/or unanticipated negative impacts?

CHAPTER 2

Literature Review on Cool Pavement Research

Contents

Pavement Materials for Heat Island Mitigation
ISBN 978-0-12-803476-7
http://dx.doi.org/10.1016/B978-0-12-803476-7.00002-7

Based on the key issues listed in Chapter 1, an extensive literature review was performed to identify potential technologies and strategies for cool pavements, key questions and what research has addressed them, and gaps in research and knowledge. The detailed literature review is presented in the corresponding chapters for each area investigated in this dissertation. This chapter summarizes the overall literature review.

While there are strengths and limitations associated with each study, this review does not systematically identify such strengths and limitations on a study-by-study basis. Instead, this review provides a comprehensive assessment of the strengths and weaknesses of the cool pavement literature for each subject, especially to identify key questions that remain unanswered. In doing so, the intent is to provide a more useful and constructive assessment of cool pavements as a field of scientific inquiry, and to highlight broad research directions that should be considered to advance the state of the art, and then to identify the most important questions in the author's opinion based on the review for investigation in this study.

2.1 COOL PAVEMENTS AND COOLING MECHANISMS

As mentioned previously in Chapter 1, pavements are viewed as an important factor that contributes to heat islands. However, pavements also could be part of the solution, not just part of the problem.

According to the EPA definition [3], "cool pavements refer to a range of established and emerging materials and technologies. These pavement materials and technologies could potentially make pavements have lower surface temperature and tend to release less heat into atmosphere compared with conventional pavements."

There are several potential strategies to make pavements cooler, with different cooling mechanisms, which can be classified into four categories as follows:

1. Modification of thermal properties of pavement materials;
2. Enhancement of evaporation from pavements;
3. Enhancement of convection; and
4. Reducing heat energy on/in pavements.

2.1.1 Modification of Thermal Properties of Pavement Materials

The thermal behavior of pavements is largely dependent on the thermal properties of pavement materials, including thermal conductivity, specific heat capacity, density, solar reflectivity (i.e., albedo), and thermal emissivity. Appropriate modification of these properties could help keep pavements and near-surface air cooler.

2.1.1.1 Reduce Pavement Thermal Conductivity

Thermal conductivity is the ability of materials to conduct or transmit heat. It determines how fast and readily the heat would be conducted from a high-temperature object or part to a low-temperature object or part. Pavements with low thermal conductivity may heat up at the surface but will not transfer that heat to the other pavement layers as quickly as pavements with higher thermal conductivity [3]. Therefore, reducing thermal conductivity of pavements could slow and reduce the heat flow into pavements under solar radiation and high air temperatures and generally lower the temperatures of pavements and near-surface air.

2.1.1.2 Increase Pavement Heat Capacity

Heat capacity is the amount of energy (or heat) required to raise the temperature of one unit weight of a substance by 1 °C without change of phase. In the case of pavements, it determines how much energy is absorbed and stored in the pavement at a certain temperature. Many man-made engineered materials, including pavements, can store more heat than natural materials such as dry soils and gravel/rock. As a result, built-up areas typically capture more of the solar energy—sometimes retaining twice as much as their rural surroundings during daytime [24]. The higher heat capacity of conventional urban materials contributes to heat islands at night, when these materials in urban areas release the heat absorbed and stored during daytime.

However, increasing the specific heat capacity as well as the density and thickness of pavement layers could increase the effective heat capacity of the whole pavement and help reduce the daytime high temperature and increase the nighttime low temperature. This is similar to the moderating effect of large water bodies as heat sinks (e.g., pool, pond, lake, sea) [25].

2.1.1.3 Increase Pavement Surface Reflectance

Solar reflectance, or albedo, is the percentage of solar energy reflected by a surface. Most existing studies on cool pavements have focused on solar reflectance, which is the primary determinant of maximum pavement surface temperature [3]. High albedo also could help to reduce pavement subsurface temperatures, because less heat is available at the surface to then be transferred into the pavement layers below the surface. Many opportunities exist to improve this property of materials, owing to the simplicity and convenience of reflectivity improvement for both new and existing surfaces of both asphalt and concrete pavements. In addition, increased albedo of pavement and other surfaces such as roofs might help offset global warming potential through radiative forcing [26–28].

Conventional asphalt pavement consists of asphalt binder mixed with dense-graded aggregate and is usually impervious. It can be used in a wide range of applications including trails, city streets, rural roads, parking lots, and highways. It can be modified with high-albedo materials, such as light-colored aggregates, which are exposed when traffic wears the asphalt off the surface aggregates, or cool-colored asphalt mixed with pigments or sealant, or by using light-colored tree resin in place of asphalt [29]. It also could be treated after installation to raise reflectance by applying light-colored coatings or light-colored chip seals, or, for existing asphalt pavement in need of rehabilitation, by whitetopping (placement of a concrete pavement over it) or ultrathin whitetopping, or, if in need of maintenance, by microsurfacing with light-colored aggregate and/or emulsified polymer resin [30]. These treatments could be applied to a wide range of functions from parking lots to highways.

Conventional concrete pavement is made mainly by mixing hydraulic cement, water, and dense-graded aggregate and is usually impervious. It can be used in a wide range of applications, including trails, city streets, rural roads, parking lots, and highways. Concrete pavement generally has a higher initial reflectance than asphalt pavement. It can be modified to increase the reflectance by using white cement or cement blended with light-colored slag [3,31]. When an asphalt overlay is placed on top of the concrete pavement, this kind of composite pavement tends to store less heat and cool down faster than conventional asphalt pavements according to the preliminary results of the existing research [21,32].

Similar to concrete pavement, concrete block pavement (also referred to as interlocking concrete paver) is also made mainly by mixing hydraulic cement, water, and dense-graded aggregate and is usually impervious. However, the gap between individual pavers can provide some path to allow water to drain, making the whole pavement section pervious to some extent. The block paver can be modified by using white cement or cement blended with light-colored slag or pigment to increase the reflectance.

As mentioned above, many opportunities exist to improve the albedo of pavements. However, the increased albedo might increase the thermal interactions between the pavement surface and the surfaces of other objects, such as buildings, vehicles, and human bodies, surrounding the pavement. The reflected heat energy from pavement surfaces has the potential to hit these surrounding surfaces and be absorbed by them, which will probably increase the temperature of these surrounding surfaces. The mechanism of thermal interactions between surfaces is addressed in detail later in this study (see Section 11.3.2).

Moreover, owing to weathering and the accumulation of dirt, the solar reflectance of conventional asphalt and concrete pavements and interlocking concrete pavers tends to change over time. Asphalt pavements consist largely of petroleum derivatives as a binder mixed with sand or stone aggregate. They tend to lighten as the binder oxidizes and more aggregate is exposed through surface wear. Concrete pavements and interlocking concrete pavers also use sand and stone aggregate, but, in contrast to asphalt pavements, typically use Portland cement as a binder. Various types of traffic (such as walking, biking, driving, etc.) generally dirty the cement, causing it to darken over time.

2.1.1.4 Increase Pavement Thermal Emissivity

A material's thermal emissivity determines how much heat it will radiate per unit area at a given temperature, that is, how readily a surface emits heat. Thermal emissivity plays a role in determining a material's contribution to heat islands. Research suggests albedo and emissivity have the greatest influence on determining how a conventional pavement cools down or heats up, with albedo having a large impact on maximum surface temperatures and emissivity affecting minimum temperatures [33]. Similar to albedo, if the increased radiation from pavements goes back out directly into the space far away with very little absorbed by the air, it might help reduce heat islands. However, if the radiation is blocked and absorbed by other surfaces (e.g., building or vehicle surfaces), it might not effectively help reduce heat islands in the urban canopy.

2.1.2 Enhancement of Evaporation from Pavements

Evaporation of water requires heat energy to achieve a *phase change* of water from liquid to gas. This process absorbs heat energy from surroundings and cools them down. The use of evaporative cooling could reduce pavement temperature and consequently air temperature through *latent heat* lost by the phase change of water (from liquid to gas) when moisture exists in the pavements or in the underlying soils or is sprinkled on hot pavement surfaces.

Two types of pavements can provide these benefits: permeable pavements (nonvegetated and vegetated) and water-retentive pavements.

2.1.2.1 Permeable Pavements

The majority of pollutants discharged to receiving water bodies are now associated with nonpoint sources. Impermeable streets, roads, and highways are among these nonpoint sources that contain large amounts of inorganic and organic pollutants [34–44]. To protect the quality of receiving waters, regulations have been established to treat the runoff prior to discharging or to reduce pollutants at the source [44–46]. In most cases, the runoff is managed or treated by constructed best management practices (BMPs) or sustainable urban development (SUDs).

Currently, many transportation agencies are employing BMPs such as sand filters, wet and dry detention basins, bioswales, and infiltration systems to comply with water quality regulations. These BMPs are impractical to construct in certain urban areas owing to space constraints, cost of construction and/or maintenance, and lack of expertise and/or equipment to provide required maintenance. In addition, collecting and treating a large volume of runoff from paved surfaces during an intense storm event is very challenging. For these reasons, transportation agencies are constantly looking for more efficient and economical methods to manage storm-water runoff. Reduction of runoff volume and any pollutants associated with it is one alternative method that is now practiced commonly in urban areas under the general term of low-impact development (LID). One method to comply with LID and eliminate the current costly practice of BMPs or SUDs is the construction of fully permeable pavement for streets, roads, or highway shoulders [47–49]. A recent simulation study performed at the University of California showed that the construction of fully permeable pavement shoulders for storm-water runoff management is technically and economically feasible [50,51]. With this proposed pavement design, no conventional BMPs are required, no additional land acquisition is needed, and, most important, no runoff will be generated if designed to handle even the most extreme storm events and hence the pollutant discharge to the environment will be significantly reduced.

Permeable pavement contains more voids than conventional impermeable pavement and is designed to allow water to drain through the surface into the sublayers and then to infiltrate into the ground below. Permeable pavements include porous asphalt pavements, pervious concrete pavements, pervious cast concrete pavement, permeable interlocking concrete pavements, and various types of permeable gravel pavements. They could potentially enhance the evaporation from pavements, and the cooling effect depends on the moisture content and evaporation rate. Permeable (a generic term used here to mean permeable, porous, or pervious) pavements can be used in city streets, parking lots, highway shoulders, etc.

In addition to these nonvegetated permeable pavements listed above, there are also some vegetated permeable pavements, such as grass pavers and concrete grid pavers, which use plastic, metal, or concrete lattices for support and allow grass or other vegetation to grow in the interstices. Unlike the nonvegetated permeable pavements, the typical use of vegetated permeable pavements is for lower traffic volumes such as alleys, parking lots, and trails; and they may be best suited to climate regions with adequate moisture to keep vegetation alive or may need irrigation systems to supply water. For vegetated permeable pavements, in addition to evaporation, the transpiration of vegetation provides an additional cooling effect to help reduce the pavement temperature.

Beyond reducing temperature, permeable pavements also could potentially reduce the air/pavement noise owing to reduced air pumping under high speed [52], if roughness and macrotexture are also kept low to avoid tire vibration, and improve high-speed driving safety by reducing splashing and hydroplaning during rainfall [53]. Also, full-depth permeable pavement could reduce storm-water runoff and improve water quality [54,55]. However, permeable pavements also could possibly increase fuel consumption of vehicles owing to higher rolling resistance if they have the rougher surfaces. In addition, current permeable pavements are generally not suitable for high-speed facilities such as highways and airfields because they are often rougher than conventional pavement and more prone to raveling. They also require thicker cross-sections than conventional pavement because of the greatly reduced structural capacity of the subgrade when the pavement is holding water.

To mitigate heat islands and reduce the associated impacts mentioned previously, use of permeable pavements may provide benefits in some situations. Permeable pavement also limits disruption of natural hydrology. This is a requirement of the LEED [1] building environmental rating system, which requires different options for project sites with imperviousness 50% or less and sites with greater than 50%. The increased use of pervious surfaces could potentially reduce other environmental impacts, including storm-water runoff and associated water pollution and the capacity required for storm-water management facilities, and can also enhance on-site infiltration for vegetation growth and recharging the underground water [56].

2.1.2.2 Water-Retentive Pavements

Some cities in Japan, such as Tokyo and Osaka, are testing the effectiveness of water-retentive pavements as part of using permeable pavements to reduce the heat island effect. These pavements can be asphalt or concrete-based and have a sublayer that consists of water-retentive materials that absorb moisture and then evaporate it through capillary action when the pavement heats up. Some of these systems involve underground water piping or surface water sprinkling to enhance the evaporation from the pavement. The sprinklers can reuse treated wastewater or stored rainwater and also can be integrated with and use the existing vegetation irrigation system, especially in arid seasons or regions. The energy needed by the sprinkling system can be provided by hybrid tower solar/wind power generators [57,58]. Results to date are promising, and the experiment shows that this pavement can effectively reduce the temperature of a road surface by up to 25 °C in midsummer, when the surface temperature can be as high as 60 °C [59]. The resulting reduction in the air temperature is 2–3 °C [59].

2.1.3 Enhancement of Convection between Pavement and Air

Pavement transfers heat to the near-surface air through convection as air moves over the warmer pavement surface. The rate of convection depends on the velocity and temperature of the air passing over the surface, the pavement roughness, and the total surface area of the pavement exposed to air [3]. Some permeable pavements (e.g., permeable asphalt pavement, pervious concrete pavement, pervious cast pavement, pervious brick or block pavers, etc.) have rougher surfaces and contain more air voids than conventional pavements, which increases their effective surface area exposed to air and creates air turbulence/circulation over and within the pavement. This will increase the convective heat exchange between pavement and air and thus help to reduce the temperatures of pavement and moving near-surface air through mass exchange and mixing due to air flow.

In particular, pervious cast pavements proposed by the University of California Pavement Research Center for storm-water management [38], which are made of dense-graded concrete and contain holes to allow water to drain through, potentially have the most effective convection and stay cooler than other pavement types owing to more ventilation paths created by the large holes. The pervious cast pavements use standard dense-graded Portland cement concrete with precast or cast-in-place holes instead of an open-graded mix and therefore also have relatively higher structural capacity per unit thickness than pervious concrete mixes. However, careful consideration needs to be given to the design of the holes to ensure sufficient strength, adequate drainage of water, and safe use for bicycle, motorcycle, motor vehicle, and possibly pedestrian traffic. Another study reported a similar idea and observed a cooling effect [60].

While surface roughness can increase convection and cooling effects, it might also have the potential to reduce the surface's net solar reflectance [3]. It is observed that in direct sunlight, dry pervious concrete pavement became hotter (on the surface) than traditional pavement, indicating the uncertainty of a cooling effect of the pervious pavement without evaporation [61].

2.1.4 Reducing Heat Energy in Pavements

Other measures to keep pavement cooler include shading pavement surfaces from solar radiation using shading trees or buildings and canopy covers (e.g., regular or solar panels) installed over the pavement. Another cooling option is the use of active mechanical cooling associated with harvesting and converting the heat energy stored in pavement. The active mechanical cooling strategies include cooling by fluid (e.g., water) circulating through pipes embedded in the pavement as one kind of heat exchanger [62,63] and cooling through thermoelectric devices embedded in the pavement [64]. The advantage of the active mechanical cooling is that it can be used for both cooling in hot seasons and heating (deicing, melting snow, etc.) in cold seasons, as well as for harvesting energy for local use. However, most approaches are still under theoretical development, and there are many technological, practical, and cost problems related to implementation. Therefore, the active mechanical cooling associated with harvesting and converting the heat energy in pavements is out of the range of this study at present.

2.1.4.1 Shading Pavements

Shading pavements could reduce the sunlight hitting on the pavement and thus directly reduce the heat source (solar energy) coming into the pavement, thus reducing the pavement temperature.

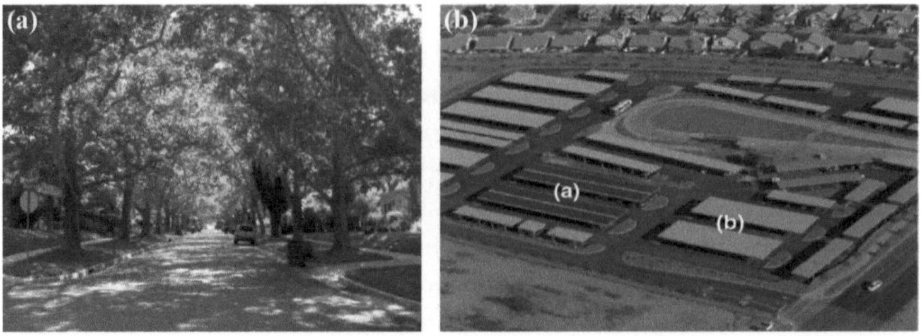

Figure 2.1 Shading pavements. (a) shading with trees. (b) shading with canopy and solar panels [65].

Pavement shading includes tree shading (Figure 2.1(a)) and vegetation shading, which especially could be used for city streets and parking lots. In addition to tree shading and vegetation shading, another emerging option considered by some local governments and private firms is to install canopies that incorporate solar panels in parking lots (Figure 2.1(b)) [65] and along highways [66]. Beyond shading pavement surfaces from incoming solar radiation to reduce the pavement temperature, these photovoltaic canopies also could generate electricity that can be used to power nearby buildings or provide energy for electric vehicles.

With respect to this cooling technology using shading, only a few site-specific case studies have been done. No specific data on the cooling effects for different seasons and different climate regions are available.

2.2 SUMMARY OF RESEARCH RELEVANT TO COOL PAVEMENTS

Based on the review of existing research and literature relevant to cool pavements, the potential cool pavement technologies and impact assessments reported in the literature are summarized in Table 2.1 and Table 2.2, respectively.

2.2.1 Pavement and Near-Surface Air Temperature

All the studies of cool pavements assess the cooling effects on surface or air or both. However, as shown in Table 2.1, no consistent cooling effects were observed in the literature by researchers. A variety of research problems are identified from the literature and listed as follows.

Table 2.1 Summary of cool pavement technologies and strategies

Cooling technology	Cooling mechanism	Eff. (°C)[a]	Co-benefits	References
1. Modify material thermal properties				
1.1 Increase albedo/emissivity	• Increase reflected heat	4/0.1[b] (s) 0.6/0.25 (a)	• Enhance illumination • Offset global warming potential	[20,29,66–75]
1.2 Increase heat capacity/density	• Increase heat capacity	—[c]	—	[66,71]
1.3 Reduce thermal conductivity	• Reduce transfer of heat into material	—	—	—
2. Evaporation/evapotranspiration				
2.1 Permeable pavements (+vegetation)	• Increase latent heat • Increase convection	10∼25 (s) 2∼8 (a)	• Reduce storm-water runoff • Reduce water pollution • Recharge groundwater • Increase greening (+vegetation)	[33,55,76–80]
2.2 Water-retentive pavements (+sprinkling)	• Increase latent heat	10∼25 (s) 1∼5 (a)	• Reuse wastewater/rainwater	[77,81–83]

Continued

Table 2.1 Summary of cool pavement technologies and strategies—cont'd

Cooling technology	Cooling mechanism	Eff. (°C)[a]	Co-benefits	References
3. Shading				
3.1 Canopy cover (+trees)	• Reduce absorbed heat	~10 (s)	• Increase greening (+trees)	[65]
3.2 Photovoltaic panels	• Reduce absorbed heat	~25 (s)	• Reduce land use dedicated for solar farms	[65]
4. Enhance convection				
4.1 Ventilation paths	• Increase convection	–	–	[38,60]
5. Harvesting energy				
5.1 Heat exchangers/converter	• Convert heat energy	–	–	[62–64]

[a]Eff., effectiveness. Temperature reduction effect: s, surface; a, air at ~2 m height.
[b]4/0.1 = 4 °C temperature reduction per 0.1 increase in albedo.
[c]No data available.

Table 2.2 Summary of literature relevant to cool pavements

Reference		Technology		Impact assessment[c]										LCA[c]	
Author (year) [Reference]	Title	Country and research unit[a]	Cool pavement type[b]	Microclimate	Energy use	Air quality	Water quality	Storm water	Thermal comfort	Lighting	Noise	Durability	Safety	Economic	Environmental
Pomerantz et al. (1997) [29]	Paving materials for heat island mitigation	US-LBNL	1.1	√	×	×	×	×	×	×	×	×	×	×	×
Pomerantz et al. (2000) [67]	The effect of pavements' temperatures on air temperatures in large cities	US-LBNL	1.1	√	×	×	×	×	×	×	×	×	×	×	×
Pomerantz et al. (2000) [20]	Cooler reflective pavements give benefits beyond energy savings: durability and illumination	US-LBNL	1.1	√	×	×	×	×	×	√	×	√	×	×	×
Ting et al. (2001) [68]	Preliminary evaluation of the lifecycle costs and market barriers of reflective pavements	US-LBNL	1.1	×	×	×	×	×	×	×	×	×	×	√	×
Levinson and Akbari (2001) [69]	Effects of composition and exposure on the solar reflectance of portland cement concrete	US-LBNL	1.1	√	×	×	×	×	×	×	×	×	×	×	×

Continued

Table 2.2 Summary of literature relevant to cool pavements—cont'd

Reference		Technology		Impact assessment[c]												LCA[c]	
Author (year) [Reference]	**Title**	**Country and research unit[a]**	**Cool pavement type[b]**	**Microclimate**	**Energy use**	**Air quality**	**Water quality**	**Storm water**	**Thermal comfort**	**Lighting**	**Noise**	**Durability**	**Safety**	**Economic**	**Environmental**
Pomerantz et al. (2003) [70]	Examples of cooler reflective streets for urban heat-island mitigation: portland cement concrete and chip seals	US-LBNL	1.1	√	×	×	×	×	×	×	×	×	×	×	×
Golden et al. (2006) [66]	Mesoscale and microscale evaluation of surface pavement impacts on the urban heat island effects	US-ASU	1.1 1.2	√	×	×	×	×	×	×	×	×	×	×	×
Golden et al. (2007) [65]	A comparative study of the thermal and radiative impacts of photovoltaic canopies on pavement surface temperatures	US-ASU	3.1 3.2	√	×	×	×	×	×	×	×	×	×	×	×

Kaloush et al. (2008) [71]	The thermal and radiative characteristics of concrete pavements in mitigating urban heat island effects	US-ASU	1.1 1.2	√	×	×	×	×	×	×	×	×	×	×	×
Kevern et al. (2009) [76]	Hot weather comparative heat balances in pervious concrete and impervious concrete pavement systems	US-ASU	2.1	√	×	×	×	×	×	×	×	×	×	×	×
Mallick et al. (2009) [63]	Harvesting energy from asphalt pavements and reducing the heat island effect	US-WPI	–	√	×	×	×	×	×	×	×	×	×	×	×
Mallick et al. (2009) [84]	Reduction of urban heat island effect through harvest of heat energy from asphalt pavements	US-WPI	–	√	×	×	×	×	×	×	×	×	×	×	×
Kubo and Kido (2006) [77]	Study on pavement technologies to mitigate the heat island effect and their effectiveness	Japan-PWRI	2.1 2.2	√	×	×	×	×	×	×	×	×	×	×	×

Table 2.2 Summary of literature relevant to cool pavements—cont'd

Reference		Technology		Impact assessment[c]										LCA[c]	
Author (year) [Reference]	Title	Country[a]	Cool pavement type[b]	Microclimate	Energy use	Air quality	Water quality	Storm water	Thermal comfort	Lighting	Noise	Durability	Safety	Economic	Environmental
Kawakami and Kubo (2008) [72]	Accelerated loading tests on the durability of cool pavement at PWRI	Japan-PWRI	1.1	√	×	×	×	×	×	×	×	×	×	×	×
Kinoshita et al. (2009) [81]	Evaporation performance analysis for water retentive material based on outdoor heat budget and transport properties	Japan	2.2	√	×	×	×	×	×	×	×	×	×	×	×
Takahashi and Yabuta (2009) [73]	Road temperature mitigation effect of "road cool," a water-retentive material using blast furnace slag	Japan	1.1	√	×	×	×	×	×	×	×	×	×	×	×
Nakayamaa and Fujita (2010) [82]	Cooling effect of water-holding pavements made of new materials on water and heat budgets in urban areas	Japan	2.2	√	×	×	×	×	×	×	×	×	×	×	×

Furumai et al. [83]	Recent application of rainwater storage and harvesting in Japan	Japan	2.2	√	×	×	×	×	×	×	×	×	×	×	×
Hasebe et al. (2006) [64]	Thermoelectric generators using solar thermal energy in heated road pavement	Japan	–	√	×	×	×	×	×	×	×	×	×	×	×
Tan et al. (1992) [78]	Influence of pavement materials on the thermal environment of outdoor spaces	Singapore	2.1	√	×	×	×	×	×	×	×	×	×	×	×
Wong et al. (2009) [74]	A study on the effectiveness of heat mitigating pavement coatings in Singapore	Singapore	1.1	√	×	×	×	×	×	×	×	√	×	×	×
Cao et al. (2011) [85]	Cooling principle analysis and performance evaluation of heat-reflective coating for asphalt pavement	China	1.1	×	×	×	×	×	×	×	×	√	×	×	×

Continued

Table 2.2 Summary of literature relevant to cool pavements—cont'd

Reference		Technology		Impact assessment[c]										LCA[c]	
Author (year) [Reference]	**Title**	**Country[a]**	**Cool pavement type[b]**	**Microclimate**	**Energy use**	**Air quality**	**Water quality**	**Storm water**	**Thermal comfort**	**Lighting**	**Noise**	**Durability**	**Safety**	**Economic**	**Environmental**
Sha (2010) [86]	Proceedings of International Workshop on Energy and Environment in the Development of Sustainable Asphalt Pavements	China	1.1	√	×	×	×	×	×	×	×	√	×	×	×
Zhang (2011) [87]	Research on heat reflection and thermal resistance technology of asphalt pavement cooling mechanism and its application	China	1.1	×	×	×	×	×	×	×	×	√	×	×	×
Lin et al. (2007) [79]	Seasonal effect of pavement on outdoor thermal environments in subtropical Taiwan	China/ Taiwan	2.1	√	×	×	×	×	×	×	×	√	×	×	×
Wang et al. (2009) [60]	Benefit analysis of permeable pavement on sidewalks	China/ Taiwan	4.1	√	×	×	×	×	×	×	×	√	×	×	×

Yilmaz et al. (2008) [80]	Determination of temperature differences between asphalt concrete, soil and grass surfaces of the City of Erzurum, Turkey	Turkey	2.1	√	×	×	×	×	×	×	×	×	×	×	×
Synnefa et al. (2009) [75]	Measurement of optical properties and thermal performance of coloured thin layer asphalt samples and evaluation of their impact on the urban environment	Greece	1.1	√	×	×	×	×	×	×	×	√	×	×	×

[a]US-LBNL, Lawrence Berkeley National Laboratory in United States; US-ASU, Arizona State University in United States; US-WPI, Worcester Polytechnic Institute in United States; Japan-PWRI, Public Works Research Institute.
[b]See Table 2.1 Summary of Cool Pavement Technologies; – not included in this study.
[c]√, with assessment; ×, without assessment; LCA, life cycle assessment.

2.2.1.1 More Attention and Effort Are Needed for Solutions other than Only Albedo

To mitigate the heat island effect, new or existing pavements can be made with higher albedo through various technologies as discussed previously. According to the review above (Table 2.2), it is found that most studies have focused on increasing albedo to reduce pavement and air temperature.

The major advantages of employing high-albedo surfaces as a heat island mitigation strategy are a large available area for implementation (e.g., all impervious streets, sidewalks, and roofs) and possibly a relatively low cost per unit area. However, light surfaces are difficult to keep clean and may lose up to one-third and even more of their reflectivity in a few years. This is due to normal staining, weathering, and soot deposition that occur on pavement surfaces. Light surfaces also scatter radiation (sensible heat) to other surfaces (e.g., buildings, etc.), resulting in a lower net effect on the energy budget than other surface cooling (through latent heat) with vegetation and evaporation. In addition, glare resulting from very high reflectivity of pavement surface materials might cause safety issues during the daytime, although it may help with lighting during the nighttime. Therefore, in addition to more research on the solution of high albedo, more solutions beyond high albedo should be further developed and applied to enhance the diversity of cooling pavement technologies and strategies. This is probably necessary to some extent to reduce the uncertain and/or unintended potentially negative consequences of using one single technology.

As mentioned previously other thermal properties of pavement materials, such as heat capacity, density, and thermal conductivity, as well as thermal emissivity, also play important roles in determining the temperatures of pavement and near-surface air. Identifying the effects of these factors and exploring effective and feasible solutions to modify them are also necessary to effectively lower temperatures.

In addition to modifying the thermal properties of pavement materials, other cooling technologies also should be given more attention from research institutes and industries. Permeable pavements, incorporated with vegetation if possible, and water-retentive pavements are promising solutions for heat island mitigation. Also, as mentioned previously, these solutions can bring multiple benefits in addition to heat island mitigation, including reducing storm-water runoff and associated pollution, recharging underground water, reducing the demand for dedicated storm-water management facilities, and reusing treated wastewater or stored rainwater for sprinkling.

In addition to enhancing evaporation from water and vegetation, improving the natural ventilation and convection to reduce temperatures of pavement and near-surface air is another promising solution. Similar to evaporation, this is another great benefit that can be gained from the power of natural processes with few or no negative impacts on the environment.

Shading the pavement surfaces (especially large parking lots) and incorporating solar panels is also a potential solution, especially as the number of solar panels that will be installed to collect clean solar energy is increasing. Incorporating solar panels into parking lots will be a win–win solution, as it reduces land demand for solar farms and also helps to mitigate heat island effect.

The effects of combinations of these various cooling technologies also need to be assessed to optimize the mitigation effects with lower overall cost and environmental impact.

2.2.1.2 Low Temperatures during Nighttime and Wintertime Also Need to Be Inspected

High temperatures of pavement and near-surface air during the daytime need to be reduced to mitigate heat island effects in hot climates. However, while reducing the daytime high temperature, it is important to consider the low temperature during the nighttime, especially for dry land climates with large diurnal temperature changes, such as the Central Valley or the desert of California. What is expected is a relatively stable and comfortable temperature range throughout the whole day. Therefore, efforts are also needed to ensure little or no reduction in nighttime low temperatures while lowering the daytime high temperatures.

Similarly, most studies focus just on the effects of mitigation measurements on lowing high temperatures during summertime. Efforts are also needed to ensure little reduction in low temperatures during wintertime in locations with cold winters while lowering the summertime high temperatures.

2.2.1.3 Temperature and Temperature Gradient in the Pavement Structure and Near-Surface Air

For pavement life, not only the pavement temperature but also the temperature gradient (both spatial and temporal) is of great importance, as it influences rutting, aging (asphalt pavement), and thermal stress (contraction, expansion, and curling).

2.2.1.4 Site-Level Effects

The existing studies mostly put emphasis on the city-wide (global mesoscale) effects of heat islands, rather than site-level (local microscale) effects.

However, for different impact assessments, different scales should be used. For example, the effect of urban heat islands on air quality is more related to the global mesoscale conditions, which can be given a more accurate assessment if assessed on a city-wide scale. In contrast, human thermal comfort, building/vehicle energy use, and pavement life are likely to be more influenced by local thermal conditions and thermal interactions between pavements and surroundings rather than global mesoscale conditions.

2.2.1.5 Effects for Different Regions Should Be Assessed

Heat island effects are different for different regions. To date, most existing studies are case studies for certain regions and specific climate conditions. Therefore, more systematic assessments for different regions are needed to determine the necessity and applicability of various cool pavement technologies and strategies.

2.2.2 Thermal Comfort

During summertime, especially in hot climates, people will experience a hot environment. This contributes to general discomfort on streets and parking lots, if outside of a range of maximum human comfort, and in buildings and vehicles (if air conditioning is not being used). The discomfort caused by heat might also result in respiratory difficulties, heat cramps and exhaustion, nonfatal heat stroke, and heat-related illness.

The reduced pavement surface and near-surface air temperature could help improve thermal comfort without increasing cooling-energy demand. In addition, improved human thermal comfort could potentially encourage more outdoor activities [88], including potentially more walking and cycling rather than driving for short-distance trips. This will help to improve the quality of life, reduce vehicle miles traveled, and consequently reduce energy consumption and reduce associated air pollution and greenhouse gas (GHG) emission.

However, very limited studies were found that assessed the effects of heat island mitigation measures on improving human thermal comfort. Especially for permeable pavements and water-retentive pavements, enhanced evaporation will increase humidity while reducing temperature, which might influence negatively the human thermal comfort if outside of the human comfort range for humidity. Therefore, comprehensive models should be used to assess the impacts of cool pavements on thermal environment and thermal comfort indices.

2.2.2.1 Thermal Comfort Index

Thermal comfort is influenced by a large number of factors, such as surrounding thermal environment (temperature, humidity, radiation flux, air flow), human activities, clothing, and perception about how hot an area is. Assessing comfort outdoors is not simply due to the complexity and methodological differences observed in the related literature, which make any comparison with available results difficult. Generally, comfort can be assessed by means of comfort indices. There are a large number of indices referred to in the literature, such as predicted mean vote, index of thermal stress, perceived temperature, operative temperature, standard effective temperature, mean radiant temperature, and physiological equivalent temperature (e.g., [89–103]). However, most of these indices have been developed from and for uniform indoor thermal environments, mostly using a single factor, and therefore might not be suitable for spatially and temporally severely nonuniform outdoor thermal environments. Hence, a rational index that combines several significant factors (e.g., temperature, humidity, radiation flux, air flow, etc.) into a single variable, which sums up their simultaneous effects on the sensory and physiological responses of the body, should be identified and chosen for the assessment of outdoor thermal comfort.

Moreover, considerable variability exists in the tolerance levels of different people (with age, health, and gender being some of the variables) [104]. Consequently, criteria for the thermal comfort index also need to be based on some form of population mean or on the tolerance levels of more sensitive members of the population.

One research paper by Ahmed [105] presents findings on defining outdoor comfort based on field investigations conducted in Dhaka, Bangladesh, a city in the tropics. Findings from a survey conducted on a large number of randomly selected people from urban spaces are presented. The findings include factors affecting comfort outdoors for Dhaka, and a comfort regime based on environmental parameters for urban outdoors is presented. The authors found that comfortable ambient climate leads to comfortable indoor environment particularly for buildings without running an air conditioner. With regard to mechanically controlled indoor environments, a comfortable outdoor condition helps reduce the demand on energy for cooling or heating. Evaluating and improving outdoor environments by using outdoor comfort index is an important step toward achieving sustainable urban environments.

2.2.3 Energy Use

2.2.3.1 Surface and Near-Surface Air Temperature Influences on Energy Demand

Surface temperature is the "skin" temperature of a surface, and near-surface air temperature is generally defined as the air temperature usually within the 2 m above the street level. Energy demand for air conditioning in buildings is affected by both surface temperature and near-surface air temperature through interrelated processes. Reductions in surface temperature on building walls and roofs directly reduce the conductive heat flow, partially driving energy demand. Rooftop summer surface temperatures are dramatically lowered by vegetated and light-colored roof surfaces, and this tends to reduce heat conduction in a downward direction. Conduction of heat through the walls of buildings is also reduced when urban tree canopies shade the walls. Reductions in near-surface air temperature can reduce the temperature of air entering buildings through ventilation and infiltration into buildings, one of the primary determinants of air-conditioning loads. Since all building surfaces are affected by near-surface air temperature, this can have a large effect on energy demand.

Currently, most building energy models calculate surface temperatures based on a given ambient air temperature. However, it is not clear how well these models calculate surface temperatures affected by sensible heat fluxes from reflective surfaces and latent heat fluxes from vegetated surfaces. To the extent that building energy models do not fully simulate sensible and latent heat fluxes, they may underestimate the impacts of reflective pavement surfaces and pavement surfaces with evaporation and vegetation on building energy demands.

2.2.3.2 Building Energy Models

Energy demand is one of the impacts of heat islands that has received the most research effort investigating mitigation measures. However, most researchers merely mention this potential benefit or have conducted qualitative analysis. A limited number of studies attempt to quantitatively analyze the effects of cool pavements on building energy saving; no studies could be found that have quantitatively analyzed the impacts of cool pavements on vehicle energy use because of increased air conditioning use. The only study found is one that was done by the Lawrence Berkeley National Laboratory (LBNL) [5,16]. The methodology (Figure 2.2) used by the LBNL to analyze effects of cool pavements and other heat island

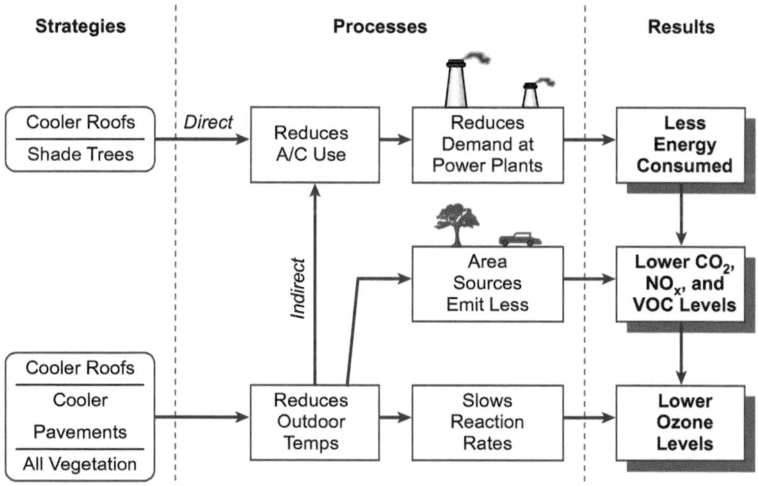

Figure 2.2 Methodology to analyze the impacts of shade trees, cool roofs, and cool pavements on energy use and air quality (smog) [5,16].

mitigation measures on building energy saving is shown in Figure 2.2 and summarized as follows:

1. Obtaining local full-year hourly weather data with and without modification as input to the DOE-2 building energy simulation program;
2. Defining prototypical buildings for each city;
3. Simulating the base heating and cooling energy use for each prototype;
4. Simulating the energy effects of reflective roofs for each prototype;
5. Estimating the total roof area for each prototype by city;
6. Estimating the energy saving by city-wide regions.

Figure 2.2 depicts the overall methodology used in analyzing the impacts of heat island mitigation measures on energy use and urban air pollution [5,16]. The DOE-2 building-energy simulation program is used to calculate the energy use and energy savings in buildings. To calculate the *direct* effects, prototypical buildings are simulated with dark- and light-colored roofs and with and without shade trees. Typical weather data for each climate region of interest are used in these calculations. To calculate the *indirect* effects, the typical weather data input to DOE-2 are first modified to account for changes in the urban climate due to heat island mitigation measures. The prototypical buildings are then simulated with the modified weather data to estimate savings in heating and cooling energy consumption.

Computer simulations for Los Angeles, California, using DOE-2 show that resurfacing about two-thirds of the pavements and rooftops with reflective surfaces and planting three trees per house could cool down Los Angeles by an average of 2–3 °C [5,16]. Every degree increase adds about 500 MW to the air-conditioning load in the Los Angeles Basin [5,16] according to the modeling. Peak urban electric demand rises by 2–4% for each 1 °C increase in daily maximum temperature above a threshold of 15–20 °C [5,16]. Thus, the additional air-conditioning use caused by this urban air temperature increase is responsible for 5–10% of urban peak electric demand [5,16].

The authors of the study [5] believe that "Cool pavements provide only indirect effects [on energy use] through lowered ambient temperatures," which is very small and could even be omitted [5]. However, as mentioned above, the interactions (especially huge reflected sensible heat and radiation due to high-reflective surfaces) between pavements and building are also of great significance to building energy use on a local microscale, which the authors of reference [5] did not consider. Recently, some researchers (e.g., [106]) used thermodynamics incorporated with the computational fluid dynamics approach to consider the interactions of the outdoor environment (including pavements) and buildings in an urban block.

2.3 LIFE CYCLE ASSESSMENT

Like most new technologies, there are some negative impacts of cool pavements associated with their benefits. To ensure a net benefit, both economically and environmentally, economic and environmental assessments should be performed over the entire life cycle of cool pavements with appropriate system boundaries.

In addition, comprehensive economic and environmental impacts of cool pavements in each phase over the entire life cycle could be identified through life cycle assessment (LCA). This also could provide a comparative LCA of these cool pavement technologies and supply an opportunity to identify potential redesign solutions to minimize the negative impacts. If designs were adapted to specific usability requirements, the environmental impact factors associated with pavements, and therefore urban areas, would be significantly optimized and reduced [107].

However, very few studies of cool pavements have conducted cost–benefit analyses and no study of cool pavements has included an environmental life cycle assessment.

2.3.1 Life Cycle Cost Analysis

There are many potential benefits from cool pavements relative to conventional pavements, such as reducing energy use. Some possible costs, however, also coexist with these benefits. For example, while cool pavements could potentially reduce building and vehicle energy use, special materials (e.g., high-reflection coatings or aggregates or cement, large-size and high-quality aggregates for open-graded materials) and special maintenance are needed to make them cooler. These factors may increase the overall costs of installing and maintaining the pavement with cooling effects. In addition, these costs will vary for different locations and applications. Therefore, the overall costs of cool pavements should be estimated through life cycle cost analysis.

2.3.2 Environmental Life Cycle Assessment

Similarly, although there may be potential environmental benefits from cool pavements in various contexts compared with conventional pavements, some possible negative environmental impacts may also coexist with these benefits. For example, cool pavements could potentially reduce building and vehicle energy use and associated GHG emissions; however, special materials (e.g., high-reflective paint or aggregates or cement, special aggregates for permeable mixes, etc.) are needed to make them cooler. Producing and processing these special materials might consume more energy and produce more associated GHG emissions. While thermal comfort might be improved during the daytime or summertime, people might experience more discomfort during the cold period.

Moreover, among those various cooling strategies, each one has unique cooling effectiveness and associated unique environmental impact (both positive and negative) as well as unique associated costs. Life cycle assessment provides a method to comprehensively compare these cooling strategies, to ensure the largest net benefit and lowest environmental impacts and costs through their entire life cycle, including material production, construction, use phase, and end of use.

2.4 SUMMARY OF RESEARCH AND KNOWLEDGE GAPS

Although some amount of research has been completed or is being conducted by various research institutes and groups around the world, there are still a number of knowledge gaps related to the environmental and engineering performance of cool pavements, as well as economic performance and implementation and policy issues. Some major research and knowledge gaps are identified from the literature review above and summarized as follows.

2.4.1 Environmental Performance

1. Thermal behavior of cool pavement materials and strategies, especially for pervious pavement under both wet and dry conditions, is still not clear.
2. Values for fundamental material properties of cool pavement materials, such as permeability, albedo, thermal conductivity, and heat capacity, especially for open graded permeable pavement materials, are very limited.
3. The seasonal effects on local microclimate of both conventional and cool pavements have not been fully investigated, especially the local effects in different seasons under different climate conditions.
4. The thermal interactions between pavement and building/vehicle surfaces and the specific impact on energy use for cooling and heating in different seasons over a year need to be estimated.
5. The effect of cool pavements on the improvement of outdoor human thermal comfort has not been investigated.

2.4.2 Life Cycle Analysis

1. Uncertainties related to LCA of cool pavements need to be fully identified and specified.
2. Life cycle cost analysis with uncertainties of cool pavements over the entire life cycle is missing.
3. Environmental life cycle assessment with uncertainties of cool pavements is missing.
4. Comparative life cycle assessment of various cool pavements and other mitigation measures (vegetation, cool roofs, etc.) of heat islands is missing.

2.4.3 Implementation Issues

Comprehensive recommendations for implementation and further research of cool pavements are missing.

CHAPTER 3

Scope, Methodologies, and Organization

Contents

3.1 PROBLEM STATEMENT

Some studies have been performed or are being conducted by various research institutes and groups around the world regarding the use of cool pavements as a strategy for mitigating heat island effects, improving outdoor thermal comfort, and potentially reducing energy use. However, there are a number of knowledge gaps related to the environmental and engineering performance of cool pavements as identified in Chapter 2.

Based on the background and literature review on heat island effects and cool pavements, and the identified gaps in the knowledge discussed previously in Chapters 1 and 2, this study will mainly address the following key issues on environmental performance of cool pavements:

1. Fundamental material properties of cool pavements, such as permeability, albedo, thermal conductivity and heat capacity, and evaporation rate, and corresponding measurement methods;
2. Seasonal thermal performance of various cool pavement technologies and strategies, in particular the performance of permeable pavement under both wet and dry conditions;
3. Thermal interactions between pavement and near-surface air;
4. Thermal interactions between pavement and other surfaces (e.g., buildings, vehicles);
5. Thermal behavior of various pavements and impacts on outdoor human thermal comfort and thermal load of cooling and heating.

Pavement Materials for Heat Island Mitigation
ISBN 978-0-12-803476-7
http://dx.doi.org/10.1016/B978-0-12-803476-7.00003-9

These issues are critical and of great significance for the evaluation and implementation of cool pavements for effectively mitigating heat islands and potentially producing maximum benefits.

3.2 STUDY GOAL AND SCOPE

The goal of this study is to evaluate the effectiveness and applicability of various cool pavement technologies for reducing local heat islands and improving the near-surface outdoor thermal environment when applied in various climate regions. The primary focus of this study will be on cool pavement technologies and design strategies and their effectiveness at reducing localized heat island effects and improving the outdoor thermal environment in areas with hot climates where the heat island is mainly a negative issue. The scope of this study comprises: (1) experimental investigation on the thermal behavior and cooling effects of various cool pavement strategies, in particular permeable pavements; (2) use of the field measurement data to validate thermal behavior modeling and application of the validated model to simulate the thermal behavior and cooling effects of different pavements under different climates; and (3) examination of their effects on outdoor human thermal comfort and building energy use, mostly for urban areas with hot climate.

3.3 STUDY OBJECTIVES

To achieve the study goal above, the specific research objectives are listed as follows;

1. Measure the fundamental material properties affecting the thermal performance of cool pavements, including permeability, albedo, thermal conductivity and heat capacity, and evaporation rate;
2. Measure seasonal thermal performance of various pavement technologies and strategies, in particular permeable pavements;
3. Develop, validate, and apply a theoretical model and simulation system to conduct sensitivity analysis and evaluate near-surface pavement thermal performance in various climate regions;
4. Investigate the thermal impacts of pavements on near-surface air and effects on human thermal comfort;
5. Investigate the thermal impacts of pavements on buildings and effects on thermal load of cooling and heating;

6. Provide recommendations for implementation of cool pavement strategies for various climate regions;
7. Identify future research needs.

3.4 TASKS AND METHODOLOGIES

An overview of the tasks and methodologies to realize the specific objectives above is briefly described by a roadmap in Figure 3.1. The details of tasks and methodologies are presented in the corresponding chapters.

3.5 ORGANIZATION OF THE FOLLOWING PARTS OF THIS BOOK

This book is organized as follows. Chapter 4 presents the design, construction, and instrumentation of test sections and methods and measurements of albedo (i.e., solar reflectivity) for reflective pavements. Chapter 5

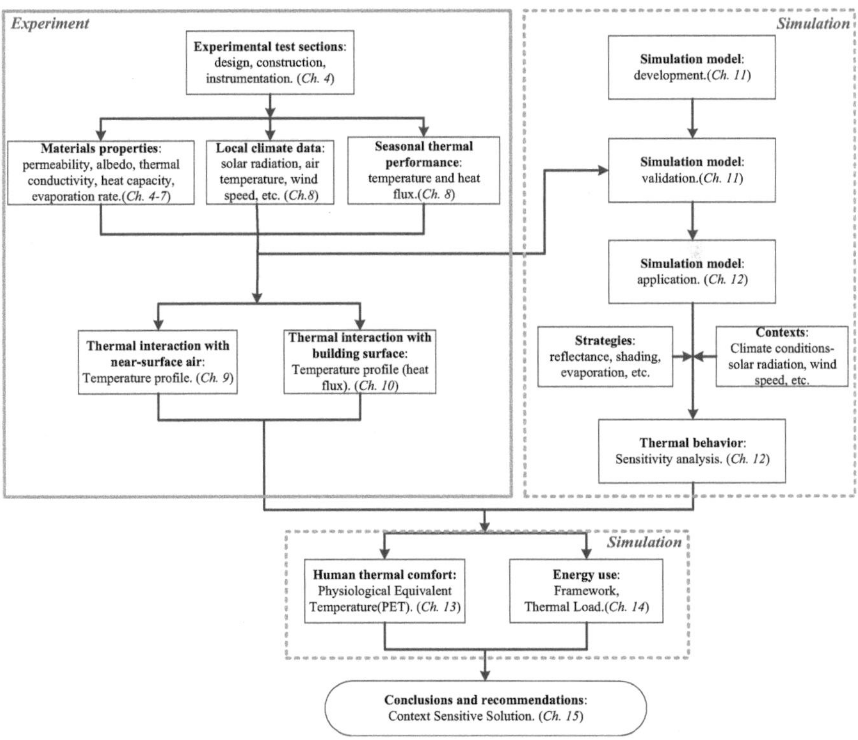

Figure 3.1 Roadmap for this study on cool pavements. Note: Ch. = Chapter.

through 7 present methods and measurements for other fundamental pavement material properties affecting thermal performance, including permeability (permeable pavements), thermal properties (i.e., thermal conductivity and heat capacity for thermal resistance pavements), and evaporation rate (evaporative cooling). Chapter 8 presents the measurement of seasonal thermal performance of various pavement design and management strategies, in particular permeable pavements. Thermal impacts of pavements on near-surface air temperatures are investigated in Chapter 9. Chapter 10 presents the thermal impacts of pavements on building surfaces. Development and validation of the pavement thermal modeling and simulation system are presented in Chapter 11. Chapter 12 uses the validated pavement thermal model to run temperature simulations for various factors to identify the significant factors affecting thermal behavior. Chapters 13 and 14 present results of an investigation of the effects of various pavement design and management strategies on human thermal comfort and building thermal load, respectively. The last chapter, Chapter 15, summarizes the entire study and gives conclusions and recommendations for implementation of cool pavement strategies for some climate regions and recommendations for future research.

CHAPTER 4

Reflective Pavements and Albedo

Contents

4.1 INTRODUCTION

Reflective pavements can be used as a cool pavement to mitigate heat island effects through reflective cooling. Albedo (or solar reflectivity) is an indicator of the reflecting power of a surface (e.g., pavement) and a key thermal characteristic that significantly influences the thermal performance of the surface. It is defined as the ratio of the reflected solar radiation to the incident solar radiation at the surface. Albedo is a dimensionless fraction and is measured on a scale from 0 to 1. An albedo of 0 means no reflecting power of a perfectly black surface (none reflected, all absorbed), an albedo of 1 means perfect reflection off a perfectly white surface (100% reflected). Solar reflectivity depends on the frequency or wavelength of the solar

Pavement Materials for Heat Island Mitigation
ISBN 978-0-12-803476-7
http://dx.doi.org/10.1016/B978-0-12-803476-7.00004-0

radiation. When albedo is reported without qualifications, it usually refers to some appropriate average across the spectrum of frequencies or wavelengths [108,109].

It is assumed that reflected radiation is both diffuse and specular in nature, meaning that it is diffuse if the reflected radiation is the same in all angular directions and specular if the surface of reflection is smooth with respect to the wavelength of the incident radiation such that the laws of reflection are satisfied. Pavement materials and most land-cover types are generally diffuse, and thus the reflected radiation is uniform or isotropic in all angular directions [109,110].

The temperatures of pavement surfaces exposed to solar radiation are generally higher than the adjacent air temperatures owing to heating by absorbed solar radiation, thus creating a surface heat island. The extent to which solar radiation influences surface temperatures depends on the solar reflectance of the exposed pavement surface. A low solar-reflectance material, such as a black stone surface, would result in a very large effect, while a high solar-reflectance material, such as fresh snow, would result in a small effect on surface temperature. Therefore, solar reflectivity is a factor of great significance for evaluating and modeling the thermal performance of pavements and other land-cover types [26,27,111].

There are some values of albedo reported in the literature for some pavement materials. Pomerantz et al. conducted some studies on more reflective pavements and their benefits and measured the albedos of some pavement materials such as Portland cement concrete and chip seals using light-colored aggregates [20,29,67,68,70]. Levinson and Akbari [69] performed a study on the effects of concrete mix composition (proportioning of cement, aggregate, and sand in the concrete) and exposure on the solar reflectance of Portland cement concrete. Synnefa et al. [75] measured the optical properties and thermal performance of asphalt samples with colored thin coatings and evaluated their impact on the urban environment. Wong et al. [74] performed a study on the effectiveness of heat-mitigating pavement coatings in Singapore and measured the albedo of different types of coatings.

However, data on albedo of different types of pavement materials is relatively limited or absent for some pavement surface types. Most existing studies refer to albedo values from a very limited number of literature sources or simply assume a value for evaluating and modeling the thermal performance of pavements and other land-cover surfaces (e.g., [33,112–116]) and their impacts on human thermal comfort, building energy use (e.g., [34,36,39,106,117–119]), and air quality. This limitation increases the barriers and uncertainty for understanding, evaluating, and modeling thermal performance and consequential environmental impacts of pavements and other land-cover types with different albedos.

4.2 OBJECTIVES

The objectives of this chapter are to: (1) design and construct experimental sections with various pavement types, (2) perform field measurements of albedo for various pavement materials on experimental test sections and other existing pavements, (3) compare the albedo for various materials, (4) examine the factors affecting the field measurement of albedo, (5) examine diurnal and seasonal changes in albedo, and (6) examine the effect of albedo on pavement temperature.

4.3 DESIGN AND CONSTRUCTION OF EXPERIMENTAL SECTIONS

Nine 4-m by 4-m (13-ft by 13-ft) experimental sections were designed and constructed. These specifically built test sections and some other existing pavements were used to measure the fundamental materials properties, including albedo, permeability, thermal properties, and evaporation rate (for some of the materials used in the experimental sections), and empirically examine the thermal behaviors of different pavement types at different seasons and under different moisture conditions and their impacts on near-surface air and building surfaces, which are presented in Chapters 4–10. The plan design and cross-sections for each experimental section are shown in Figure 4.1. The experimental sections include three different pavement surfacing materials, namely interlocking concrete paver (surfacing type A), open-graded asphalt concrete (surfacing type B), and pervious concrete (surfacing type C). For each pavement surfacing type, one impermeable pavement design (design 1) and two permeable pavement designs (designs 2 and 3) were designed. Both of the permeable interlocking concrete paver pavements have the same cross-sections, the difference is in the solar reflectivity of the pavers. Both of the permeable asphalt sections have the same surface material, the difference is in the thickness of the layers. The two permeable concrete sections have different thickness and concrete surface material mix designs, meaning that

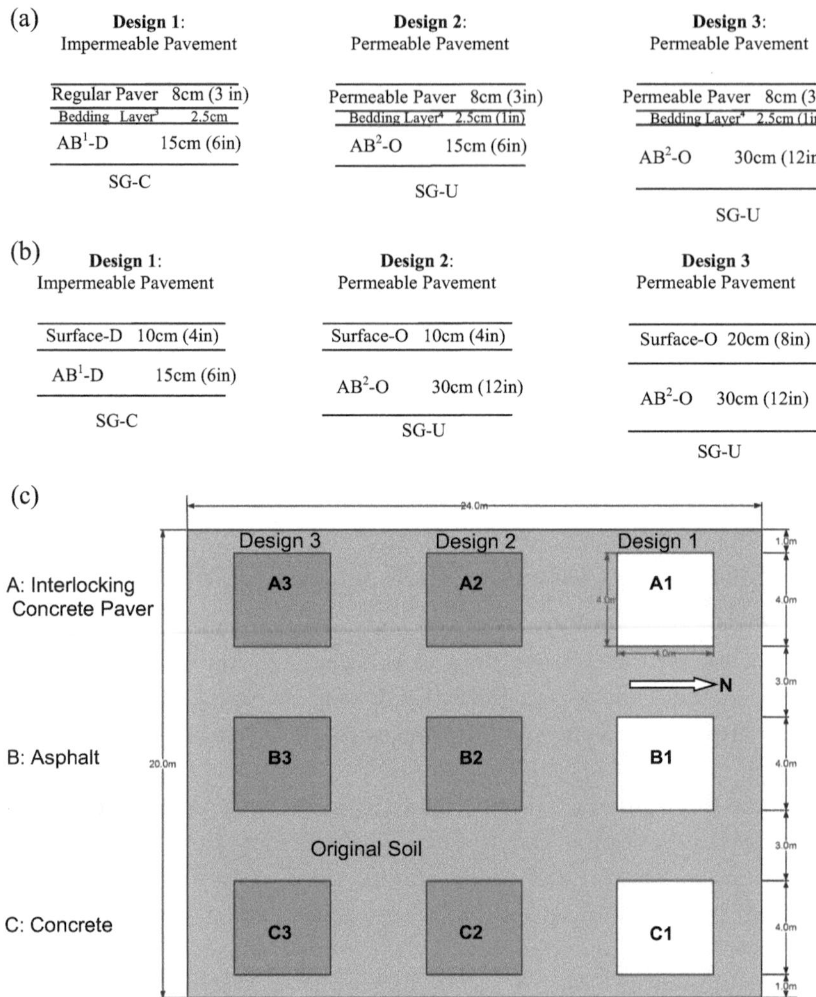

Figure 4.1 Designs of experimental sections for the cool pavement study. (a) Cross-sections for interlocking concrete paver pavements (A). (b) Cross-sections for asphalt (B) and concrete (C) pavements. (c) Schematic plan view (six permeable pavements shown in shaded area, i.e., left two columns).

the aggregate gradations, cement contents, and other ingredient proportions are different. The permeable concrete section C2 is darker than C3, which will give different albedos.

The nine experimental sections were constructed as designed during the summer of 2011 at the University of California Pavement Research Center

(UCPRC) test facilities in Davis, California. The construction materials used for the test sections are listed in Table 4.1.

Some example photos of the construction process for each layer of the pavement structures are shown in Figure 4.2. The soil of each section was first excavated to the design elevation to create the space for the base and surface layers (Figure 4.2(a)). The surface of the subgrade was prepared to create a level surface. For the six permeable sections, the subgrade was not compacted to ensure a high permeability of subgrade. The subgrade was compacted for the three impermeable sections. The permeability of the subgrade was measured on at least three locations for each section before the aggregate base was constructed. As shown in Figure 4.2(b), the aggregate was then poured into the prepared space and compacted three times using a jumping jack compactor. The thickness and level of the aggregate base were checked to ensure the design thickness and the surface level. After the aggregate bases were in place, the surface layers of paver, asphalt, and concrete were paved as shown in Figure 4.2(c) and (d). A photo view of the complete construction process for the experimental sections is presented in Figure 4.3.

4.4 MEASUREMENT METHODOLOGY FOR ALBEDO

4.4.1 Measurement Method and Equipment

There are two American Society for Testing and Materials (ASTM) standard testing methods for determining solar reflectance of a surface: (1) ASTM C1549, Standard Test Method for Determination of Solar Reflectance Near Ambient Temperature Using a Portable Solar Reflectometer [46], and (2) ASTM E1918, Standard Test Method for Measuring Solar Reflectance of Horizontal and Low-Sloped Surfaces in the Field [47].

4.4.1.1 ASTM C1549 Test Method

Solar spectrum reflectance measurement with this method relies on a testing instrument with an integrated radiation source and four detectors with filters for four specific wavelength ranges. This test method is best suited to use on flat and homogeneous smooth surfaces, such as single-ply membranes and smooth modified-bitumen membranes. The test method also requires that a surface to be tested is dry. However, it is not suitable for rough surfaces such as gravel and some other pavement surfaces [46].

Table 4.1 Construction material quantity needed for test sections

Material	Type	Thickness	Quantity per section[c]	No. of sections	Total quantity[c]
Base layer	O[a] (ASTM No. 57)	30 cm (12 in)	9.5 tons	6	57 tons
Base layer	D[b] (Class 2)	15 cm (6 in)	5.5 tons	3	16.5 tons
Bedding layer	O[a] (ASTM No. 8)	2.5 cm (1 in)	1 ton	2	2 tons
Bedding layer	Sand (ASTM C33)	2.5 cm (1 in)	1 ton	1	1 ton
HMA surface	O[a]	10 cm (4 in)	3.8 tons	1	11.4 tons
		20 cm (8 in)	7.6 tons	1	
	D[b]	10 cm (4 in)	4.4 tons	1	4.4 tons
PCC surface	O[a]	10 cm (4 in)	1.9 m^3 (2.5 yd^3)	1	5.7 m^3 (7.5 yd^3)
		20 cm (8 in)	3.8 m^3 (5.0 yd^3)	1	
	D[b]	10 cm (4 in)	1.9 m^3 (2.5 yd^3)	1	1.9 m^3 (2.5 yd^3)
Paver surface	Permeable	8 cm ($3^1/_8$ in)	16 m^2 (~180 ft^2)	2	32 m^2 (~360 ft^2)
Paver surface	Impermeable	8 cm ($3^1/_8$ in)	16 m^2 (~180 ft^2)	1	16 m^2 (~180 ft^2)
Edge curb	Wood	5 × 10 cm (2 × 4 in)	16 m (~55 ft)	9	150 m (~495 ft)

[a]Open-graded.
[b]Dense-graded.
[c]Metric ton (1000 kg).

(a)

Figure 4.2 Construction of experimental sections for the cool pavement study. (a) Preparation of the subgrade layers. (b) Construction of base layers. (c) Construction of surface layers (paver pavements). (d) Construction of surface layers (concrete pavements). (e) Construction of surface layer (asphalt pavement).

4.4.1.2 Pyranometer Test Method (ASTM E1918)

The device employed in this test method allows for calculation of solar reflectance based on alternate readings of incoming solar radiation and reflected solar radiation on a surface using only one pyranometer. The test procedure is weather-sensitive. It requires cloudless weather and a sun angle to the normal from the test surface of less than 45° to obtain valid and repeatable solar reflectance values [47].

This test method is suited to measurements over all types of flat surfaces, including textured or irregular surfaces such as gravel surfacing.

(b)

Figure 4.2 Cont'd

However, it has only one pyranometer to measure both the incoming solar radiation and the reflected solar radiation on a surface. After the incoming solar radiation is measured, the pyranometer has to be flipped over to measure the reflected solar radiation [47]. This is not convenient and might increase measurement error since the incoming solar radiation and reflected solar radiation are not measured at the same time, especially on cloudy days.

4.4.1.3 Dual-Pyranometer Test Method

To improve the convenience and reduce the measurement error of ASTM E1918, a dual pyranometer was selected and used to perform the

(c)

Figure 4.2 Cont'd

measurement of solar reflectivity in this study. A dual pyranometer (also called albedometer) is composed of two star pyranometers. One pyranometer faces upward and the other faces downward. Incident global solar radiation (diffuse and direct solar radiation) with wavelengths of 0.3–3 μm is measured by the upward-facing pyranometer, while reflected solar radiation from surfaces is measured by the downward-facing pyranometer [120]. Separate outputs are provided for each pyranometer, which can be read from an indicator or recorded together automatically using a data logger.

Testing in this study was conducted in accordance with ASTM E1918, following the standard method except using a dual pyranometer. The

Figure 4.2 Cont'd

Model 240-8140 dual pyranometer was purchased from NovaLynx Corporation® in July 2011, with a calibration certificate and showing a measurement error of less than 1 W/m^2. The data were recorded automatically using a data acquisition system (DAS) composed of a CR 10X data logger (from Campbell Scientific, Inc.®) powered by a battery and connected to a computer. This allows monitoring of the solar reflectivity of a surface over long time periods. The whole measurement system is shown in Figure 4.4.

4.4.2 Pavement Materials for Measurement

The albedo measurements for this study were performed mainly on the nine 4-m by 4-m (13-ft by 13-ft) test sections, which were specifically

(e)

Figure 4.2 Cont'd

Figure 4.3 Photo view of all experimental sections at UCPRC facility.

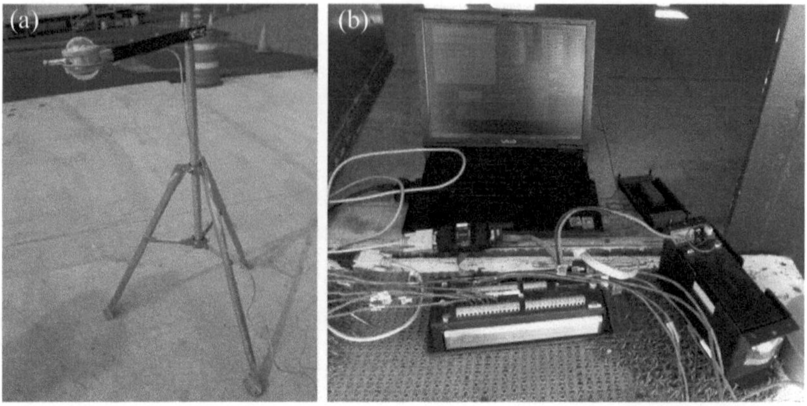

Figure 4.4 Albedo measurement system with a dual pyranometer. (a) Dual pyranometer (albedometer). (b) DAS: data logger (CR10X), battery, and computer.

constructed during the summer of 2011 at the UCPRC test facilities in Davis, California (as shown in Figure 4.3). The summary for each test section along with surface mix design and other characteristics is provided in Table 4.2. As noted previously, both of the permeable interlocking concrete paver pavements have the same cross-section, the difference is in the solar reflectivity of the pavers. Both of the permeable asphalt sections have the same surface material, the difference is in the thickness of the layers. The two permeable concrete sections have different thicknesses and different concrete surface material mix designs, meaning that the aggregate gradations, cement contents, and other ingredient proportions are different. The permeable concrete section C2 is darker than C3, resulting in different albedos.

In addition to these nine sections, some extra pavement sections with conventional impermeable asphalt and concrete surfacing were also included in the study for field measurement of albedo. In addition, albedo was measured on some other land-cover materials, including gravel, soil, and grass, for comparison. Some of these materials were different ages when the measurement of solar reflectivity was conducted on them. The overall summary of materials used for albedo measurement in this study and the dates of measurement are summarized in Table 4.3.

4.4.3 Plan for Measurement

The dual pyranometer with DAS was used to measure the albedo of all pavement test sections in the experimental sections (pavers, asphalt, and

Table 4.2 Experimental plan for field albedo measurement on nine test sections

Section	Pavement layer		Base layer		Test locations[b]	Test method
	Type[a]	Thickness (cm)	Type[a]	Thickness (cm)		
A1	ICP-I	10[c]	AB-I	15	SE, NE, NW, SW, CT	Dual pyranometer
A2	ICP-P	10[c]	AB-P	15	SE, NE, NW, SW, CT	Dual pyranometer
A3	ICP-P	10[d]	AB-P	30	SE, NE, NW, SW, CT	Dual pyranometer
B1	AC-I	10	AB-I	15	SE, NE, NW, SW, CT	Dual pyranometer
B2	AC-P	10	AB-P	30	SE, NE, NW, SW, CT	Dual pyranometer
B3	AC-P	20	AB-P	30	SE, NE, NW, SW, CT	Dual pyranometer
C1	PCC-I	10	AB-I	15	SE, NE, NW, SW, CT	Dual pyranometer
C2	PCC-P1[e]	10	AB-P	30	SE, NE, NW, SW, CT	Dual pyranometer
C3	PCC-P2[e]	20	AB-P	30	SE, NE, NW, SW, CT	Dual pyranometer

[a]ICP, interlocking concrete paver; AC, asphalt concrete; PCC, Portland cement concrete; AB, aggregate base; I, impermeable; P, permeable.
[b]SE, southeast; NE, northeast; NW, northwest; SW, southwest; CT, center.
[c]Includes the thickness of paver (6 cm) and underlying bedding layer (4 cm).
[d]Includes the thickness of paver (8 cm) and underlying bedding layer (2 cm).
[e]The two pervious concrete materials have very different mix designs, cement and aggregate types.

Table 4.3 Summary of materials and plan for albedo measurement

Surface category	Permeability	Binder/cement/color type	Code[a]	Measurement date (month/day/year)
Asphalt	Impermeable	Conventional	B1	9/19/2011, 10/13/2011, 2/15/2012, 5/2/2012
	Permeable	Polymer modified	B2[b]	9/19/2011, 10/13/2011, 2/15/2012, 5/2/2012
	Permeable	Polymer modified	B3	9/19/2011, 10/13/2011, 2/15/2012, 5/2/2012
	Impermeable	Polymer modified	PMA	10/13/2011
	Impermeable	Rubberized	RHMA	10/13/2011
	Impermeable	Warm mixed asphalt	WMA	10/13/2011
	Permeable	Polymer modified	OGFC	10/13/2011
	Impermeable	Conventional	Aged AC	10/13/2011
Concrete	Impermeable	Conventional	C1	9/19/2011, 10/13/2011, 2/15/2012, 5/2/2012
	Permeable	Conventional	C2[c]	9/19/2011, 10/13/2011, 2/15/2012, 5/2/2012
	Permeable	White	C3	9/19/2011, 10/13/2011, 2/15/2012, 5/2/2012
	Impermeable	Conventional	PCC	10/13/2011
Interlocking	Impermeable	Conventional–orange	A1	9/19/2011, 10/13/2011, 2/15/2012, 5/2/2012
Concrete paver	Permeable	Conventional–champagne	A2	9/19/2011, 10/13/2011, 2/15/2012, 5/2/2012
	Permeable	Conventional–orange	A3	9/19/2011, 10/13/2011, 2/15/2012, 5/2/2012
Gravel	Permeable	–	Gravel	10/13/2011
Soil	Permeable	–	Soil	10/13/2011
Grass	Permeable	–	Grass	10/13/2011

[a]Code: A1–A3, B1–B3, C1–C3 are experimental test sections described in Figure 4.1. PMA, polymer modified asphalt; RHMA, rubberized hot mixed asphalt; WMA, warm mixed asphalt; OGFC, open graded friction course; AC, asphalt concrete; PCC, Portland cement concrete.
[b]Monitored continuously over time.
[c]Monitored continuously in one day.

concrete). For each test section, the albedo measurement was conducted at five different locations: the southeast corner, northeast corner, northwest corner, southwest corner, and center. At least six measurements were performed at each location. A summary of the experimental test plan for the nine test sections is shown in Table 4.2.

The measurements on the nine experimental test sections (A1–A3, B1–B3, and C1–C3) were repeated on four different dates in one year (September 2011 to May 2012). The measurements on the other materials were conducted at two or three different locations with at least six measurements for each location. All these measurements were performed at midday (mostly 12:00 PM–2:00 PM). For some materials (mainly for B2), the solar reflectivity was monitored continuously over time.

The weather data (including air temperature, relative humidity, solar radiation, wind speed, and rainfall) during the measurement period were also monitored using a nearby mobile weather station from Campbell Scientific, Inc.®. The measurement plan in this study is summarized in Table 4.3.

The pavement temperatures of the nine experimental test sections (A1–A3, B1–B3, and C1–C3) were continuously monitored from the fall of 2011 through the summer of 2012 using type T thermocouple sensors (from Omega Engineering, Inc.® and with measurement error of less than 0.5 °C) and a CR10X data logger. Some temperature data were used to examine the seasonal effects of albedo on the pavement thermal performance.

4.5 RESULTS AND DISCUSSION ON ALBEDO

4.5.1 Albedo of the Nine Test Sections

The albedos of the nine test sections (A1–A3, B1–B3, and C1–C3) were initially measured at five different locations (four corners and one center) at around 1:00 PM on 19 September 2011. The measured albedos are illustrated in Figure 4.5 for each location on all nine sections using a box plot. As shown in the plot, the albedo measured at one location on a particular plot is relatively constant, which verifies the high accuracy of the albedo measurement equipment used in this study. However, the albedos measured at different locations on one section show some level of variation; this implies that the pavement surface is not uniform in albedo or color on each section, especially for concrete section C3. The summary box plot of albedos at five locations on each section, Figure 4.6, gives an even clearer

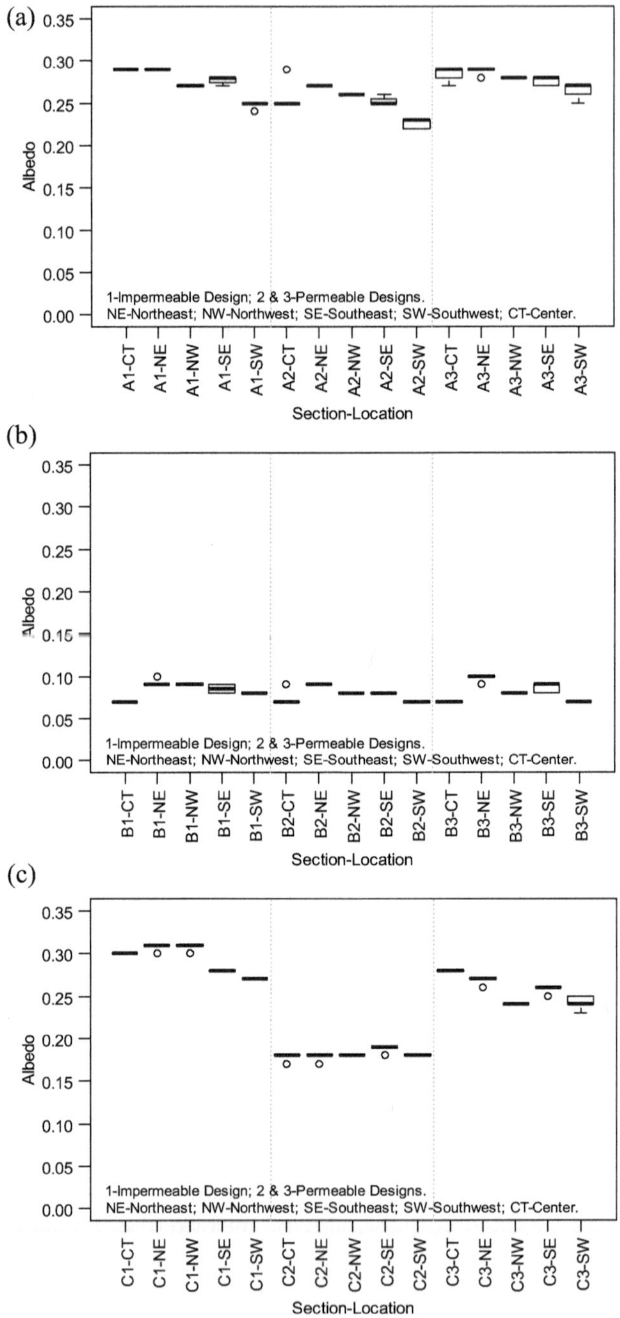

Figure 4.5 Albedo of different materials at different locations measured on 19 September 2011. (a) Interlocking concrete paver sections A1–A3. (b) Asphalt pavement sections B1–B3. (c) Concrete pavement sections C1–C3. CT, center; NE, northeast; NW, northwest; SE, southeast; SW, southwest.

illustration of this variation of albedo across these nine sections. Sections B2 and C2 are the most uniform, with only small variations across the 4-m-wide pavements. Sections C1 and C3 show the largest variation. This suggests that for any pavement the albedo should be measured at different locations to obtain a representative sample of albedo value, since one point may not be representative of the whole paved area.

The summary statistics of albedo for the nine sections measured on 19 September 2011 are listed in Table 4.4. As expected, the asphalt sections, which are black, have lower albedos (0.09 for B1 and 0.08 for B2 and B3). As mentioned previously, sections B2 and B3 have the same surface material (only thicknesses are different) and have the same albedo (0.08), as expected. The three concrete sections (C1–C3) have a range of mean albedo of 0.18–0.29. The darker concrete section C2 has a lower albedo of 0.18 compared to the other two concrete sections (0.26 for C3 and 0.29 for C1). The paver sections (A1–A3) have albedos close to the more reflective concrete sections (C1 and C3), which are in the range of 0.25–0.28. The relatively low albedo of asphalt pavements will absorb more incident solar radiation and produce a high temperature. In contrast, the concrete and paver pavements generally have a higher albedo, which will reflect more incident solar radiation and produce a lower pavement temperature.

4.5.2 Albedo of Other Pavement Materials and Some Other Land Covers

The albedo also was measured on pavements with asphalt and concrete surfacing materials other than the above nine experimental sections to

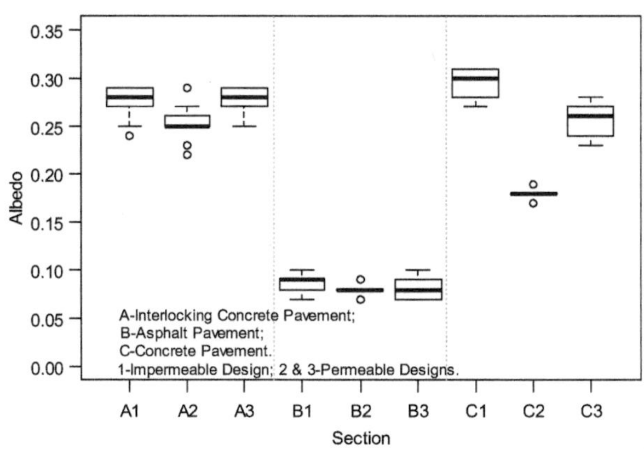

Figure 4.6 Overall albedos of nine test sections measured on 19 September 2011.

Table 4.4 Summary statistics of albedo for various materials measured on 19 September 2011

Section no.	Surface category	Permeability	Number of tests	Mean	SD	Median	Min	Max	Max − min
A1	Paver	Impermeable	40	0.28	0.01	0.28	0.24	0.29	0.05
A2	Paver	Permeable	40	0.25	0.02	0.25	0.22	0.29	0.07
A3	Paver	Permeable	33	0.28	0.01	0.28	0.25	0.29	0.04
B1	Asphalt	Impermeable	58	0.09	0.01	0.09	0.07	0.10	0.03
B2	Asphalt	Permeable	58	0.08	0.01	0.08	0.07	0.09	0.02
B3	Asphalt	Permeable	33	0.08	0.01	0.08	0.07	0.10	0.03
C1	Concrete	Impermeable	38	0.29	0.02	0.30	0.27	0.31	0.04
C2	Concrete	Permeable	44	0.18	0.00	0.18	0.17	0.19	0.02
C3	Concrete	Permeable	70	0.26	0.02	0.26	0.23	0.28	0.05

include additional pavement surface materials for comparison. These pavements have different mix designs for materials, with different binder or cement and aggregate types compared to those used in the nine test sections. Moreover, these pavements have ages between 1 and 5 years, while the nine experimental test sections had ages of less than 3 months. Despite their older age, the additional pavement surfaces had not been subjected to much traffic, so aging processes related to traffic were not visible. For example, the asphalt covering the aggregate on the pavement surface had not been worn off. Albedos measured on gravel, bare soil, and grass are also included for comparison. The results are shown in Figure 4.7.

The additional types of asphalt pavement materials tested include a polymer-modified asphalt (PMA; should not affect the albedo), gap-graded rubberized hot mix asphalt (RHMA; which contains a higher asphalt binder content and recycled tire rubber, which is expected to affect the albedo), rubberized warm mix asphalt (RWMA; a different RHMA mix with a "warm mix" additive that should not affect the albedo), an open-graded asphalt friction course (OGFC), and an aged asphalt concrete (AC). The PMA section showed a slightly higher mean albedo of 0.12 compared to 0.08 for the RHMA and 0.06 for the

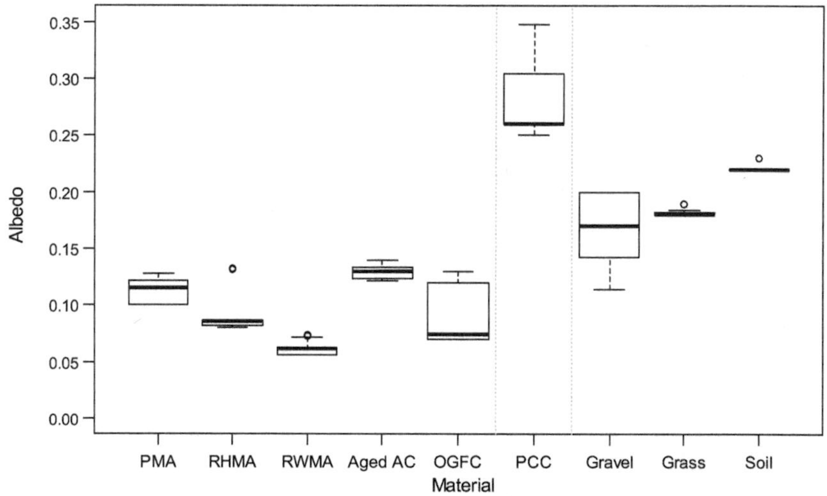

Figure 4.7 Albedo of other pavement and land-cover materials. PMA, polymer modified asphalt; RHMA, rubberized hot mixed asphalt; RWMA, rubberized warm mixed asphalt; Aged AC, aged asphalt concrete; OGFC, open graded friction course; PCC, Portland cement concrete.

RWMA. The aged asphalt pavement (aged AC) of 4 years resulted in a higher albedo of 0.14. The newly paved OGFC layer had a low albedo of 0.07. The lower albedos of the RHMA, RWMA, and OGFC were expected because of their higher asphalt content and the inclusion of tire rubber in the binder of the RHMA and RWMA. The higher albedo of the aged asphalt was also expected because of the oxidation of the asphalt.

The extra concrete pavements (PCC) had albedos in the range of 0.25–0.35 with an average of 0.26, which is close to that of concrete pavements in the nine test sections.

The gravel measured in this study was crushed open-graded basalt paved on the yard of the UCPRC facility. It had a maximum aggregate size of 19 mm and was blue/gray in color. This gravel had albedos in the range of 0.12–0.22 with an average of 0.18. The bare soil (native clay between the nine test sections) had an albedo of 0.22. The grass (green lawn) had an albedo of 0.19.

4.5.3 Diurnal Variation of Albedo

To examine the diurnal variation of albedo, the solar reflectivity was continuously monitored over time on one clear day on a concrete pavement (C2) and on another clear day on an asphalt pavement (B2) and is plotted in Figure 4.8. Regarding the measured albedo over time during the day, it is high in the early morning and then low and relatively constant around the middle of the day (9:00–15:00 h) and then low again in the late afternoon when there is a low incident angle of solar radiation as in the early morning. The diurnal variation of solar reflectivity of an asphalt pavement (B2) over three days is plotted in Figure 4.9. The same trend was observed for the concrete pavement results. This implies that the albedo should be measured in the middle of the day (9:00–15:00 h) to obtain a constant and conservative value. Otherwise, if the albedo is measured in the early morning and late afternoon, the value obtained will tend to be larger than that measured in the middle of the day.

4.5.4 Seasonal Variation of Albedo

The solar reflectivity was continuously monitored over one year (fall 2011 to summer 2012) on a permeable asphalt pavement (B2). The albedo on three days of each season is plotted in Figure 4.10. As seen from Figure 4.10, there was no significant change in the measured midday albedo

(a)

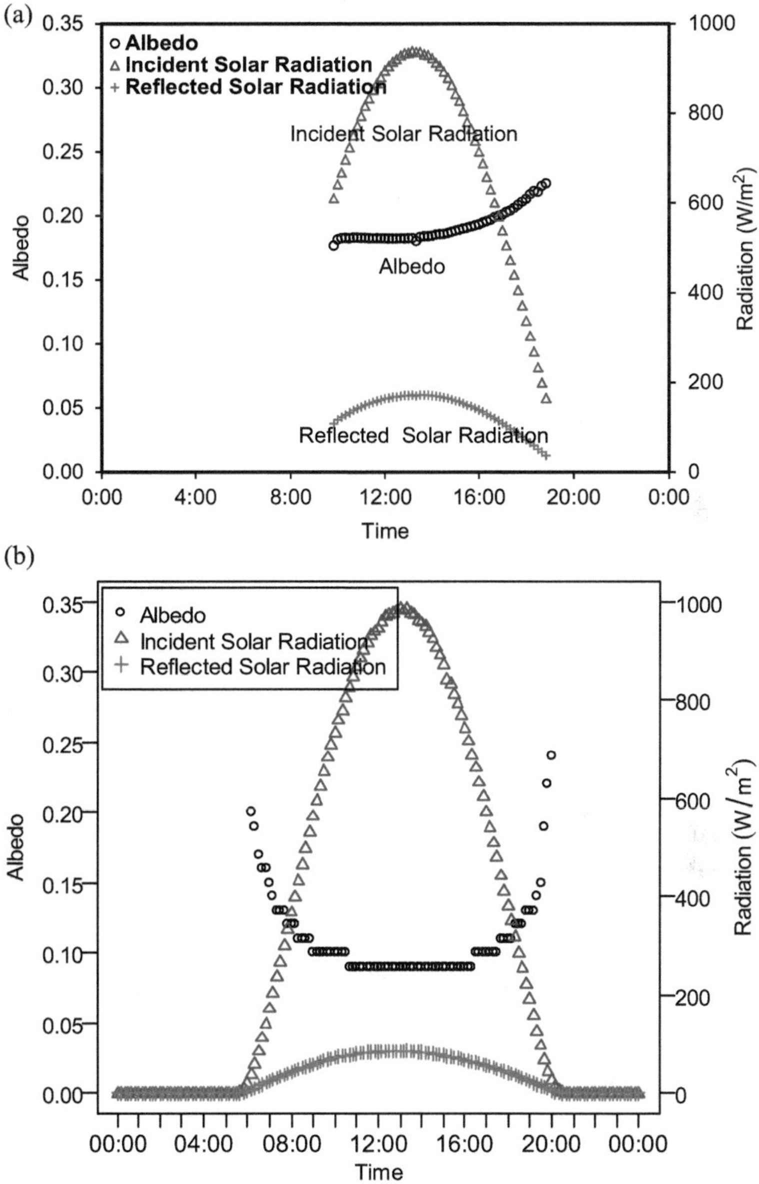

Figure 4.8 Diurnal variation of solar reflectivity during one day. (a) Concrete pavement (C2). (b) Asphalt pavement (B2).

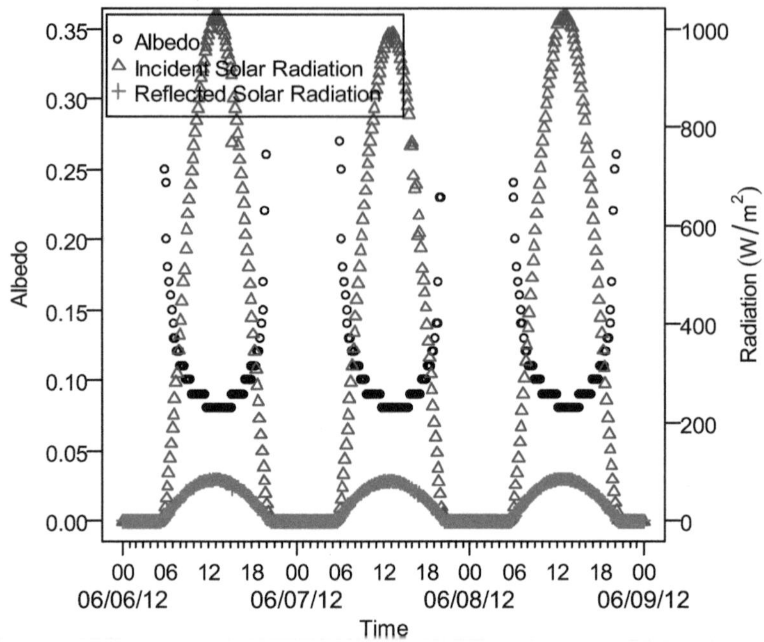

Figure 4.9 Diurnal variation of solar reflectivity over three days (B2). Time is shown as month/day/year.

over a season in a year, except a slightly higher albedo in winter (Figure 4.10(b)), when the angle of the sun was lower.

To verify the impact of season on albedo, the albedos at various times of day on one clear day for each month in one year are plotted against time in Figure 4.11. As found previously, the albedo measured in early morning and late afternoon was higher than that of midday. The midday albedo (11:00–15:00 h in summer, 12:00–15:00 h in winter) was lower and constant over time (falling in one curve in Figure 4.11). As observed from Figure 4.10, it is noted from Figure 4.11 that the midday albedo did not significantly change over season, other than being slightly higher in wintertime.

4.5.5 Change of Albedo over Time

The color of the pavement surface tends to change over time owing to weathering and traffic. Figure 4.12 shows the albedos of nine test sections at various times after construction. These nine test sections were constructed for testing and were not open to any type of traffic. As noted, the albedos of concrete pavements (C1–C3, although albedos of C1 and C3 increased

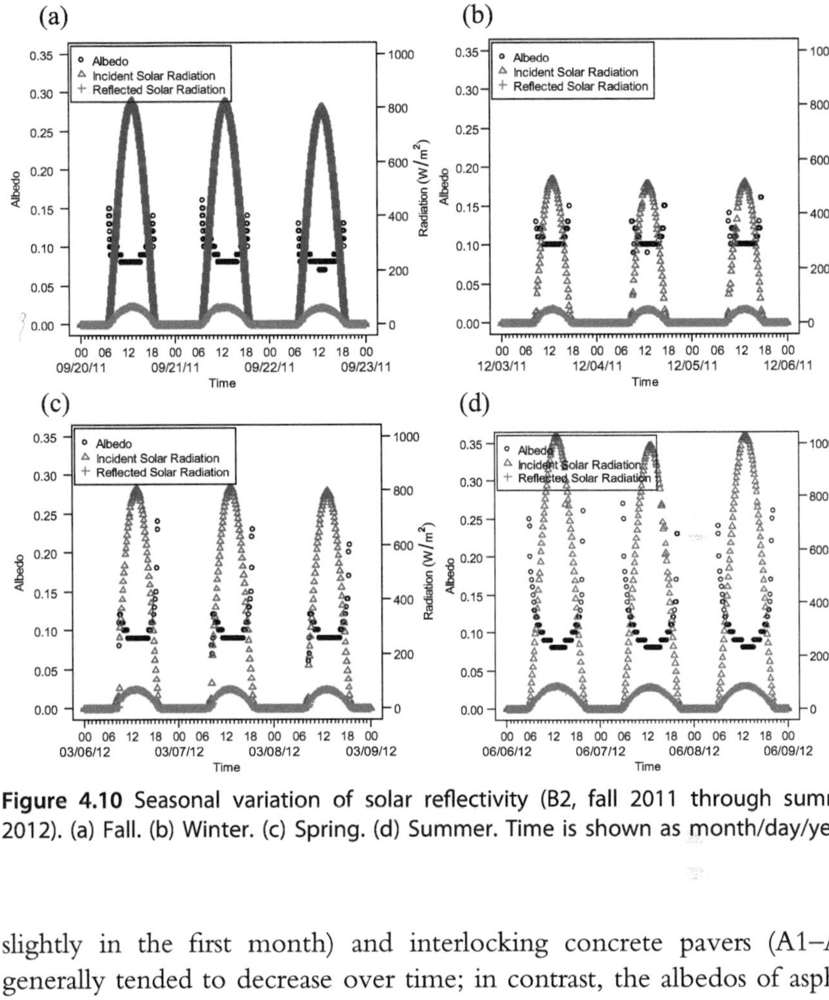

Figure 4.10 Seasonal variation of solar reflectivity (B2, fall 2011 through summer 2012). (a) Fall. (b) Winter. (c) Spring. (d) Summer. Time is shown as month/day/year.

slightly in the first month) and interlocking concrete pavers (A1–A3) generally tended to decrease over time; in contrast, the albedos of asphalt pavements (B1–B3) increased slightly over time. The change in albedo mostly happened in the first month just after the construction owing to weathering. Under both continued weathering and trafficking, the change in albedo is expected to be larger, and for concrete and especially asphalt, traffic will wear the binder (cement or asphalt) off of the surface aggregate, which will result in albedo being influenced by the reflectivity of the aggregate.

4.5.6 Influence of Cloudiness on Albedo

Cloudiness has an important influence on the incident solar radiation, reducing the amount of solar radiation incident to the pavement or other

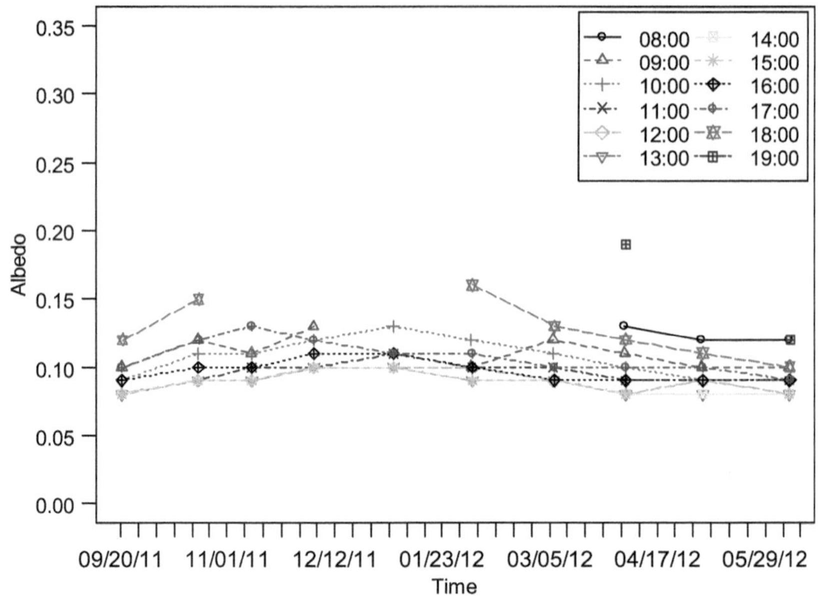

Figure 4.11 Seasonal variation of solar reflectivity at various times of the day (B2). *Note: the data with very low solar radiation (<5 W/m²) but a very high albedo in early morning and late afternoon were neglected.* Time is shown as month/day/year.

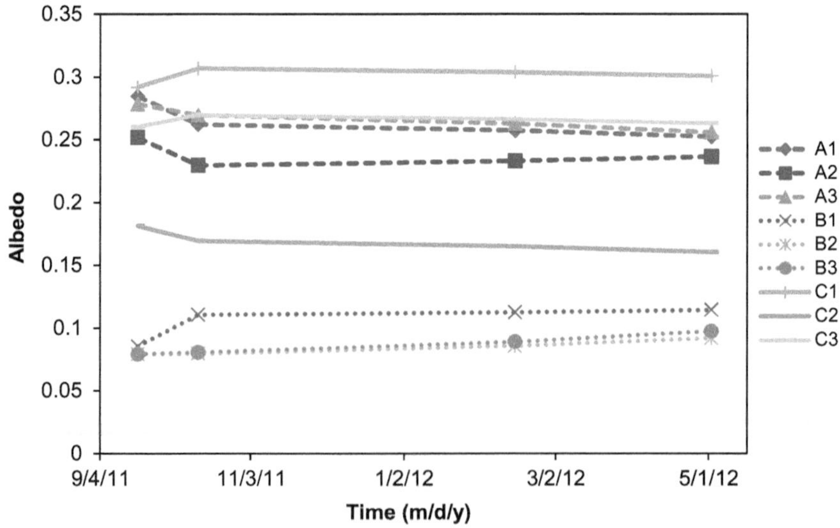

Figure 4.12 Change of solar reflectivity over time. *Nine test sections, only weathered, pavers A1–A3, asphalt pavements B1–B3, and concrete pavements C1–C3.*

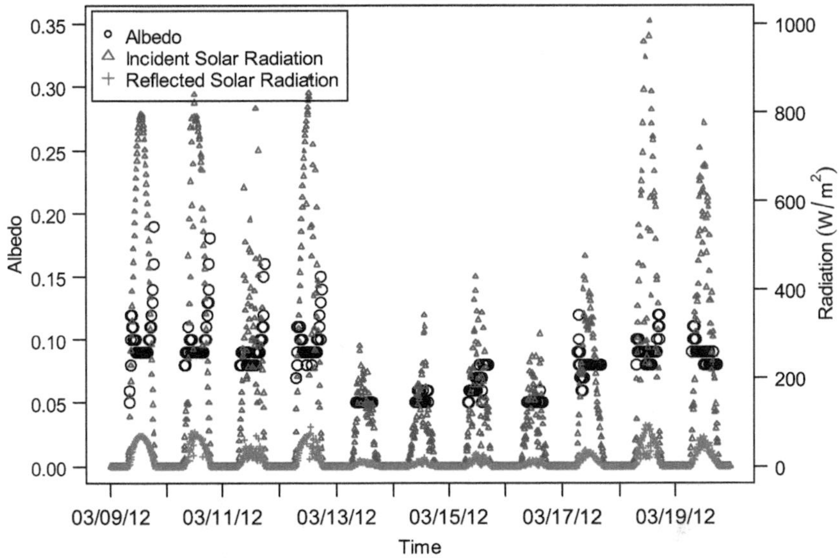

Figure 4.13 Influence of clouds on solar reflectivity (B2). Time is shown as month/day/year.

ground surfaces. To examine the influence of clouds on the albedo of pavements, the albedo of asphalt pavement (B2) on days with different cloud levels was measured for comparison. As shown in Figure 4.13, the cloud levels over the days 9–19 March 2012 were different. The cloud levels on 13–17 March 2012 were much higher than all the other days during this period, during which the incident solar radiation was quite low. The midday albedos ($\sim$0.05) during the cloudy days and resultant low incident solar radiation were much lower than those (0.08) on days with few or no clouds. This implies that the albedo should be measured on a clear day. Otherwise, a lower albedo will be given, even measured at midday.

4.5.7 Influence of Wind Speed and Air Temperature on Albedo

Albedo is expected to be constant regardless of wind speed or air temperature. To verify this concept, the albedo of the asphalt pavement (B2) on some days with different wind speeds and air temperatures was measured for comparison. As shown in Figure 4.14(a), the wind speed and air temperature on 23–25 February 2012 were quite different. The wind speed on 23 February 2012 was much higher than on the other two days.

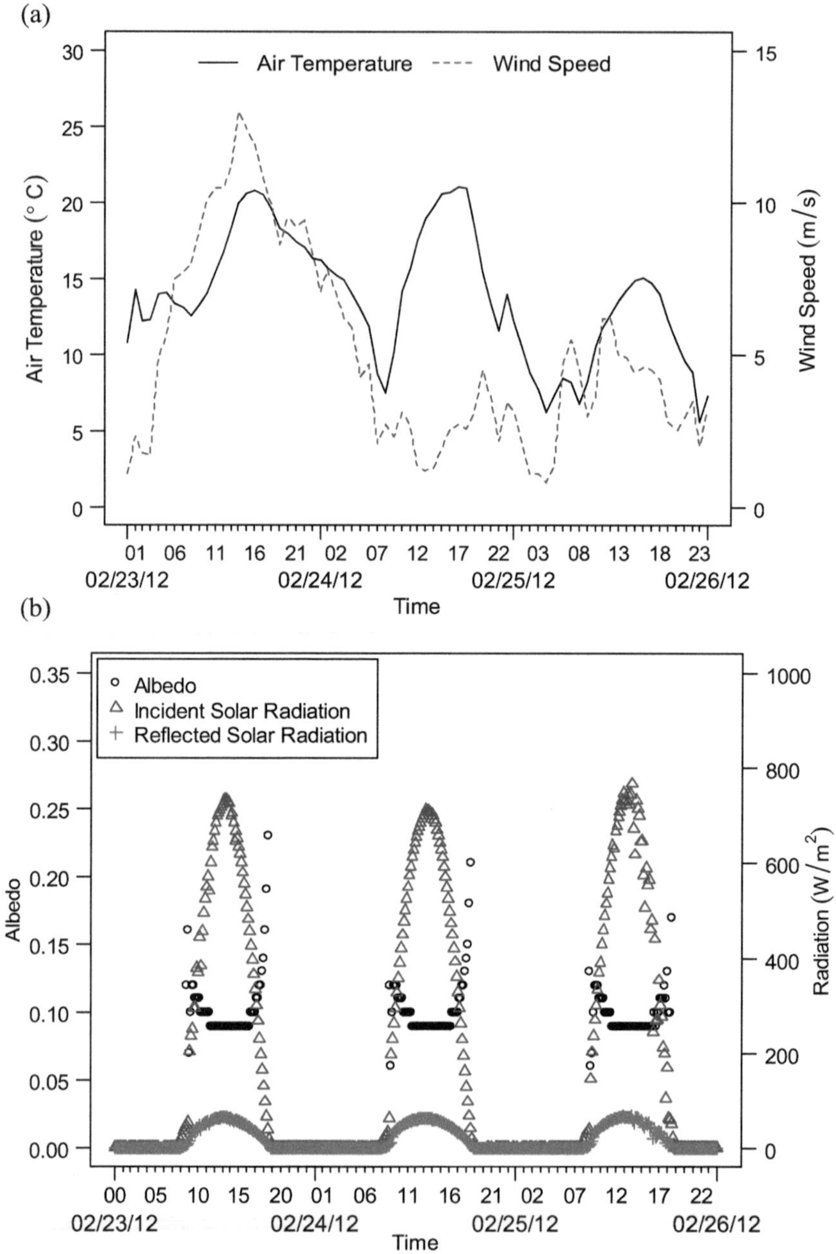

Figure 4.14 Influence of wind speed on solar reflectivity (B2). (a) Wind speed and air temperature on 23–25 February 2012. (b) Albedo and solar radiation on 23–25 February 2012.

The air temperatures on 23 and 24 February 2012 were higher than that on 25 February 2012. The albedo and solar radiation on 23–25 February 2012 are plotted in Figure 4.14(b). The albedos around noon were all around 0.08 on these three days. No significant variation in midday albedo was observed in this period with quite different wind speeds or air temperatures.

4.5.8 Effect of Albedo on Pavement Temperature

As mentioned previously, albedo plays an important role in pavement temperature. To examine the influence of the albedo on the pavement temperatures, the high (T_{max} at 3:00 PM) and low (T_{min} at 6:00 AM) temperatures of the nine experimental pavements with different albedos were measured from fall 2011 through summer 2012 at Davis, California. The temperatures for one day of summer (1 July 2012) and winter (15 January 2012) were selected and plotted against their albedos in Figure 4.15. This figure illustrates that the albedo has a large influence on the high temperature in both summer and winter. Under the same weather conditions in summer, the high temperature of pavement with low albedo (0.08 for the asphalt) was about 15 °C (65 vs 50 °C) higher than that

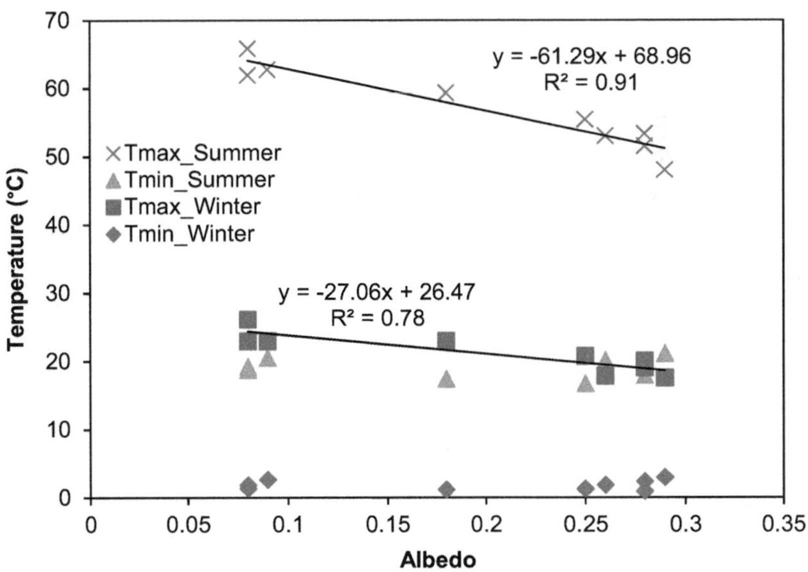

Figure 4.15 Influence of solar reflectivity on pavement surface temperature. *Summer on 1 July 2012 with daytime peak solar radiation approximately 1000 W/m² and winter on 15 January 2012 with daytime peak solar radiation approximately 500 W/m².*

of pavement with high albedo (0.28 for concrete). In winter, this high temperature difference was approximately 8 °C (26 vs 18 °C).

An increase in albedo of 0.1 can reduce the high surface temperature in summer by approximately 6 °C (i.e., −6 °C/+0.1 albedo, which is approximately the slope of the fitting line for T_{max}_Summer in Figure 4.15), and by approximately 3 °C (i.e., −3 °C/+0.1 albedo, which is approximately the slope of the fitting line for T_{max}_Winter in Figure 4.15) in winter.

This reveals that albedo has different effects on pavement temperature in summer and winter with different peak solar radiations. The peak solar radiation intensity at around 1:00 PM is approximately 1000 W/m^2 in summer in Davis, California. It is approximately halved to 500 W/m^2 in winter. The temperature effect of the albedo in winter is also approximately half of that in summer (−6 °C/+0.1 albedo in summer and −3 °C/+0.1 albedo in winter). This implies that solar radiation positively influences the cooling effect of increased albedo (CE_{albedo}), and the cooling effect of increased albedo in hot climates with high solar radiation is larger than that in climates with low solar radiation. This also can be seen from the insignificant effect of albedo (∼0 °C/+0.1 albedo) on the low temperature during nighttime with no solar radiation for both summer and winter.

To further illustrate the influence of solar radiation on the cooling effect of an increased albedo, the cooling effects of increased albedo on high temperatures in each month of a year were calculated from measured temperature data (just obtained from the slopes of the fitting lines between surface temperature and surface albedo for each month as shown in Figure 4.16). These calculated cooling effects of increased albedo on high temperatures in each month of a year are presented in Figure 4.16(a) along with the peak solar radiation intensity in each month. It is revealed that the cooling effect of increased albedo does change over seasons and thus changes with the peak solar radiation intensity. The cooling effect is higher in summer, with high peak solar radiation intensity, and lower during winter, with low peak solar radiation. The correlation between the cooling effect of increased albedo and solar radiation is shown in Figure 4.16(b) and with a best-fit linear relation in Eq. (4.1) below. An increase in peak solar radiation intensity of 100 W/m^2 will produce an increase in cooling effect on the surface temperature of albedo of 0.6 °C/0.1 change in albedo.

$$CE_{albedo} = 6q \quad (R^2 = 0.9) \tag{4.1}$$

In Eq. (4.1) q is the peak global solar radiation flux hitting the pavement (kW/m^2) and CE_{albedo} is the cooling effect of albedo increased by 0.1 on

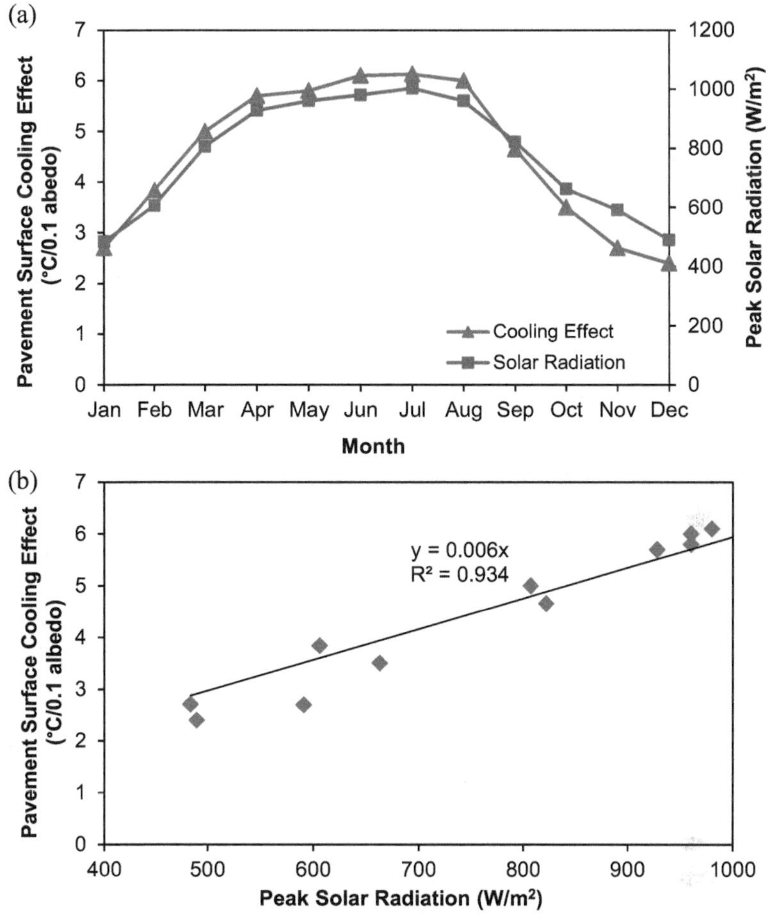

Figure 4.16 Influence of solar radiation on cooling effect of increased albedo. (a) Cooling effect of increased albedo and peak solar radiation intensity in different months. (b) Correlation between cooling effect of increased albedo and solar radiation.

pavement temperature (°C/0.1 albedo). This empirical relationship can be used to roughly estimate the cooling effect of increased albedo or heating effect of reduced albedo on pavement for various climates and seasons with different solar radiation.

4.6 SUMMARY AND CONCLUSIONS

Nine 4-m by 4-m (13-ft by 13-ft) experimental sections were designed and constructed at Davis, California, including asphalt pavement, concrete pavement, and interlocking concrete paver, with a focus on permeable structures with impermeable pavements as controls. These specifically built test sections (with some other existing pavements) were used to measure the fundamental materials properties, including albedo, permeability, thermal properties, and evaporation rate (for some of the materials used in the experimental sections), and empirically examine the thermal behavior of various pavement types in different seasons and under different moisture conditions and their impacts on near-surface air and building surfaces, the results of which are presented in Chapters 4–10.

Field measurements of albedo for various pavement materials were performed on experimental test sections and other existing pavements. The albedos for various materials were compared, and the factors affecting the field measurement of albedo and the diurnal and seasonal changes in albedo were examined. The effect of albedo on pavement temperature was also examined using measured data of albedo and pavement surface temperatures. This study contributes three developments to research on pavement albedo and thermal performance: (1) a new albedo measurement system, (2) new data documenting differences in albedo across pavement types and over pavement surfaces, and (3) a correlation between the cooling effect of increased albedo and solar radiation. Main findings include:

1. A new albedo measurement system using a dual pyranometer and an automatic data collection system was developed. It can be used to conveniently measure albedo in the field and perform long-term monitoring of albedo when connected to a data acquisition system.
2. Albedo was measured for commonly used paving materials including asphalt, concrete, and interlocking concrete paver surfacing materials, with different designs. These new data enhance basic knowledge on albedo values for pavements as well as the other land-cover types measured (gravel, soil, and lawn), which can help reduce the uncertainty in understanding, evaluating, and modeling their thermal behavior and their consequences for human thermal comfort and building energy use.
3. This study found that the measured albedo of pavement materials is high in the early morning and in the late afternoon; it is low and constant over time during midday. This suggests that the albedo should be measured during midday on a clear day to obtain a stable and conservative value. No significant seasonal variation in albedo was found. Pavement albedo will change over time under weathering and trafficking, especially in the first month after construction. Cloud cover will

negatively influence the value of albedo measured. No impact of wind speed or air temperature on albedo was observed.

4. Albedo has a great influence on the pavement surface high temperatures in the daytime (6 °C per 0.1 albedo change in a hot sunny day with solar radiation of 1000 W/m^2) and no significant impact on pavement low temperatures in the nighttime. An empirical relationship between the cooling effect of increased albedo on a pavement high temperature and solar radiation was developed. The cooling effect has a positive correlation with the peak solar radiation intensity in the daytime. This simple correlation can help estimate the cooling effect of increased albedo or heating effect of reduced albedo on pavement for various climates and seasons with different solar radiation.
5. An increase in albedo indeed can reduce the pavement surface temperature and thus might help mitigate the surface, near-surface, and atmospheric heat island effects, potentially improve outdoor thermal comfort, and reduce building cooling energy use during hot periods, depending on the characteristics of the location and the overall urban area (see the list of additional factors in the introduction). While the urban heat island effect is a large concern for many cities in hot climates, and of course is a great concern in the summer, winter effects of pavement albedo and temperature are also important to consider. The reduced temperature during winter might increase the outdoor thermal discomfort and building energy use for heating in some climates and locations. Building energy use is also highly dependent on whether air conditioners are widely used. There is a chance that this penalty (i.e., colder pavements) in winter might outweigh the benefits gained in summer in some regions.
6. This study looked at pavement surface temperature with a focus on the effects of solar radiation and albedo. It must be noted that near-surface temperature is also influenced by wind speed, with higher wind speeds reducing near-surface temperatures. However, compared to solar radiation the effect of wind speed would be much smaller. The effect of wind speed was assumed to be identical for each case and not explicitly included in the simple model of cooling effect proposed in this study. For developing a more comprehensive model, however, the wind speed can be included, as well as thermal proprieties of pavement materials (i.e., thermal conductivity and heat capacity).
7. One additional factor that was not measured as part of this research is the increased reflected solar radiation caused by high albedo, which might

hit and be absorbed by surrounding people or building/vehicle surfaces. This could increase the human thermal discomfort and building cooling energy use during hot periods. Therefore, attention should be given to the complete assessment of both the benefit and penalty over one entire year to ensure that a positive net benefit will be obtained. The optimal strategy for changing albedo in a hot climate might be to increase the albedo during summer and reduce or not change it in winter for certain places. This intention is to maximize the benefit in summer and minimize the penalty in winter.

CHAPTER 5

Permeable Pavements and Permeability

Contents

5.1 INTRODUCTION

As mentioned previously, permeable pavements have many environmental benefits beyond those of conventional impermeable pavements, including reducing storm-water runoff, improving water quality, and recharging underground water [54,55]. In addition to these already identified benefits, permeable pavements, as one potential cool pavement type, also could be an effective solution for improving outdoor thermal environment and mitigating the heat island effect in hot climates, thereby reducing the associated impacts. The major potential cooling mechanism for permeable pavements is evaporative cooling.

The study of cool pavements of this dissertation focuses on permeable pavements, including various surfacing types (paver, concrete, and asphalt). Permeability is a very important property for permeable pavement, which influences the hydraulic performance of the permeable pavement as well as

Pavement Materials for Heat Island Mitigation
ISBN 978-0-12-803476-7
http://dx.doi.org/10.1016/B978-0-12-803476-7.00005-2

the thermal performance related to evaporation (see Chapter 7 for more details on evaporation rate). This chapter is devoted to the measurement of permeability for the permeable pavements of the experimental sections.

5.1.1 Review and Background

As the application of fully permeable pavement increases, the need for a proper tool to measure the functionality of permeable pavement, especially with respect to clogging, is also increasing [1,53,115,121–124]. One way to assess the infiltration capacity of pavement and evaluate the hydraulic performance without surface overflow is through permeability measurement. Because there is at this time no single standard method, multiple methods have been employed to measure the permeability of porous asphalt, pervious concrete materials, and pavements in existing studies [38].

Most of the existing studies were based on a single permeability measurement method on multiple pavements or multiple permeability devices used on a single pavement surface. In one study Prowell and Dudley [123] compared the National Center for Asphalt Technology (NCAT) field permeameter with a laboratory permeameter using the American Society for Testing and Materials (ASTM) standard method PS129-01 for asphalt pavement. Differences in results were observed because the laboratory measurements were performed on core specimens that were disturbed, while the field measurements were done on an intact pavement. Separate comparative studies performed by Maupin [125,126] and by Gogula [200] showed inconsistency between the field and the laboratory results, and generally the permeability values for asphalt pavements with higher void ratios were higher under laboratory measurement compared to measurement of the same materials in the field.

The difference in permeability results is mostly due to unrestricted horizontal flow in field permeability tests, which is not present in the laboratory tests. In one more recent study, Williams [124] tested three field permeameters and compared the results. The three field permeameters compared were the NCAT field permeameter, the Kuss field permeameter, and the Kuss vacuum permeameter. The NCAT field permeameter uses the falling-head method, while the Kuss field permeameter uses a constant-head approach. The permeability results from the different permeameters did not always agree for most pavement sections. In general, however, the performance of the NCAT permeameter for permeability measurement of porous asphalt in the field was found to be more practical and produced results with low variability compared to the other devices.

In late 2009, a new standard method, ASTM C1701, was developed to specifically measure the permeability of pervious concrete in the field. At present, no published peer-reviewed article is available to show any laboratory or field data using this new standard method. There also does not appear to be any comparison of the results obtained by ASTM C1701 and the NCAT method in the literature for either porous asphalt or pervious concrete pavements or permeable interlocking concrete pavers. It is important for pavement designers and maintenance engineers to be able to compare the permeability of various pavement types and to be able to translate results between the different methods for input into hydraulic models and hydraulic design equations.

5.1.2 Objectives

This study was undertaken with the following objectives:

1. To measure the permeability of test sections of porous asphalt, pervious concrete, and permeable interlocking concrete paver pavements using both the ASTM C1701 and the NCAT methods.
2. To evaluate the relative reliability of both measurement techniques.
3. To determine the correlation between the measured values by the ASTM C1701 and NCAT methods.
4. To evaluate the implications of using results from the different test methods for design and performance evaluation of permeable pavement.

5.2 METHODS

5.2.1 Description of the National Center for Asphalt Technology Method

The NCAT permeameter was developed by the NCAT in late 1990 [127]. While this method has not been standardized, it is widely used and their application for pervious asphalt and thin layers of open-graded asphalt on impermeable pavement is very common. A photo of the NCAT permeameter during field measurement is shown in Figure 5.1. It can be seen that this permeameter comprises four tiers with different inside diameters. Ordinarily, if the pavement is highly permeable, then tier 4 is sufficient. Other tiers are added when the permeability is low and a longer time is needed to complete the test.

The permeability test is performed based on the falling-head method by filling the cylinder in tier 4 or in tier 3, 2, or 1 with water and measuring the time required for water to fall the height of the permeameter tiers. The

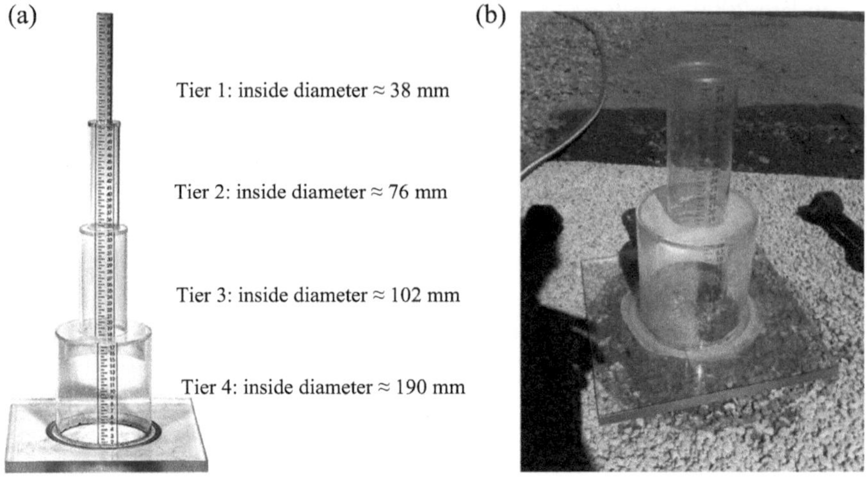

Figure 5.1 NCAT field permeameter. (a) Four cylindrical tiers with different inside diameters. (b) Photo view of the permeameter with bottom two tiers used under field operation for this study.

collected data are then used to calculate the coefficient of permeability, *K*, also known as the saturated hydraulic conductivity, using the following relationship:

$$K_s = \frac{al}{At} \ln\left(\frac{h_1}{h_2}\right) \tag{5.1}$$

where

K_s is the saturated hydraulic conductivity, in cm/s;
a is the inside cross-sectional area of the inlet standpipe, in cm^2;
l is the thickness of the permeable (porous asphalt or pervious concrete) pavement, in cm;
A is the cross-sectional area of tested pavement, in cm^2;
t is the average elapsed time of water flow between timing marks $(t_1 - t_2)$, in s;
h_1 is the hydraulic head on the pavement at time t_1, in cm; and
h_2 is the hydraulic head on the specimen at time t_2, in cm.

One important aspect of an accurate permeability measurement under field conditions is to ensure that there is no water leaking between the pavement surface and the base of the permeameter. The NCAT method specifies the use of plumbing putty to prevent water leakage. However, previous and current field measurements performed in this study verified that putty does not adequately prevent water leakage [122]. To do a better job of sealing the permeameter to the pavement and to effectively prevent leaking, the alternative used was Ecoflex 5 silicone rubber by Smooth-On®. To correctly seal the NCAT permeameter to the pavement, a caulking gun was used to fill a ring at the base of the permeameter and the permeameter was quickly turned over and placed flat on the pavement surface. Small weights were placed around the base of the permeameter to help create a uniform seal. Depending on the air and pavement temperatures, after roughly 3–5 min, a quick check was performed by tapping the base of the permeameter to see if it would move. If the permeameter did not move, that meant the sealant was ready for permeability measurement. If water leaks were present for any reason, the test was terminated.

5.2.2 Description of the American Society for Testing and Materials Method C1701

The ASTM C1701 test method was developed under the jurisdiction of ASTM Technical Committee C09 on Concrete and Concrete Aggregates and is the direct responsibility of Subcommittee C09.49 on Pervious Concrete. The test was approved in August 2009 and published in September 2009. No device for conducting ASTM C1701 is currently commercially available, unlike the NCAT permeameter.

The principal specifications of the ASTM C1701 permeameter and the device used for this study are shown in the Figure 5.2. The permeability test was performed based on the constant-head method. The procedure of the test is as follows: (1) secure the permeameter on the pavement with sealant (plumber's putty is recommended) to prevent water leakage; (2) prewet the pavement by pouring about 3.6 kg of water into the ring at a rate sufficient to maintain a head between the two marked lines at a distance of 10 and 15 mm from the bottom, respectively, until all 3.6 kg of water has been

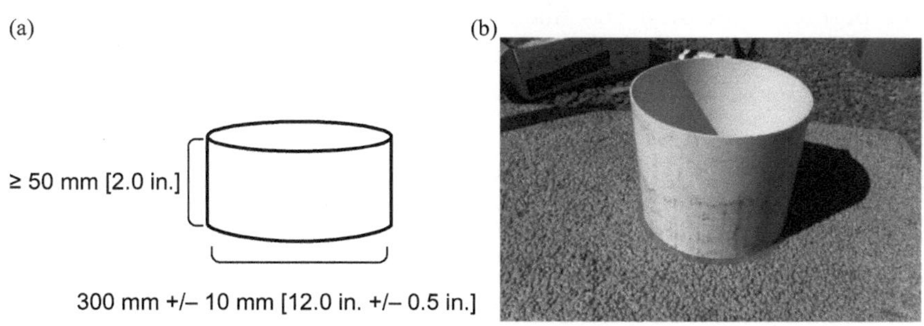

Figure 5.2 ASTM C1701 field permeameter. (a) Specified dimensions and (b) photo view of the permeameter used under field operation for this study.

used; (3) begin timing as soon as the water contacts the pervious pavement surface and stop timing when free water is no longer present on the pervious surface, and record the elapsed time for prewetting to the nearest 0.1 s; (4) run the actual test (same as items (2) and (3) above) by using 3.6 kg water if the prewetting elapsed time is not less than 30 s, otherwise use 18 kg water and record the appropriate weight of water *M* and the elapsed time *t*; and finally (5) calculate the infiltration rate (aka coefficient of permeability, hydraulic conductivity) by using the following formula:

$$I = \frac{KM}{D^2 t} \tag{5.2}$$

where

I is the infiltration rate, in mm/h;

M is the mass of infiltrating water, in kg;

D is the inner diameter of the infiltration ring, in mm;

t is the time required for the measured amount of water to infiltrate the pavement, in s; and

K is the constant factor, 4,583,666,000, in SI units.

As indicated earlier, the proper sealing of the permeameter is extremely important for accurate measurement of the infiltration rate or permeability coefficient. The experience during the initial measurements for this study revealed that the use of plumber's putty as a sealant did not fully prevent water leakage and also left some residue on the pavement that could not be easily removed. To prevent these problems the same silicone material that was used in the NCAT method was used for the ASTM C1701 testing. To do this correctly, the ASTM permeameter was placed on the surface of the pavement and a caulking gun was used to fill the outer ring at the base of the permeameter, and the silicone was allowed to dry for approximately 3–5 min, depending on the air and pavement temperatures. The permeameter then was filled with about 5 kg of water and checked for any obvious water leakage. If no water leakage was observed, the actual permeability measurement was performed.

5.2.3 Experimental Plan and Data Collection

The permeability tests for this study were performed on the six 4-m by 4-m permeable test sections constructed at the University of California Pavement Research Center test facilities in Davis, California. The summary of each test section along with pavement design mix and other characteristics is provided in Table 5.1 and previously in Figure 4.3.

Table 5.1 Experimental test plan for field permeability measurements

Section	Pavement layer Type	Pavement layer Thickness (cm)	Base layer Type	Base layer Thickness (cm)	Test locations	Test method
A2	IB-P	10[a]	AB-O	15	SE, NE, NW, SW, CT	ASTM, NCAT
A3	IB-P	10[b]	AB-O	30	SE, NE, NW, SW, CT	ASTM, NCAT
B2	AC-O	10	AB-O	30	SE, NE, NW, SW, CT	ASTM, NCAT
B3	AC-O	20	AB-O	30	SE, NE, NW, SW, CT	ASTM, NCAT
C2	PCC-O	10	AB-O	30	SE, NE, NW, SW, CT	ASTM, NCAT
C3	PCC-O	20	AB-O	30	SE, NE, NW, SW, CT	ASTM, NCAT

IB-P, interlocking block paver–permeable; AC-O, asphalt concrete–open graded; PCC-O, portland cement concrete–open graded (two types); AB, aggregate base–open graded. SE, southeast; NE, northeast; NW, northwest; SW, southwest; CT, center.
[a]Includes the thickness of the paver (6 cm) and underlying bedding layer (4 cm).
[b]Includes the thickness of paver (8 cm) and underlying bedding layer (2 cm).

Both test methods (ASTM and NCAT) were used to measure the permeability of all permeable pavement test sections (interlocking pavers, asphalt, and concrete). For each test section, the permeability measurement was made at five different locations: the southeast corner (SE), northeast corner (NE), northwest corner (NW), southwest corner (SW), and center (CT). Triplicate measurements were performed at each location for both ASTM and NCAT methods. A summary of the experimental test plan for the various pavement types and designs is shown in Table 5.1.

5.2.4 Statistical Analysis

Box-and-whisker plots were used as a descriptive statistical method to analyze and interpolate the permeability data. This method is a convenient way of graphically depicting groups of numerical data through their five-number summaries: the smallest measurement (sample minimum), lower quartile (25%, Q1), median (50%, Q2), upper quartile (75%, Q3), and largest measurement (sample maximum). A box-and-whisker plot may also indicate which observations, if any, might be considered outliers. In addition, a descriptive statistics analysis was performed for each test section and the coefficient of variation (CV) was used to compare the relative repeatability of the two test methods. Finally, linear regression analysis was performed to evaluate the correlation between the ASTM and the NCAT methods.

5.3 RESULTS

The statistical summary results for all locations within each test section are presented in Figure 5.3. As shown, the summary results include the number of measurements, mean, standard deviation, minimum, lower quartile (25%, Q1), median (50%, Q2), upper quartile (75%, Q3), and maximum. The entire permeability results measured by both methods for each test section are shown in Table 5.2. Generally, the mean was higher for the NCAT permeameter compared with ASTM C1701. In addition, larger variability was noticed with the NCAT permeameter measurements

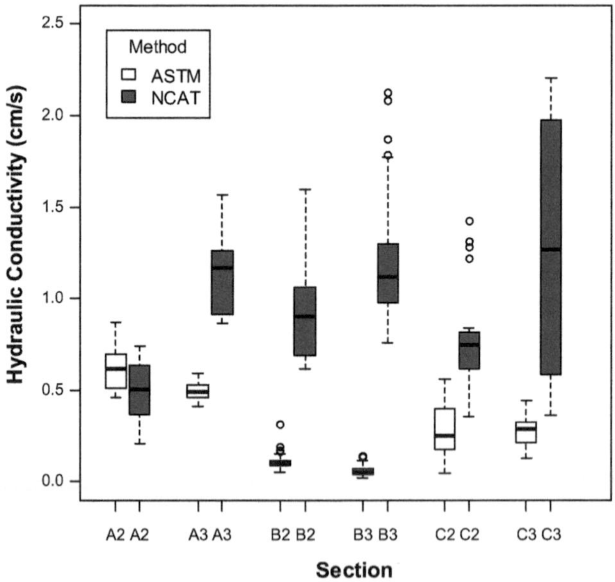

Figure 5.3 Box and whisker plot of permeability data for all test section measurements using the ASTM and NCAT methods.

Table 5.2 Summary statistics of all permeability measurements by ASTM and NCAT methods

	ASTM C1701 (cm/s)						NCAT permeameter (cm/s)						
Test ID	**N**	**Mean**	**SD**	**Min**	**Median**	**Max**	**Test ID**	**N**	**Mean**	**SD**	**Min**	**Median**	**Max**
A2-CT	7	0.53	0.03	0.49	0.50	0.54	A2-CT	4	0.22	0.01	0.21	0.22	0.23
A2-NE	7	0.65	0.03	0.61	0.63	0.65	A2-NE	4	0.37	0.01	0.37	0.37	0.37
A2-NW	7	0.63	0.05	0.57	0.59	0.64	A2-NW	4	0.50	0.01	0.49	0.50	0.51
A2-SE	8	0.80	0.05	0.71	0.76	0.82	A2-SE	4	0.62	0.02	0.59	0.62	0.63
A2-SW	7	0.49	0.02	0.46	0.48	0.51	A2-SW	4	0.73	0.01	0.72	0.72	0.73
A2 (all)	**36**	**0.62**	**0.12**	**0.46**	**0.52**	**0.62**	**A2 (all)**	**20**	**0.49**	**0.18**	**0.21**	**0.37**	**0.51**
A3-CT	9	0.47	0.04	0.41	0.46	0.48	A3-CT	9	1.13	0.23	0.89	0.91	1.16
A3-NE	3	0.46	0.03	0.44	0.44	0.45	A3-NE	4	0.89	0.02	0.87	0.88	0.89
A3-NW	3	0.54	0.02	0.53	0.53	0.54	A3-NW	4	1.19	0.05	1.13	1.16	1.19
A3-SE	3	0.55	0.04	0.53	0.53	0.54	A3-SE	5	1.14	0.08	1.04	1.06	1.17
A3-SW	3	0.51	0.06	0.45	0.49	0.53	A3-SW	4	1.42	0.12	1.32	1.32	1.39
A3 (all)	**21**	**0.50**	**0.05**	**0.41**	**0.46**	**0.49**	**A3 (all)**	**26**	**1.15**	**0.21**	**0.87**	**0.92**	**1.17**
B2-CT	8	0.15	0.07	0.09	0.11	0.11	B2-CT	4	0.84	0.15	0.73	0.76	0.78
B2-NE	6	0.10	0.01	0.09	0.10	0.10	B2-NE	4	1.44	0.10	1.36	1.40	1.41
B2-NW	5	0.11	0.04	0.08	0.08	0.10	B2-NW	5	0.98	0.14	0.90	0.91	0.93
B2-SE	5	0.07	0.02	0.05	0.06	0.06	B2-SE	4	0.70	0.12	0.61	0.62	0.66
B2-SW	5	0.12	0.04	0.09	0.09	0.10	B2-SW	4	0.72	0.14	0.63	0.63	0.66
B2 (all)	**29**	**0.11**	**0.05**	**0.05**	**0.09**	**0.10**	**B2 (all)**	**21**	**0.94**	**0.30**	**0.61**	**0.69**	**0.90**
B3-CT	6	0.07	0.02	0.05	0.05	0.06	B3-CT	7	1.11	0.06	0.99	1.11	1.13
B3-NE	4	0.06	0.02	0.05	0.05	0.06	B3-NE	7	1.65	0.32	1.28	1.35	1.77
B3-NW	7	0.03	0.01	0.02	0.02	0.03	B3-NW	10	0.93	0.16	0.76	0.80	0.90

Continued

Table 5.2 Summary statistics of all permeability measurements by ASTM and NCAT methods—cont'd

	ASTM C1701 (cm/s)							NCAT permeameter (cm/s)					
Test ID	**N**	**Mean**	**SD**	**Min**	**Median**	**Max**	**Test ID**	**N**	**Mean**	**SD**	**Min**	**Median**	**Max**
B3-SE	4	0.13	0.01	0.12	0.13	0.13	B3-SE	4	1.49	0.40	1.23	1.29	1.32
B3-SW	4	0.05	0.01	0.05	0.05	0.05	B3-SW	8	1.06	0.11	0.93	0.98	1.06
B3 (all)	**25**	**0.06**	**0.04**	**0.02**	**0.04**	**0.06**	**B3 (all)**	**36**	**1.20**	**0.34**	**0.76**	**0.98**	**1.12**
C2-CT	3	0.23	0.09	0.16	0.17	0.19	C2-CT	8	0.74	0.06	0.64	0.69	0.75
C2-NE	3	0.30	0.10	0.23	0.24	0.25	C2-NE	4	0.71	0.23	0.36	0.70	0.82
C2-NW	3	0.10	0.07	0.05	0.07	0.09	C2-NW	4	0.55	0.08	0.50	0.50	0.51
C2-SE	5	0.44	0.07	0.38	0.40	0.42	C2-SE	4	1.31	0.09	1.22	1.27	1.30
C2-SW	3	0.19	0.11	0.09	0.13	0.17	C2-SW	4	0.64	0.08	0.57	0.60	0.62
C2 (all)	**17**	**0.28**	**0.15**	**0.05**	**0.17**	**0.25**	**C2 (all)**	**24**	**0.78**	**0.27**	**0.36**	**0.62**	**0.75**
C3-CT	4	0.14	0.01	0.13	0.13	0.14	C3-CT	8	0.47	0.10	0.36	0.38	0.46
C3-NE	6	0.43	0.01	0.41	0.42	0.43	C3-NE	5	2.04	0.10	1.93	1.99	2.02
C3-NW	8	0.22	0.01	0.21	0.22	0.22	C3-NW	4	1.27	0.02	1.24	1.26	1.27
C3-SE	7	0.32	0.01	0.32	0.32	0.32	C3-SE	4	2.01	0.03	1.98	1.98	2.00
C3-SW	8	0.28	0.02	0.26	0.27	0.29	C3-SW	8	1.23	0.13	1.11	1.12	1.19
C3 (all)	**33**	**0.29**	**0.09**	**0.13**	**0.22**	**0.29**	**C3 (all)**	**29**	**1.27**	**0.61**	**0.36**	**0.59**	**1.27**

compared with those from ASTM C1701. While in general the variability remained within a close range, for test section C3 the variability was much higher for the NCAT method. For example, the hydraulic conductivity measured by the NCAT permeameter at the center of section C3 (C3-CT) was 0.468 cm/s. However, the hydraulic conductivities measured by the NCAT permeameter at the southeast (C3-SE) and northeast (C3-NE) sections of the same test section were 2.005 and 2.039 cm/s, respectively.

5.4 DISCUSSION

This section is devoted to the discussion of questions related to: (1) repeatability of measurements made using both methods, (2) influence of the operator on the measurement values, and (3) correlation between the measurement values obtained by the two methods.

5.4.1 Measurement Variability

The CV was used to evaluate the measurement variability of both the NCAT permeameter and the ASTM C1701 methods. The relationship between CV and mean permeability values for all locations within the test sections is shown in Figure 5.4.

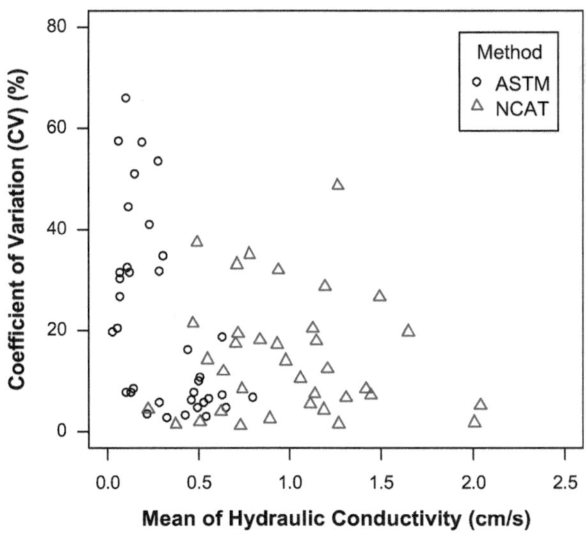

Figure 5.4 Measurement repeatability based on coefficient of variation for permeability values measured using the ASTM and NCAT methods.

Generally, the mean was higher for the NCAT method compared with ASTM C1701. While in general the variability remained within a close range, for test section C3 the variability was much higher for the NCAT method. Variability can also be looked at in terms of the CV. The relationship between the CV and the mean permeability values for all locations on all test sections is shown in Figure 5.4. As shown, the CV is generally higher for the ASTM C1701 method compared with the NCAT permeameter method, although the standard deviation is lower. The CV is higher for the ASTM C1701 method because the mean permeabilities are lower than the means for the NCAT permeameter. The average CV for the ASTM C1701 method was around 30%, while the average CV value for the NCAT permeameter was around 10%. Also, as a general trend, the CV decreased as the mean permeability increased. When the mean permeability was higher than 0.5 cm/s, the CV values for both methods were comparable. From this it appears that the variability of measurement by the NCAT method was slightly superior compared with the ASTM C1701 method. However, using the standard deviation as the basis for accuracy, then the ASTM C1701 method produces less variable results.

5.4.2 Influence of Operator on Permeability Measurement

The repeatability or variability of permeability values with respect to multiple operators is another important parameter for evaluating the measurements made using the ASTM C1701 and the NCAT permeameter methods. To perform this comparative evaluation, the permeability measurements were performed by three operators on test section A3 at the center location (A3-CT). The results of these measurements are shown Figure 5.5. As shown, it is clear that the measurement values are comparable and the variability between measurements by all operators is relatively low for both methods. Generally, however, the variability of measurements made by different operators for ASTM C1701 is slightly lower than that of the NCAT permeameter method. This lower variability in measurement may be partially due to the larger permeameter diameter and the use of constant head.

5.4.3 Correlation between ASTM and NCAT Methods

A linear regression analysis was performed to assess a possible correlation that may exist between the permeability measurements made using the ASTM and NCAT methods. To do this analysis, the mean permeability measured from all test sections was used. The result of the analysis is shown

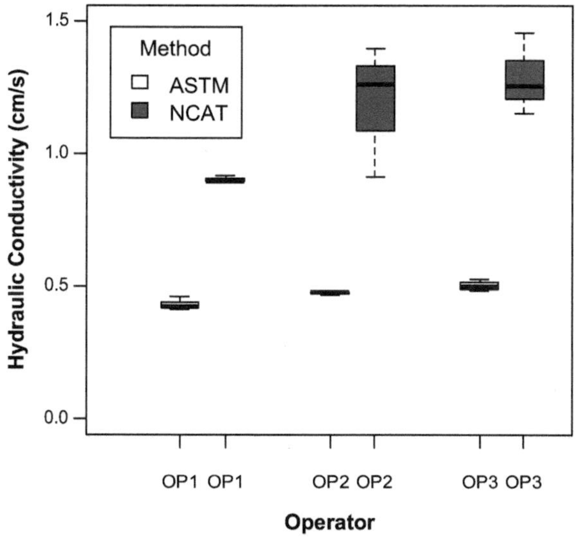

Figure 5.5 Influence of operator on permeability measurements made using the ASTM and NCAT methods.

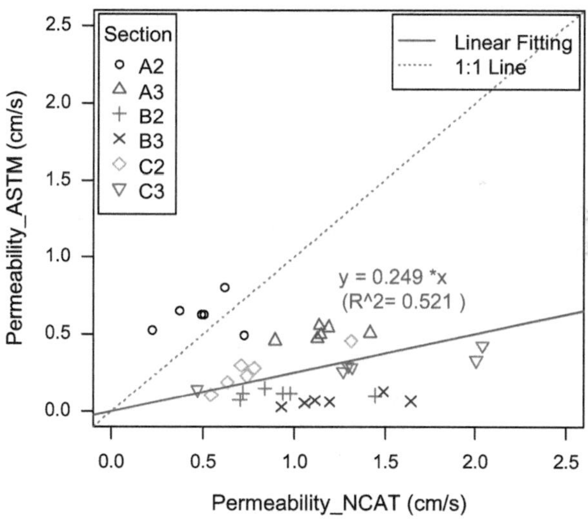

Figure 5.6 Correlation between the permeability value measurements using the ASTM and NCAT methods.

in Figure 5.6. As noted, a relatively weak correlation ($R^2 = 0.521$) was found between measurement values made using the NCAT and ASTM methods. From this relationship, the permeability measurement values obtained with the ASTM C1701 method are lower, generally about 25% of

Table 5.3 Correlation between the permeability values measured using the ASTM C1701 and NCAT permeameter methods for various pavement materials

Material	Test section	Coefficient α	Std. error	t value	Pr(>|t|)	R^2
All pavement types	All	0.25	0.04	6.17	4.60E-07	0.52
Interlocking paver	A2	1.15	0.18	6.49	1.29E-03	0.89
	A3	0.43	0.02	18.94	7.57E-06	0.97
	A (all)	0.55	0.09	5.80	1.19E-04	0.75
Asphalt	B2	0.11	0.02	6.55	1.24E-03	0.90
	B3	0.06	0.01	6.32	1.46E-03	0.89
	B (all)	0.08	0.01	6.57	4.02E-05	0.80
Concrete	C2	0.34	0.02	14.10	3.22E-05	0.98
	C3	0.20	0.01	17.05	1.27E-05	0.98
	C (all)	0.23	0.02	11.35	2.05E-07	0.92

Correlation relationship: $k_{ASTM} = \alpha \times k_{NCAT}$.

the values measured with the NCAT permeameter method. However, when the regression analysis was conducted on each type of pavement material and on individual sections, strong correlations (R^2 from 0.754 to 0.986) were found, and the estimates of coefficients are all statistically significant at the level of 0.01, with most statistically significant at 0.001, as shown in Table 5.3.

5.5 IMPLICATIONS OF THE RESULTS AND RECOMMENDATIONS

There are a number of fundamental differences between the ASTM and the NCAT methods. Two notable differences are the size of the permeameter ring and whether the operation is under constant head or falling head. Further investigation to determine whether the ring size is a major factor influencing the accurate measurement of permeability was therefore initiated, with a focus on determining if measurement accuracy is related to the ring size or number of rings. Several experiments were performed to quantify the effects of these modifications and the results are shown in Figure 5.8(a) and (b). Results shown in Figure 5.8(a) indicate that the permeability values measured with the single ring are generally higher than those measured with the double ring (one outer ring and one inner ring for testing, as shown in Figure 5.7). This can probably be attributed in part to the outer ring in the double-ring test, which constrains the water flow to

Figure 5.7 Double ring used for testing (compared to the single ring as shown in Figure 5.2).

the inner ring by reducing the lateral flow and forces the water flow closer to a one-dimensional flow as assumed by Darcy's Law.

In addition, when the ASTM C1701 permeameter ring was reduced to a smaller size, the permeability values increased (see Figure 5.8(b)). It is also important to note that higher permeability was measured with the smaller ring size even when the double ring was used. This also can be explained by the relative increase in lateral water flow area on the pavement caused by smaller ring size. Therefore, the larger ring size, as recommended by the ASTM C1701 method, is reasonable since it will cover a larger pavement surface and make the flow closer to one dimensional and reduce the lateral flow movement. In addition, from Figure 5.9 it can be noted that operating the recommended ASTM C1701 permeameter under falling head (similar to the NCAT permeameter method) will result in a higher permeability value compared with operating the same permeameter under constant head. Therefore, based on this finding it is possible to obtain comparable results when the bottom ring size in the NCAT permeameter is increased closer to the 12-inch diameter of the ASTM C1701 method. With this modification, it is possible to obtain a much better correlation between the NCAT permeameter falling-head method and the ASTM C1701 method, as can be seen in Figure 5.9.

One additional constraint associated with the NCAT permeameter method is related to the advance knowledge of pavement thickness that is a required input to the calculation (see Eq. (5.2)). The pavement thickness

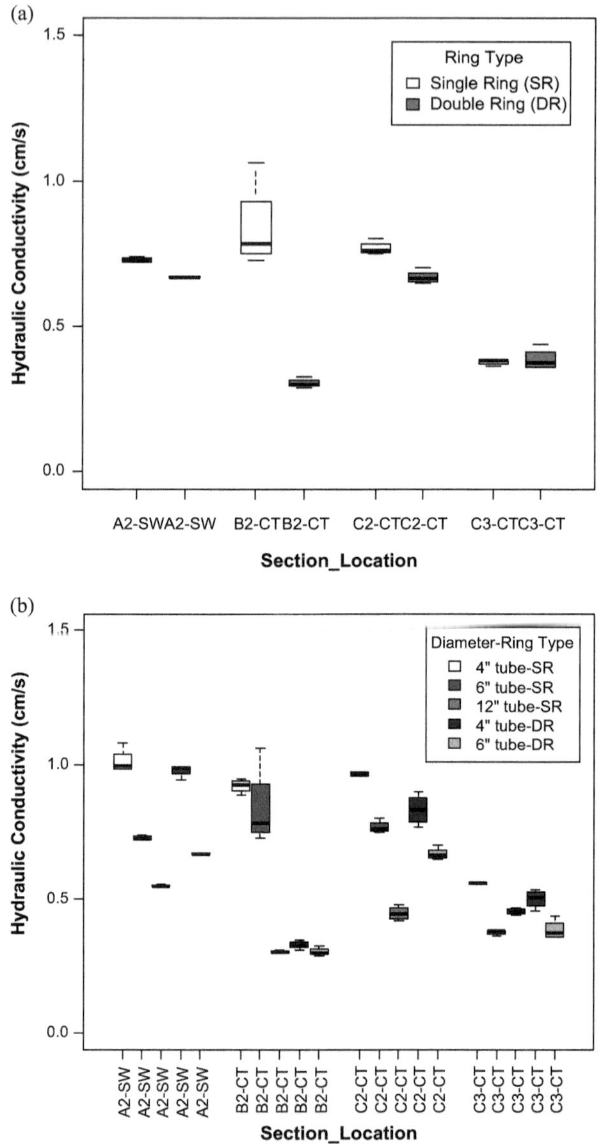

Figure 5.8 (a) Influence of ring type and (b) ring size (SR = single ring; DR = double ring) on permeability measurement applicable to ASTM C1701 constant-head method.

may not always be known or practical to measure without seriously disturbing the pavement. Even if the thickness is known, experience has shown that the pavement thickness may not always be uniform and hence the permeabilities measured within a specified pavement area may not be

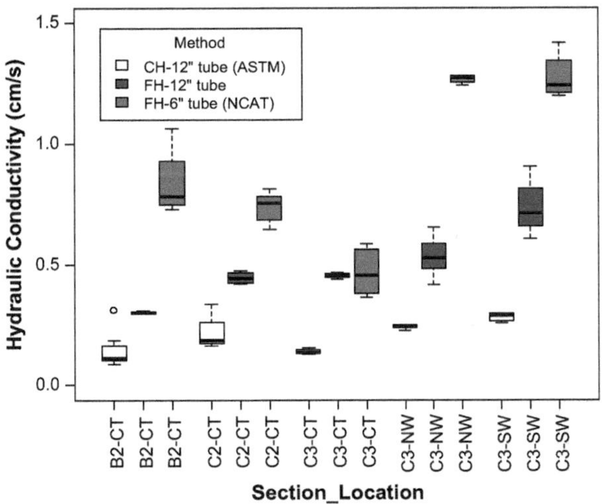

Figure 5.9 Influence of ring type and size on permeability measurement based on operation as falling head (FH) or constant head (CH). *Note: FH-12″ tube size was used for falling-head permeability measurements to compare ASTM C1701 and NCAT permeameters.*

the same owing to variations in pavement thickness. This problem is not an issue with the ASTM C1701 method.

While the NCAT permeameter method is based on the well-established falling-head theory that is derived from Darcy's law, no specific theory has been presented for the derivation of the equation presented in the ASTM C1701 method. For example, it is not known how coefficient K in Eq. (5.2) was derived (Section 5.2.2). More important, the constant value of K used in Eq. (5.2) can be confused with saturated hydraulic conductivity, which is also expressed as K. It is recommended that another symbol (e.g., C) be used to eliminate the confusion.

One other important permeability measurement constraint is related to the use of nonstandard terminology. For instance, inconsistent terminology, such as infiltration rate and hydraulic conductivity, is used for permeability measurements in the ASTM C1701 and NCAT permeameter methods, respectively. In addition, several terms have been used interchangeably in the literature, including saturated hydraulic conductivity, infiltration rate, permeability coefficient, or simply permeability. For the purpose of design and performance evaluation and comparative analysis, it may be more appropriate to use a consistent and standardized terminology regardless of method of measurement.

5.6 SUMMARY AND CONCLUSIONS

This chapter comparatively measured the permeability of test sections of porous asphalt, pervious concrete, and permeable interlocking concrete paver pavements using both the ASTM C1701 and the NCAT methods. The conclusions drawn from the study presented in this chapter are:

1. For accurate permeability measurement, regardless of method of measurement, water leakage must be prevented. It was found that silicone gel is superior for water sealing compared with the plumber's putty recommended in both methods.
2. Both the ASTM C1701 and the NCAT permeameter measurement methods can effectively be used to measure the permeability of all surface pavement types, and their mix design will not have a significant impact on the measurement accuracy.
3. Weak correlation (with $R^2 = 0.52$) was observed between permeability measurements made by the NCAT permeameter and ASTM C1701 methods across all pavement surface types. The correlation was stronger ($R^2 = 0.72$ through 0.9) for the measured permeability values for a single type of pavement surface material.
4. The permeability measured using the ASTM C1701 method was more conservative (lower than that from the NCAT method) and on average about 25% of the values measured using the NCAT permeameter method.
5. The larger ring size used in the ASTM C1701 method or the double-ring method reduced the variability of the permeability measurements.

CHAPTER 6

Thermal Resistance Pavements and Thermal Properties

Contents

Pavement Materials for Heat Island Mitigation
ISBN 978-0-12-803476-7
http://dx.doi.org/10.1016/B978-0-12-803476-7.00006-4

6.1 INTRODUCTION

6.1.1 Background

In addition to reflective pavements and permeable pavements, thermal resistance pavements are another promising type of cool pavement to mitigate the heat island effect through engineered thermal properties of pavement materials.

Thermal properties are fundamental parameters that influence the distribution and variation of pavement and other building material temperatures and consequently affect the thermal performance of the built environment [26,27,78,128–132]. These properties are required as inputs for understanding, evaluating, and modeling the thermal behavior and consequent environmental impact of cool roofs and cool pavements [33,78,112–116,128,129,131–135].

Moreover, temperature is a critical factor affecting building materials' deterioration speed and durability, especially for pavement materials. High temperature will increase the risk of rutting (permanent deformation) of asphalt pavement [136–138]. On the other hand, low temperatures and the adverse thermal gradient at low temperatures make both asphalt and concrete pavements more susceptible to thermal cracking [139,140]. From this point of view, it is also of great significance to measure thermal properties for predicting and optimally designing the thermal behavior of pavement structures and materials.

Conventionally, the laboratory testing procedure for thermal properties of building materials is based on American Society for Testing and Materials (ASTM) C177, which employs a one-dimensional (1-D) steady-state method to measure the thermal conductivity. To ensure a 1-D heat-flow condition, this procedure is limited to a flat slab specimen with thickness not exceeding one-third of the maximum linear dimension of the metered region. The test also requires that the temperature gradient within the test specimen be small enough to ensure reasonable approximation of differential terms in the Fourier equation for heat conduction. It is practically difficult to meet these requirements of slab size and temperature gradient. In addition, it is difficult to reduce and consider the heat loss from the edges of the testing specimen, which will influence the accuracy. Moreover, this steady-state method needs a separate test to measure the heat capacity or thermal diffusivity.

Carlson et al. [141] also employed a 1-D steady-state method to measure thermal conductivity. Conventional cylinder specimens were proposed to be used to reduce the practical difficulty of obtaining a thin slab of asphalt in the laboratory. It is still difficult to reduce and consider the heat loss from the top and bottom of the cylinder specimen or make it thermally long enough to ensure a 1-D heat flow, and this will influence the testing accuracy. In addition, since it is also using a steady-state method, a separate test is needed to measure the heat capacity or thermal diffusivity.

Some studies [112,142,143] have employed a transient method to determine the thermal conductivity and heat capacity (or thermal diffusivity) of asphalt or concrete slab specimens from a single test. However, since they used 1-D heat-transfer theory, a 1-D heat-flow condition was still required in their method. This requires that the specimen be a thermally thin slab (1-D heat transfer) as do the other 1-D methods discussed above.

Xu and Solaimanian [144] employed a multidimensional transient method to measure the thermal properties of asphalt concrete. This method relaxes the requirements on specimen shape and size. However, the authors used a one-term approximation of the series solutions of specimen temperature, which increased the error of the model. In addition, the procedure of back-calculation of thermal properties developed by the authors caused an issue on the uniqueness of the back-calculated thermal properties. These will be detailed in the following sections of this chapter.

Nguyen et al. [145] investigated the change in temperature of asphalt mixtures during cyclic tests on cylindrical specimens, which is created in the sample by the viscous dissipated energy that is completely transformed into heat. Temperature is measured at the surface and inside the specimen. From the analysis of the experimental results using the 1-D transient heat-transfer method with internal heat generation, the thermal properties were obtained. However, the thermocouple embedded into the specimen might weaken the specimen for mechanical testing.

Table 6.1 summarizes some studies on the thermal properties of asphalt and concrete materials, the thermal property values from the literature, and the corresponding measurement methods [112,141–149]. As noted from the table and the discussion on the literature, most of the existing methods of measuring thermal properties of asphalt or concrete materials are based on 1-D steady-state heat-transfer theory. The critical challenge for these methods has been how to achieve a 1-D heat-flow condition for testing the

Table 6.1 Thermal properties and test methods from literature

Study	Density ρ (kg m^{-3})	Specific heat capacity c (J kg^{-1} °C^{-1})	Conductivity k (W m^{-1} °C^{-1})	Diffusivity $\alpha = k/(\rho c)$ (m^2 s^{-1}) $\times 10^{-7}$	Material[a]	Measurement method
Carlson et al. [141]	—	987	—	—	HMA	1-D steady state
	—	977	—	—	GGAC	
	—	875	—	—	AR OGFC	
	—	1016	—	—	PCC	
	—	1055	—	—	CR PCC (80 lb rubber per yd^3)	
	—	992	—	—	CR PCC (160 lb rubber per yd^3)	
	—	956	—	—	CR PCC (240 lb rubber per yd^3)	
	—	964	—	—	PF PCC (0% fiber content)	
	—	997	—	—	PF PCC (3% fiber content)	
	—	977	—	—	PF PCC (5% fiber content)	
	—	971	—	—	PF PCC (8% fiber content)	
	—	—	0.896	—	HMA	
	—	—	1.719	—	PCC	

Mrawira and Luca [112]	2440	766.6	1.75	9.36	AC	1-D transient
Luca and Mrawira [143]	2297–2450	1475–1853	1.623–2.060	4.3–5.5	AC	
Mrawira and Luca [142]	2410	1630–2000	1.96–2.01	4.1–5.3	HMA with gravel, AV 4%	
	2420	1480–1890	1.91–1.94	4.2–5.4	HMA with hornfels, AV 4%	
Xu and Solaimanian [144]	2313	880	2.88	14.2	AC, AV 5.8%	2-D or 3-D transient
Nguyen et al. [145]	–	820–910	1.35	5.86–6.51	AC, AV 0.8%	1-D steady/transient with internal heat generation
Wolfe et al. [146]	–	879–963	1.003–1.747	5.16–8.26	AC	1-D steady state
Highter and Wall [147]	–	800–1600	0.800–1.600	3.50–7.50	AC	1-D steady state/transient
Tan et al. [148]	–	–	1.300–1.420	5.36–5.80	AC	1-D transient
Solaimanian and Bolzan [149]	–	–	0.744–2.889	–	AC	–
Range (overall)	2313–2450	767–2000	0.74–2.89	3.50–14.2	AC/PCC	–

[a]HMA, hot mixed asphalt; GGAC, gap-graded asphalt concrete; AR OGFC, asphalt rubberized open-graded friction course; PCC, Portland cement concrete; CR PCC, crum rubber Portland cement concrete; PF PCC, polypropylene fiber Portland cement concrete; AC, asphalt concrete; AV, air void.

specimen. It is difficult or even impossible for these methods to meet thermally thin slab or thermally long cylinder criteria, especially for the common beam or cylinder specimens prepared in the laboratory or extracted from field in-service pavements.

Therefore, an improved multidimensional (3-D for slab/beam specimen and 2-D for cylinder specimen) transient method is needed to reduce the challenge and requirements on the testing specimen size and shape and make it possible to accurately measure the thermal properties from one single test.

6.1.2 Objectives

The first objective of this chapter is to develop a multidimensional transient model and a practical tool to simulate the transient temperature at any location on a beam or cylinder specimen of various sizes subject to the convective heat transfer between the specimen and the surrounding airflow. This model and the practical tool developed can be used to, if the thermal properties are known, simulate the transient temperature and predict the time it takes to reach a specified target temperature at any location for specimens of various shapes and sizes, which are preheated or precooled in the forced convection oven or temperature chamber for mechanical and other temperature-related laboratory testing.

The second objective is to develop and validate a procedure for back-calculating the thermal properties of specimens of various shapes and sizes from the measured transient temperatures profile of the specimens, based on the temperature simulation model developed in the first objective.

The third objective is to use the developed and validated procedure to measure the thermal properties of pavement materials, specifically some surface materials used for the experimental sections in this dissertation.

6.2 THEORETICAL MODEL FOR SIMULATION OF TEMPERATURE

The multidimensional (3-D for slab/beam specimen and 2-D for cylinder specimen) model to simulate the transient temperature at any location on beam or cylinder specimens of various sizes subject to convective heat transfer is developed based on the 1-D transient heat-transfer theory for infinite plate and infinite long cylinder and geometric intersection of these 1-D models [35,43,110,150].

6.2.1 Governing Equations

The first law of thermodynamics (energy conservation) for a 3-D infinitesimal control volume of size $\Delta x\Delta y\Delta z$ in a Cartesian coordinate system with the origin in the center of the body can be written as

$$\begin{aligned}&\left(q''_x - q''_{x+\Delta x}\right)\Delta y\Delta z + \left(q''_y - q''_{y+\Delta y}\right)\Delta x\Delta z + \left(q''_z - q''_{z+\Delta z}\right)\Delta x\Delta y\\&+ \dot{q}\Delta y\Delta y\Delta z = \rho c\Delta y\Delta y\Delta z\frac{\partial T}{\partial t}\end{aligned} \tag{6.1}$$

where each q'' term represents a heat flux, or heat-transfer rate per unit area ($\mathrm{W/m^2}$). $\dot{q}$ is the volumetric rate of internal heat generation in the volume ($\mathrm{W/m^3}$). ρ is the density; c is the specific heat; T is the temperature, and t is the time.

According to the Fourier law (the second law of thermodynamics), each of the heat fluxes above can be written as

$$\begin{aligned}&q''_x = -k_x\frac{\partial T}{\partial x};\ q''_y = -k_y\frac{\partial T}{\partial y};\ q''_z = -k_z\frac{\partial T}{\partial z};\\&q''_{x+\Delta x} = q''_x + \frac{\partial q''_x}{\partial x}\Delta x;\ q''_{y+\Delta y} = q''_y + \frac{\partial q''_y}{\partial y}\Delta y;\ q''_{z+\Delta z} = q''_z + \frac{\partial q''_z}{\partial z}\Delta z\end{aligned} \tag{6.2}$$

Then Eq. (6.2) is substituted into Eq. (6.1). Dividing the resulting equation by $\Delta x\Delta y\Delta z$ and invoking the limits $\Delta x \to 0$, $\Delta y \to 0$, and $\Delta z \to 0$, we obtain the general equation for energy conservation at a point in a Cartesian frame:

$$\frac{\partial}{\partial x}\left(k_x\frac{\partial T}{\partial x}\right) + \frac{\partial}{\partial y}\left(k_y\frac{\partial T}{\partial y}\right) + \frac{\partial}{\partial z}\left(k_z\frac{\partial T}{\partial z}\right) + \dot{q} = \rho c\frac{\partial T}{\partial t} \tag{6.3}$$

where k_x, k_y, and k_z are thermal conductivity at x, y, and z spatial coordinate directions, respectively. The above equation is just the general three-dimensional transient conduction equation. For the case of constant conductivity and without internal heat generation, Eq. (6.3) can be simplified as

$$k_x\frac{\partial^2 T}{\partial x^2} + k_y\frac{\partial^2 T}{\partial y^2} + k_z\frac{\partial^2 T}{\partial z^2} = \rho c\frac{\partial T}{\partial t} \tag{6.4}$$

If the thermal conductivity is assumed to be uniform and homogeneous ($k_x = k_y = k_z$), then Eq. (6.3) can be further simplified as

$$k\frac{\partial^2 T}{\partial x^2} + k\frac{\partial^2 T}{\partial y^2} + k\frac{\partial^2 T}{\partial z^2} = \rho c\frac{\partial T}{\partial t} \tag{6.5}$$

or

$$\frac{\partial^2 T}{\partial x^2} + \frac{\partial^2 T}{\partial y^2} + \frac{\partial^2 T}{\partial z^2} = \frac{\rho c}{k}\frac{\partial T}{\partial t} = \frac{1}{\alpha}\frac{\partial T}{\partial t} \tag{6.6}$$

where $\alpha = k/\rho c$ and is the thermal diffusivity. For the infinite plate (plane wall), the transfer of heat is assumed to take place only in the longitudinal direction (say z). Then Eq. (6.6) can be simplified as 1-D heat transfer:

$$\frac{\partial^2 T}{\partial z^2} = \frac{\rho c}{k}\frac{\partial T}{\partial t} = \frac{1}{\alpha}\frac{\partial T}{\partial t} \tag{6.7}$$

Analogously, for the cylindrical coordinate system with the origin in the center of the body, the 3-D heat transfer with constant thermal conductivity can be described as:

$$k_r \frac{1}{r}\frac{\partial}{\partial r}\left(r\frac{\partial T}{\partial r}\right) + k_\varphi \frac{1}{r^2}\frac{\partial^2 T}{\partial \varphi^2} + k_z \frac{\partial^2 T}{\partial z^2} + \dot{q} = \rho c \frac{\partial T}{\partial t} \tag{6.8}$$

For an infinitely long cylinder with uniform and homogeneous thermal conductivity, the transfer of heat is assumed to take place only in the radial direction (say r). Then Eq. (6.8) can be simplified as 1-D heat transfer:

$$\frac{1}{r}\frac{\partial}{\partial r}\left(r\frac{\partial T}{\partial r}\right) = \frac{\partial^2 T}{\partial r^2} + \frac{1}{r}\frac{\partial T}{\partial r} = \frac{1}{\alpha}\frac{\partial T}{\partial t} \tag{6.9}$$

For simplification, the dimensionless temperature θ and dimensionless coordinate Z and R are defined as following,

$$\theta(Z, Fo) = \frac{T(z,t) - T_a}{T_0 - T_a};\ \theta(R, Fo) = \frac{T(r,t) - T_a}{T_0 - T_a};\ Z = z/d_0; R = r/r_0.$$

where d_0 is the half-thickness of the plate; r_0 is the radius of the cylinder; and $Fo = \frac{\alpha t}{d_0^2}$ is the Fourier number.

The 1-D heat transfer (Eqs (6.7) and (6.9)) would be changed to Eqs (6.10) and (6.11) for an infinite plate and an infinitely long cylinder, respectively.

$$\frac{\partial^2 \theta}{\partial Z^2} = \frac{\partial \theta}{\partial Fo} \tag{6.10}$$

$$\frac{1}{R}\frac{\partial}{\partial R}\left(R\frac{\partial \theta}{\partial R}\right) = \frac{\partial^2 \theta}{\partial R^2} + \frac{1}{R}\frac{\partial \theta}{\partial R} = \frac{\partial \theta}{\partial Fo} \tag{6.11}$$

6.2.2 Initial and Boundary Conditions

The specimen with a uniform initial temperature T_0 is conducting convective heat exchange with the surrounding fluid (air) of a constant temperature of T_a. The convective heat transfer coefficient (h) can be determined as [37,43,150]

$$h = \frac{k_{\text{air}}}{L} Nu \tag{6.12}$$

where the Nusselt number $Nu = CRe^m Pr^{1/3}$, the Reynolds number $Re = \frac{U_{\text{air}}L}{\nu_{\text{air}}}$, the Prandtl number $Pr = \frac{\alpha_{\text{air}}}{\nu_{\text{air}}}$, the thermal diffusivity of air $\alpha_{\text{air}} = \frac{k_{\text{air}}}{\rho_{\text{air}} c_{\text{air}}}$, the kinematic viscosity $\nu_{\text{air}} = \frac{\mu_{\text{air}}}{\rho_{\text{air}}}$, and k_{air}, c_{air}, and ρ_{air} are the thermal conductivity, heat capacity, and density of air, respectively. μ_{air} is the dynamic viscosity of air and L is the characteristic length.

Therefore,

$$h = C \cdot \frac{k_{\text{air}} Pr^{1/3}}{{\nu_{\text{air}}}^m} \cdot L^{m-1} \cdot {U_{\text{air}}}^m. \tag{6.13}$$

6.2.2.1 *Infinite Plate (Plane Wall)*

The initial condition is $T(z,0) = T_0$. The boundary conditions are $\frac{\partial T(0,t)}{\partial z} = 0$, and $-k\frac{\partial T(d_0,t)}{\partial z} = h[T(d_0,t) - T_a]$.

Under the dimensionless temperature θ and dimensionless coordinate Z, the initial condition and boundary conditions would be

$$\theta(Z,0) = 1; \quad \frac{\partial \theta(0,Fo)}{\partial Z} = 0; \quad -k\frac{\partial \theta(1,Fo)}{\partial Z} = h\theta(1,Fo). \tag{6.14}$$

6.2.2.2 *Infinite Long Cylinder*

The initial condition is $T(r,0) = T_0$. The boundary conditions are $\frac{\partial T(0,t)}{\partial r} = 0$, and $-k\frac{\partial T(r_0,t)}{\partial r} = h[T(r_0,t) - T_a]$.

Under the dimensionless temperature θ and dimensionless coordinate R, the initial condition and boundary conditions would be

$$\theta(R,0) = 1; \quad \frac{\partial \theta(0,Fo)}{\partial R} = 0; \quad -k\frac{\partial \theta(1,Fo)}{\partial R} = h\theta(1,Fo). \tag{6.15}$$

6.3 ANALYTICAL SOLUTION FOR SIMULATION OF TEMPERATURE DISTRIBUTION

The analytical solutions for infinite plate and infinitely long cylinder could be obtained using the separation-of-variable method, which can be found in most textbooks on heat transfer (e.g., [35,43,110,150]) and is not detailed here.

6.3.1 Infinite Plate (Plane Wall)

The analytical solution for infinite plate is

$$\theta(Z, Fo)_{\text{plate}} = \sum_{i=1}^{\infty} C_i \cos(\xi_i Z) e^{-\xi_i^2 Fo} \tag{6.16}$$

where $\xi_i \tan \xi_i = Bi$; $C_i = \frac{4\sin \xi_i}{2\xi_i + \sin(2\xi_i)}$; $Bi = \frac{hd_0}{k}$ and is the Biot number.

6.3.2 Infinitely Long Cylinder

The analytical solution for infinitely long cylinder is

$$\theta(R, Fo)_{\text{cylinder}} = \sum_{i=1}^{\infty} C_i J_0(\xi_i R) e^{-\xi_i^2 Fo} \tag{6.17}$$

where $\xi_i \frac{J_1(\xi_i)}{J_0(\xi_i)} = Bi$; $C_i = \frac{2J_1(\xi_i)}{\xi_i[J_0^2(\xi_i) + J_1^2(\xi_i)]}$; $Bi = \frac{hr_0}{k}$ and is the Biot number.

6.3.3 Short Cylinder

The short cylinder can be viewed as the intersection of an infinite plate and an infinite long cylinder that are perpendicular. The solution for a two-dimensional short cylinder is equal to the product of the one-dimensional solutions of infinite plate and infinitely long cylinder [43,150]:

$$\theta(Z, R, Fo)_{\text{short cylinder}} = \theta(Z, Fo)_{\text{plate}} \times \theta(R, Fo)_{\text{cylinder}} \tag{6.18}$$

The dimensionless normalized temperature θ of a short cylinder at location (z, r) and at time t will be

$$\theta(z, r, t)_{\text{short cylinder}} = \theta(z, t)_{\text{plate}} \times \theta(r, t)_{\text{cylinder}} \tag{6.19}$$

The temperature T of a short cylinder at location (z, r) at time t can be calculated as

$$T(z, r, t)_{\text{short cylinder}} = \theta(z, r, t)_{\text{short cylinder}} \cdot (T_0 - T_a) + T_a \tag{6.20}$$

6.3.4 Short Beam

Analogously, the short cylinder can be viewed as the intersection of three infinite plates that are mutually perpendicular [43,150]. The solution for the three-dimensional short beam could be calculated as,

$$\theta(X, Y, Z, Fo)_{\text{short beam}} = \theta(X, Fo)_{\text{plate}} \times \theta(Y, Fo)_{\text{plate}} \times \theta(Z, Fo)_{\text{plate}} \tag{6.21}$$

The dimensionless normalized temperature θ of a short beam at location (x, y, z) at time t will be

$$\theta(x, y, z, t)_{\text{short beam}} = \theta(x, t)_{\text{plate}} \times \theta(y, t)_{\text{plate}} \times \theta(z, t)_{\text{plate}} \tag{6.22}$$

The temperature T of a short beam at location (x, y, z) at time t can be calculated as

$$T(x, y, z, t)_{\text{short beam}} = \theta(x, y, z, t)_{\text{short beam}} \cdot (T_0 - T_a) + T_a \tag{6.23}$$

6.4 CASE STUDY FOR SIMULATION OF TEMPERATURE AND SENSITIVITY ANALYSIS

6.4.1 Procedure for Simulation of Temperature Profile

According to the theoretical model described in the previous sections, the procedure for the simulation of a temperature profile for cylinder and beam specimens is shown as a flowchart in Figure 6.1. The procedure was implemented by programming using the open-source R language.

6.4.2 Input Parameters for the Case Study

Taking a short cylinder specimen as a study case, the related parameters are assumed to be the following:

$D = 100$ mm, $H = 150$ mm, $k = 2.5$ W/(m °C), $c = 920$ J/(kg °C), $\rho = 2300$ kg/m^3, $T_0 = 25$ °C, $T_a = 55$ °C, $h = 15$ W/(m^2 °C).

Under the convective heat transfer from the surrounding air at a constant temperature of $T_a = 55$ °C described in the previous section for the short cylinder, the specimen will be heated up from the initial temperature of $T_0 = 25$ °C and reaches equilibrium up to the air temperature of $T_a = 55$ °C in some time t.

Using the theoretical model, we can calculate the specimen temperature $T(z, r, t)$ for the following locations and time frames: (1) center, $z = 0$ mm, $r = 0$ mm, $t = 0$–5 h; (2) surface, $z = 0$ mm, $r = 50$ mm, $t = 0$–5 h.

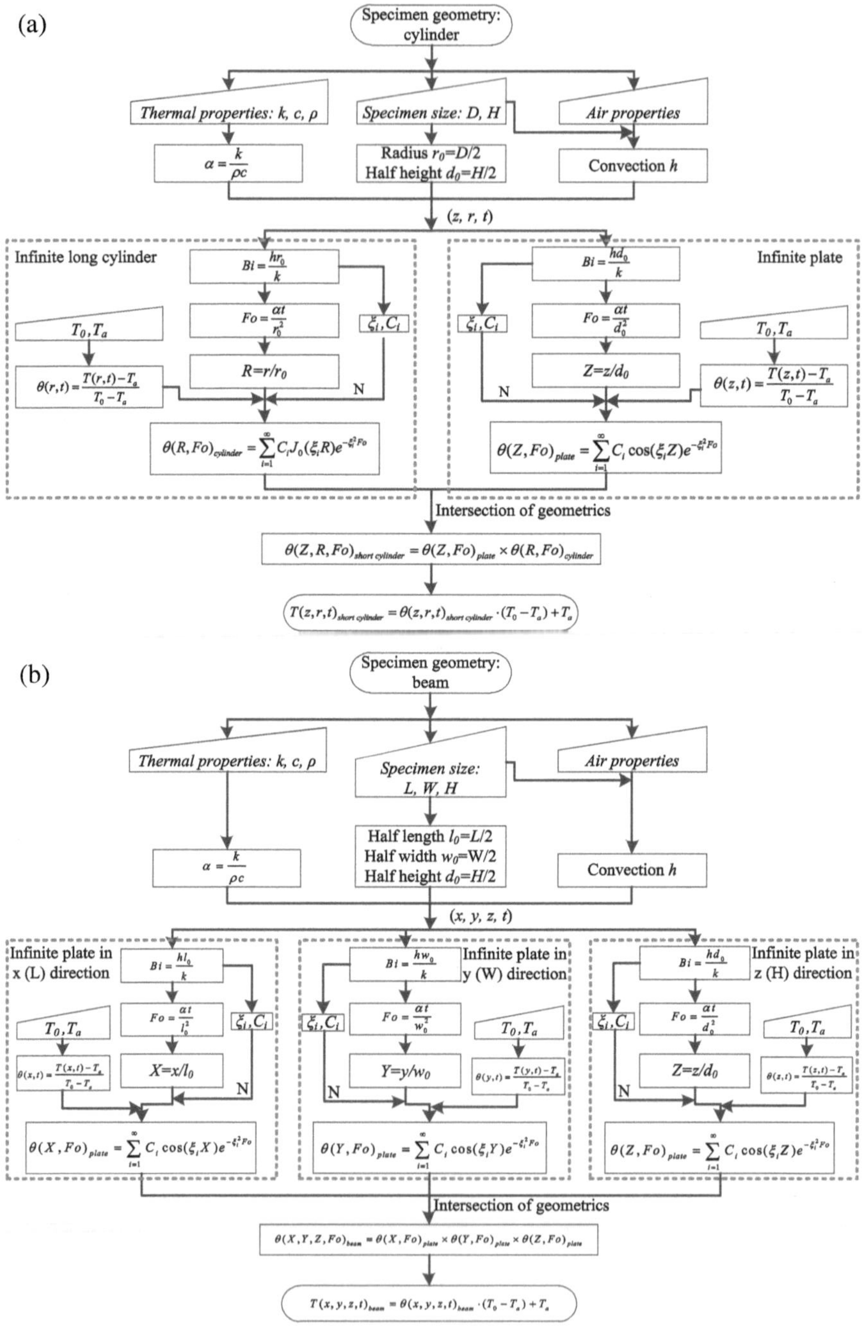

Figure 6.1 Flowchart for temperature simulation for (a) cylinder and (b) beam specimens.

6.4.3 Roots Finding for Eigenvalue Function

The analytical solutions (in Eqs (6.16) and (6.17)) for the temperature for given location z and r (or Z and R) and time t (or Fo) depend on the eigenvalue ξ_i and constant C_i, which are both determined by the eigenvalue functions ($\xi_i \tan \xi_i = Bi$ for infinite plate; $\xi_i \frac{J_1(\xi_i)}{J_0(\xi_i)} = Bi$ for infinite long cylinder). Therefore, finding the roots of the eigenvalue functions for any given Bi to obtain the series of eigenvalue ξ_i is one critical step.

Since it is difficult or even impossible to obtain the analytical solutions for the roots of the eigenvalue functions, numerical methods were employed. Numerical root finding methods include the bisection method, Newton's method, and secant method [41,42,151]. Newton's method (also called the Newton–Raphson method) and the secant method are derivative or finite-difference-based methods that converge faster but depend highly on the initial value used. They are not appropriate for finding a series of roots. The bisection method is the simplest root-finding algorithm. It works when f is a continuous function and does not need any derivative or finite difference. It requires only previous knowledge of two initial guesses, a and b, such that $f(a)$ and $f(b)$ have opposite signs. If the intervals $[a, b]$ of each root are known, we can find all the roots easily using the bisection method. Although it is reliable, it converges slowly, gaining 1 bit of accuracy with each iteration. Therefore, a hybrid method of combined bisection and Newton's method was employed to balance the accuracy and the speed of root finding. The bisection method was used to obtain the preliminary roots of lower accuracy in a small number of iterations; these roots were then used as initial guesses for Newton's method, which would give the roots with high accuracy.

To find the root intervals, the transformed forms ($g(\xi)$ and $h(\xi)$) of the eigenvalue functions are illustrated in the same plot (Figure 6.2(a) and (b) for infinite plate and infinite cylinder, respectively). From the plots, we can see the ith root (or eigenvalue) falls in the interval $((i-1)\pi, i\pi)$ for both infinite plate and infinite cylinder. According to these findings, the hybrid method of root finding was used to find the first N terms of eigenvalue. The first 10 terms of ξ and C for Biot = 5 were obtained using the hybrid method of root finding and listed in Table 6.2. The values of transformed eigenvalue functions at each eigenvalue, $f(\xi_i)$, are also listed in Table 6.2. The absolute values of $f(\xi_i)$ are all less than 1×10^{-10}, illustrating that the solved roots have high accuracy. The first 10 terms of ξ for some different

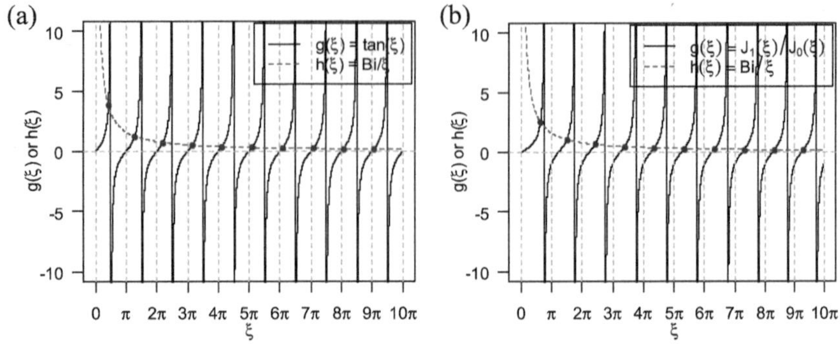

Figure 6.2 Plots for illustrating root interval for eigenvalue functions ($g(\xi) = h(\xi)$, $h(\xi) = Bi/\xi$). (a) For infinite plate ($g(\xi) = \tan \xi$) and (b) for infinite cylinder ($g(\xi) = J_1(\xi)/J_0(\xi)$).

Biot numbers are found by the hybrid method and listed in **Appendix** A for reference. These values were compared with those available from some heat transfer textbooks (e.g., [35,43,110,150]) and found correct. This verifies that the hybrid root finding method and the corresponding R program developed and used in this study are both valid and effective for root finding. This ensures a universal, convenient, and fast method of root finding for the eigenvalue functions for any given B_i (changing with h/k and specimen size) to obtain the series of eigenvalues ξ_i, which is critical for the following sections.

6.4.4 Influence of the Number of Terms *N* on the Solution

To examine the influence of the number of terms N on the solution of temperature, the temperature solutions obtained from various terms ($N = 1$, 3, 5, 10, 30, 50, and 100) in the solution were calculated and plotted in Figure 6.3 for comparison.

It is noted that there is no significant influence of the number of terms on the temperature profile for the long time of 5 h (Figure 6.3(a)). However, as shown in Figure 6.3(b), there is a significant influence of the number of terms on the temperature at the beginning, especially for only one term ($N = 1$) in the solution. One-term approximation ($N = 1$) might be used for predicting the temperature after some time from the beginning ($Fo > 0.2$ as recommend by most literature such as [35,43]) without large errors. It indeed will cause big errors for the temperature at the beginning.

Table 6.2 First 10 terms of ξ and C for Biot = 5

Infinite plate	i	1	2	3	4	5	6	7	8	9	10
	ξ_i	1.3138	4.0336	6.910	9.8928	12.935	16.01	19.106	22.213	25.328	28.448
	ξ_i/π	0.42	1.28	2.20	3.15	4.12	5.10	6.08	7.07	8.06	9.06
	$f(\xi_i)$[a]	1.6E-11	1.2E-12	1.6E-13	8.9E-15	9.3E-14	3.1E-14	−3.0E-14	2.4E-14	−2.4E-14	2.8E-14
	C_i	1.2402	−0.3442	0.1588	−0.0876	0.0543	−0.0366	0.0262	−0.0196	0.0152	−0.0121
Infinite cylinder	i	1	2	3	4	5	6	7	8	9	10
	ξ	1.9898	4.7131	7.6177	10.622	13.679	16.763	19.8640	22.975	26.0937	29.217
	ξ/π	0.63	1.50	2.42	3.38	4.35	5.34	6.32	7.31	8.31	9.30
	$f(\xi)$[b]	4.0E-12	2.0E-13	1.4E-13	8.0E-14	4.7E-14	3.6E-14	−8.9E-16	2.2E-14	−3.4E-14	3.6E-14
	C_i	1.5029	−0.7973	0.4842	−0.3220	0.2301	−0.1737	0.1365	−0.1107	0.0920	−0.0780

[a] $f(\xi) = \xi \tan \xi - Bi$.

[b] $f(\xi) = \xi \frac{J_1(\xi)}{J_0(\xi)} - Bi$.

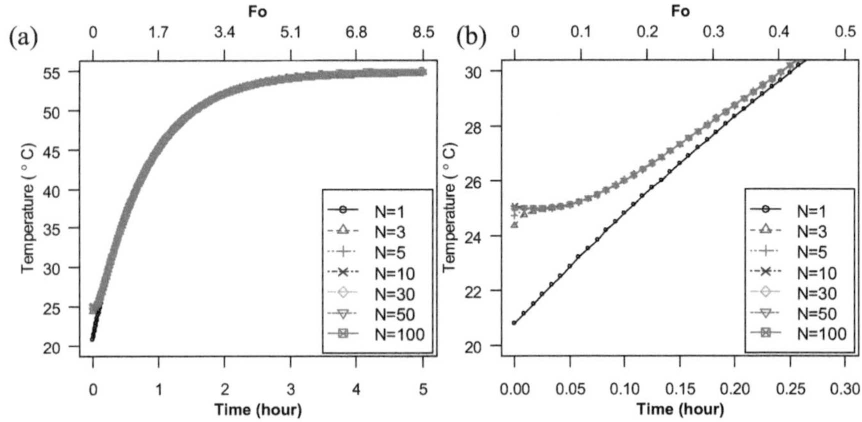

Figure 6.3 Simulated temperature with various numbers of term N in the solution (center, $z = 0$ mm, $r = 0$ mm). (a) 0–5 h, (b) 0–0.3 h.

In the study conducted by Xu and Solaimanian [144], only one-term approximation was used to back-calculate thermal properties. This one-term approximation caused some error, as noticed by the authors (the simulated temperatures at the beginning were far away from the initial temperature). Therefore, more than one term should be used in the solution to obtain a whole temperature profile that is accurate over both the beginning and the following long time, from which the thermal properties will be back-calculated and described in the following sections. As shown in Figure 6.3(b), there is no significant difference when the number of terms in the solution is larger than 10. Therefore, it is recommended that 10 terms ($N = 10$) be used in the solution to obtain a temperature profile with sufficient accuracy.

6.4.5 Simulation Results of Temperature Profiles

For the short cylinder described in the previous section, the temperature profiles at the center, middle, and surface of the short cylinder were calculated according to the producer described in the previous sections and shown in Figure 6.4 together with the normalized temperature profiles. The normalized temperature and temperature profiles of a 1-D infinite plate and 1-D infinite long cylinder are also plotted in Figure 6.4 for comparison. It is shown that the solutions from the 1-D infinite plate and 1-D infinite long cylinder both cause large errors compared to the multi-dimensional solution for a short cylinder. The temperature profiles at three

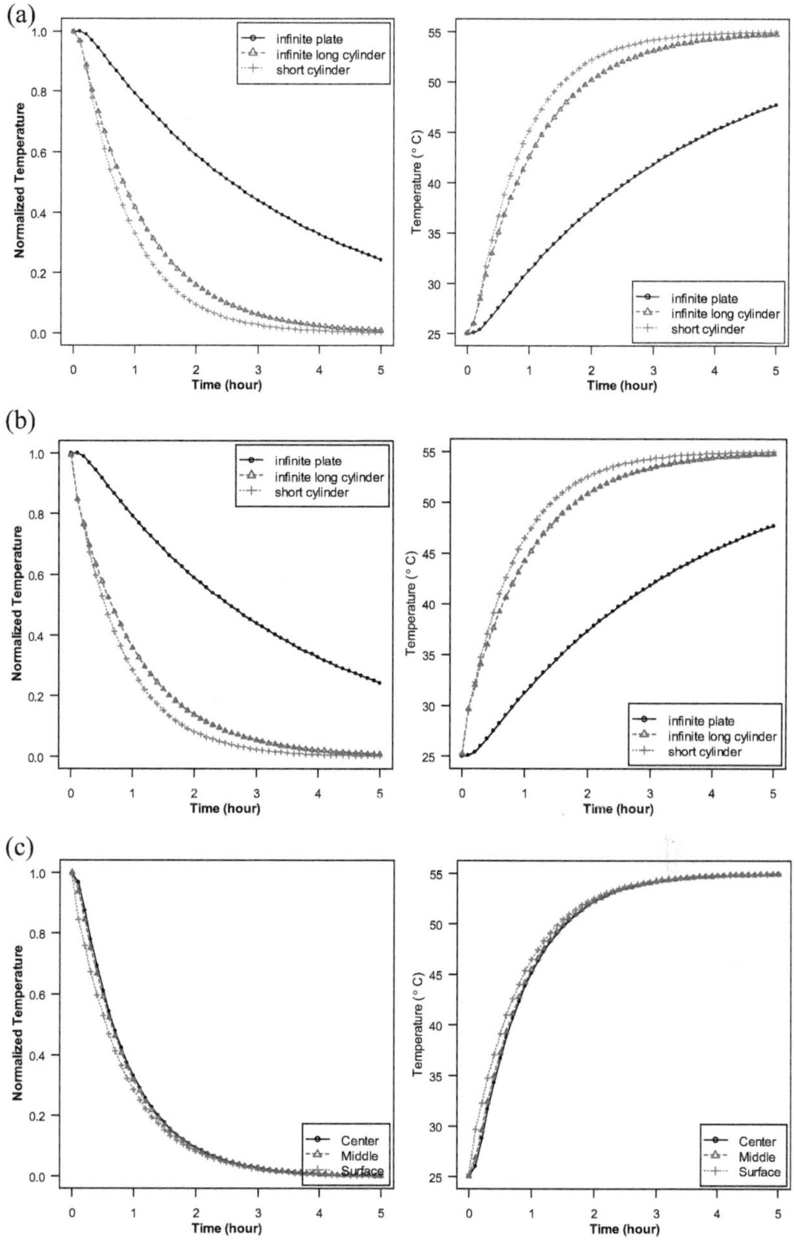

Figure 6.4 Simulated temperatures for a short cylinder at different locations. (a) Center ($z = 0$ mm, $r = 0$ mm), along with comparison to solutions from 1-D infinite plate and 1-D infinite long cylinder. (b) Surface ($z = 0$ mm, $r = 50$ mm), along with comparison to solutions from 1-D infinite plate and 1-D infinite long cylinder. (c) Comparison of temperatures for three different locations.

different locations (center, middle, and surface) are different, especially for the surface. As expected, the center location needs more time to reach the equilibrium temperature. This implies that the inside center temperature of a specimen should be considered as a thermal indictor for some laboratory testing to a certain target temperature.

6.4.6 Sensitivity Analysis of Thermal Property Parameters on the Solution

To examine the sensitivity of thermal property parameters on the solution, the center ($z = 0$ mm, $r = 0$ mm) temperature of the short cylinder under different values for each thermal property parameter was calculated and is shown in Figure 6.5. When the value of one thermal property parameter varies, the values of the other thermal property parameters stay constant at the default values as given in the previous section (i.e., $k = 2.5$ W/(m °C), $c = 920$ J/(kg °C), $\rho = 2300$ kg/m^3; $h = 15$ W/(m^2 °C)).

As mentioned previously, the short cylinder specimen will be heated up from the initial temperature of $T_0 = 25$ °C and ultimately reaches equilibrium up to the air temperature of $T_a - 55$ °C in some time t_e, through the convective heat transfer with the surrounding air at a constant temperature of $T_a = 55$ °C. Therefore, in this case the thermal property parameters will influence the shape of the temperature profile and thus the time t_e that is needed to reach equilibrium, as is shown in Figure 6.5.

The influences of thermal conductivity k, heat capacity c, density ρ, and convection coefficient h on the solution of temperature are shown in Figure 6.5(a)–(d), respectively. It is clearly shown and also easily understood that the larger values of thermal conductivity k and convection coefficient h will reduce the time t_e that is needed to reach equilibrium. In contrast, the increase in values of heat capacity c and density ρ will produce a longer time t_e. The shape of the temperature profile and the time t_e are both sensitive to the values of thermal conductivity k, heat capacity c, density ρ, and convection coefficient h.

There seem to be four fundamental thermal parameters, thermal conductivity k, heat capacity c, density ρ, and convection coefficient h, that determine the solution of temperature in the model and for the case described in this study. However, they are not completely independent parameters; there are some relationships between them for determining the solution of temperature. From the model and the procedure given previously, the solution for short cylinder temperature $T(z, r, t)$ for a given location (r, z) and a given time t depends on the Fourier number Fo and the

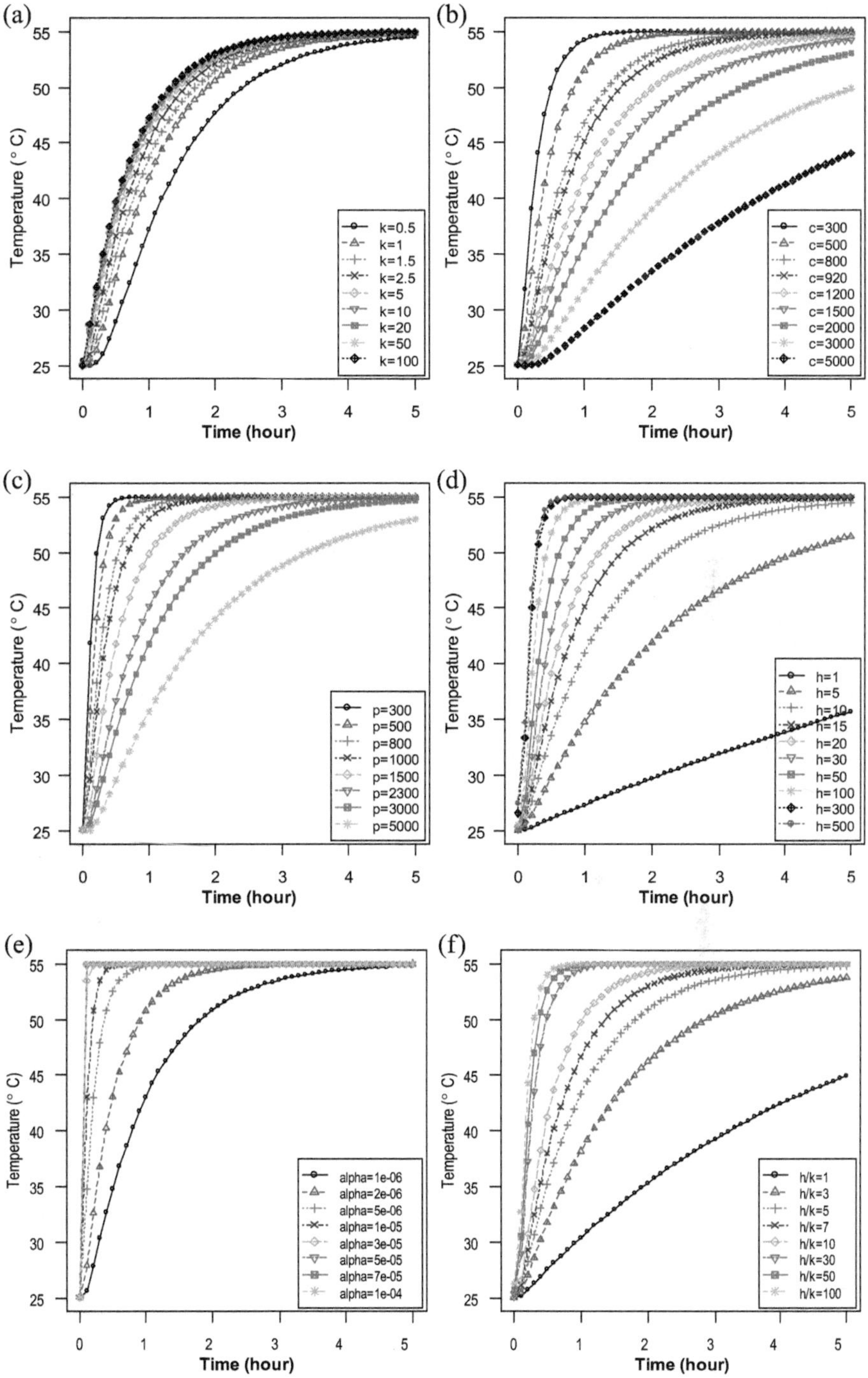

Figure 6.5 Sensitivity of thermal property parameters on the solution. (a) Thermal conductivity k (W/(m °C)), (b) heat capacity c (J/(kg °C)), (c) density ρ (kg/m^3), (d) convection coefficient h (W/(m^2 °C)), (e) thermal diffusivity α (m^2/s), (f) h/k (1/m).

eigenvalue ξ_i and constant C_i. The Fourier number Fo is determined by thermal diffusivity α ($=k/\rho c$); the eigenvalue ξ_i and constant C_i are both obtained from the eigenvalue functions, which are determined by only the Biot numbers Bi for both infinite plate and cylinder. The Biot numbers Bi for both infinite plate and cylinder are commonly determined by the ratio of convection coefficient h to the thermal conductivity k, h/k. This implies that, if you change the values of the k, c, ρ, and h in such a way that α and h/k stay constant, the solution for temperature will stay unchanged. Therefore, there are only two independent parameters (α and h/k) for determining the solution for temperature for both infinite plate and cylinder and thus for both short cylinder and beam, for which the temperatures are calculated from the solution for temperature for both infinite plate and cylinder. These give an important implication for the back-calculation method and uniqueness of thermal parameters, which is discussed in detail in the following section.

The influences of thermal diffusivity α and the ratio of convection coefficient h to thermal conductivity k, h/k, on the solution for temperature are shown in Figure 6.5(c) and (f), respectively. The larger thermal diffusivity will increase the diffusion speed of heat in the specimen and thus reduce the equilibrium time t_e. Similarly, the larger ratio of convection coefficient h to thermal conductivity k, h/k, will enhance the convection heat exchange between the specimen and the surrounding air and thus reduce the equilibrium time t_e, too.

6.4.7 Influence of Specimen Shape and Size on the Solution

To examine the influence of specimen shape and size on the solution, the predicted temperature profiles for cylinders of three sizes and beams of three sizes at the center and surface of specimens are shown in Figure 6.6. The three sizes of cylinder are (height × diameter) 100 × 50, 100 × 100, and 50 × 100 mm; the three sizes of beams are (length × width × height) 380 × 63 × 50 (standard fatigue test beam), 50 × 63 × 50, and 100 × 100 × 50 mm.

The large cylinder of 100 × 100 mm takes a longer time to reach the equilibrium of 55 °C than the small cylinders of 50 × 100 and 100 × 50 mm. The diameter D has a larger influence on the temperature profile and the equilibrium time than the height H for a cylinder. Similarly, the small beam of 50 × 63 × 50 mm takes a shorter time to reach the equilibrium of 55 °C than the large beams of 380 × 63 × 50 and 100 × 100 × 50 mm. The surface temperature of specimens of cylinders or

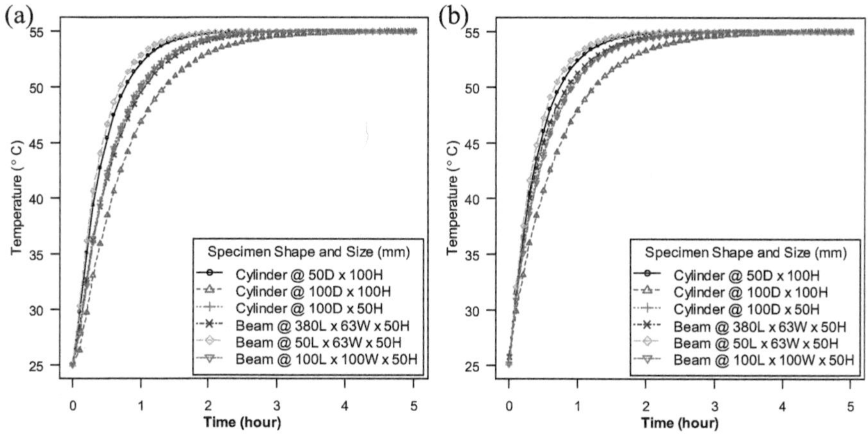

Figure 6.6 Predicted temperature profiles for various specimen shapes and sizes. (a) Center, (b) surface.

beams reaches equilibrium faster than the center temperature of a specimen with the same shape and size, as expected. Therefore, the model and the procedure for predicting temperature are also sensitive to the specimen shape and size.

6.5 PROCEDURE FOR BACK-CALCULATION OF THERMAL PROPERTIES

In the previous section, the model and procedure developed for predicting the temperature of a cylinder or beam specimen that has convective heat exchange with the surrounding air of a constant temperature were discussed. The model is sensitive to the thermal property parameters of the specimen as well as the shape and size of the specimen. Therefore, it is feasible to employ the model and the procedure as a base to develop a method and procedure for back-calculating the thermal properties of a specimen from the measured temperature profile of the specimen.

6.5.1 Optimization Method

The concept of curve fitting is used for back-calculating the thermal properties of a specimen, making the predicted temperature profile match the measured profile as well as possible. The thermal properties are optimized by minimizing the root mean squared error (RMSE), to minimize the overall difference between the predicted results and the measurements of temperature as shown in the objective function in Eq. (6.24):

$$\min_{\{\rho,\ c,\ k,\ h\}} \text{RMSE} = \sqrt{\frac{\sum_{i=1}^{N}\left(T_{m_i} - T_{p_i}\right)^2}{N}} \tag{6.24}$$

where T_{m_i} is the measured temperature at the ith time point, T_{p_i} is the theoretically predicated temperature at the ith time point based on the method described in the previous sections, N is the total number of measurements, and c, k, ρ, and h are thermal properties to be optimized.

6.5.2 Uniqueness of the Back-Calculated Thermal Properties

As discussed previously, the theoretically predicated temperature of a given specimen, at a given location and time, under the given conditions of T_0 and T_a, is ultimately determined only by the two independent thermal parameters: thermal diffusivity α ($=k/\rho c$) and the ratio of convection coefficient h to the thermal conductivity k, h/k. Therefore, if the thermal properties ρ, c, k, and h, which are not completely independent for the predicted temperature, are used for the optimization, as used by Xu and Solaimanian [144] in their study, the optimized thermal properties ρ, c, k, and h might not be unique. In other words, if the values of the ρ, c, k, and h are combined in such a way that α ($=k/\rho c$) and h/k remain constant at the optimized values, the predicted temperature profile will stay unchanged as the optimized temperature profile closest to the measured one. From this point of view, we change the optimization variables from ρ, c, k, and h to the independent parameters α and h/k, as shown in the new objective function in Eq. (6.25):

$$\min_{\{\alpha,\ h/k\}} \text{RMSE} = \sqrt{\frac{\sum_{i=1}^{N}\left(T_{m_i} - T_{p_i}\right)^2}{N}} \tag{6.25}$$

where α and h/k are thermal properties to be optimized. The others are same as in the previous definition.

Once the α ($=k/\rho c$) and h/k are determined through the optimization process, two of the four parameters ρ, c, k, and h can be calculated if only the other two are known. The density ρ of a specimen can be easily obtained if the weight and the volume of the specimen are known. The convection coefficient h can be calculated according to the model shown in Eq. (6.12).

Since it is difficult or even impossible to analytically solve this optimization problem, a numerical method of trial–comparison was used instead to find the optimized solution and is implemented through R programming. During the numerical method of trial–comparison for optimization, all the possible combinations of the two independent parameters α and h/k in the feasible ranges are selected with appropriate step lengths to run the temperature simulations and calculate the corresponding RMSEs of the predicted temperatures. The combination that gives a minimal RMSE is the optimized solution of the two parameters. The whole procedure of back-calculation of thermal properties based on the new objective function is shown in a flowchart in Figure 6.7.

6.5.3 Initial Range and Step Length of Independent Parameters

Now we have the objective function for this optimization problem shown in Eq. (6.25). However, the constraints of the variables (independent

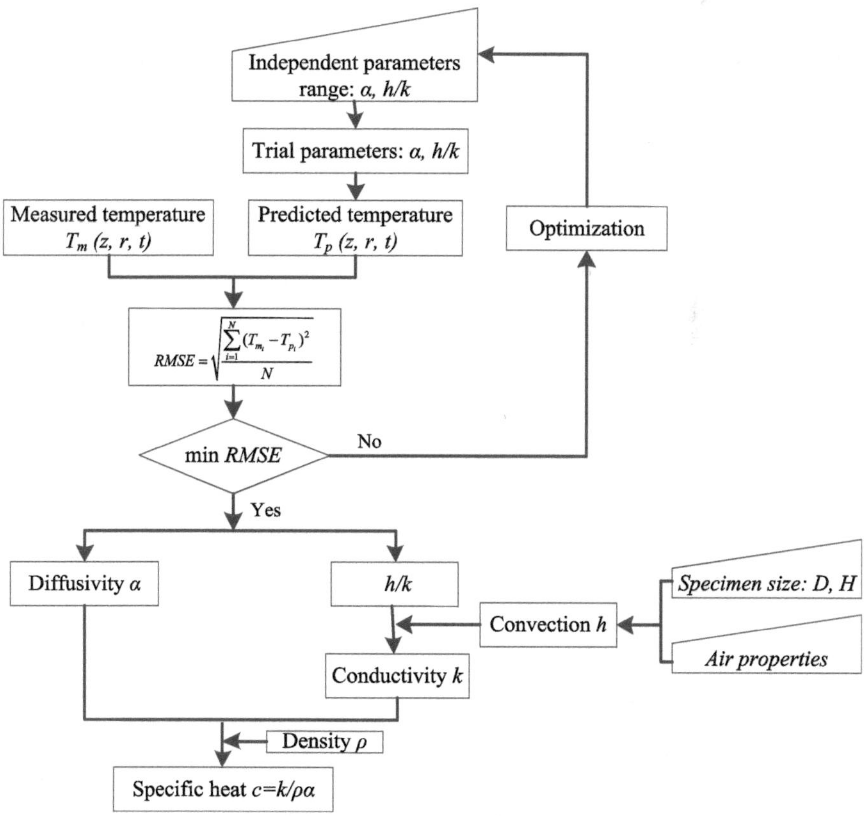

Figure 6.7 Flowchart for back-calculation of thermal properties.

parameters α and h/k) are still missing for the optimization problem. Theoretically, all positive values are possible intervals (or ranges) for both variables. However, these infinite intervals are not practical, and will be impossible for the real calculation. To reduce the calculation load (the number trial iteration), the practical interval of each independent parameter is roughly determined based on the literature (see Table 6.1). For thermal diffusivity α, the initial range is set as $[1 \times 10^{-7}, 1 \times 10^{-5}]$ m^2/s; for the ratio h/k, it is set as [1, 100] 1/m. The initial step lengths used to pick up the trial values of both parameters are set as 2×10^{-7} m^2/s and 5 1/m for α and h/k, respectively. Using these initial intervals and step lengths, the optimized parameter values can be determined, which give the minimal RMSE for all trial parameter values.

6.5.4 Adaptive Range and Step Length Method

From the initial range and step length, a set of optimized parameter values can be determined. However, the accuracy of the obtained optimized parameter values is highly dependent on the step lengths used. A small step length will produce an optimized result of high accuracy, but it will also take a long time to compute. Therefore, an adaptive range and step length (ARS) method, which is similar to the bisection method for root finding, was proposed to balance the accuracy and computation. This ARS method will have adaptive range and step length during the whole optimization process. After the optimized results are obtained using the initial range and step length, the optimization will go to the next level of optimization in which both the new range and the step length are one-half of the previous range and step length. The iteration optimization will be continued until either the current range or the step length goes down to a value small enough, say 1×10^{-9} m^2/s and 0.1 1/m for α and h/k, respectively.

6.6 CASE STUDY FOR BACK-CALCULATION OF THERMAL PROPERTIES

6.6.1 Laboratory Test Results of Temperature

To illustrate the model and the procedure for back-calculating thermal properties developed previously in this study, two cylinder specimens, one asphalt (A0) and one concrete (C0), were used to run the test (Figure 6.8). These two specimens were cored from in-service road pavements in California. The materials are standard dense-graded materials commonly used on highways, streets, and parking lot pavements in California and other

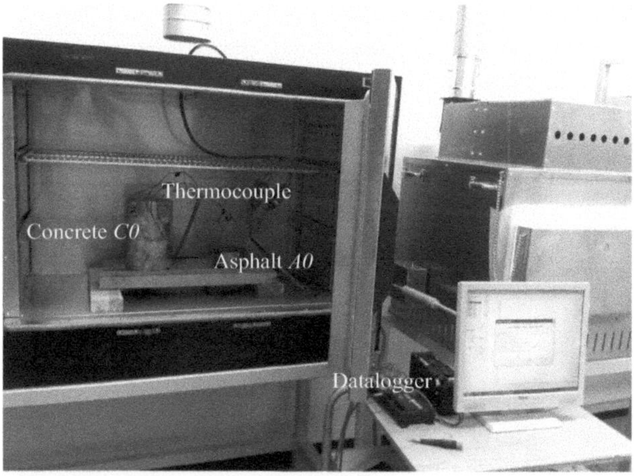

Figure 6.8 Test setup for measurement of thermal properties.

states in United States. The test setup was as shown in Figure 6.8. The specimens were first heated to a uniform temperature of 38 °C in a temperature chamber. From this initial temperature, the specimens were heated up to 70 °C through forced convection heat exchange between the specimens and the surrounding airflow of a constant temperature of 70 °C. The temperature profiles of the specimens at various locations were measured using thermocouples, and the data were recorded by a data logger. The detailed specimen parameters and testing condition are listed in Table 6.3.

Table 6.3 Specimen parameters and testing condition

Specimen No.	Height (mm)	Diameter (mm)	Mass (kg)	Density (kg/m^3)	Air void (%)
A0	63	102	1.238	2405	3.8
C0	222	145	9.095	2481	0.8

Testing temperature condition

Initial specimen temperature T_0 (°C)	38	Air temperature T_a (°C)	70

Temperature sensor location

Specimen No.	Sensor No.	1	2	3	4
A0[c]	z^a (mm)	3.5	0	0	–
	r^b (mm)	45	37	17	–
C0	z^a (mm)	4	−58	82	13
	r^b (mm)	46	35	37	20

[a]Axial position.
[b]Radial distance from the specimen center.
[c]Only three sensors were installed on the small specimen A0.

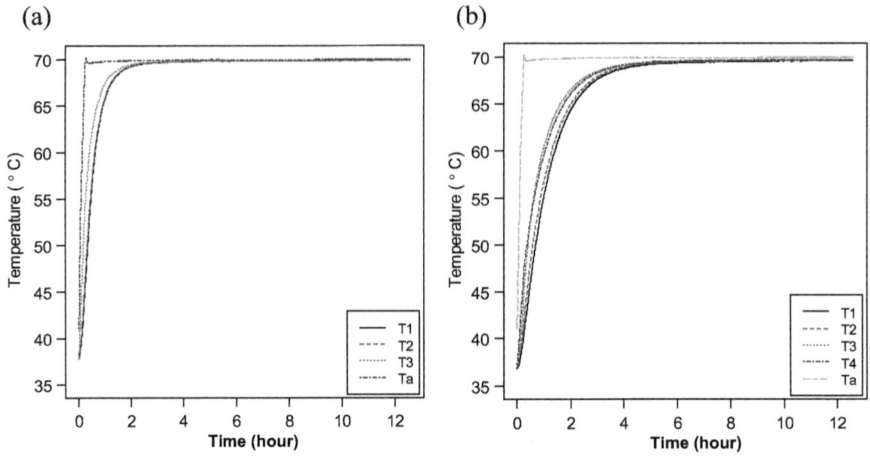

Figure 6.9 Measured temperatures at various locations for asphalt and concrete specimens. (a) Asphalt specimen A0; (b) concrete specimen C0.

The measured temperature profiles (T1 to T4) of the asphalt specimen A0 and the concrete specimen C0 are shown in Figure 6.9, as well as the temperature profile of air (T_a) in the chamber.

6.6.2 Optimized Thermal Properties and Predicted Temperature with the Optimized Parameters

The thermal properties were back-calculated according to the ARS optimization method. To illustrate the ARS optimization method, the adaptive range and step length, optimized parameters, and RMSE for various levels of optimization of the asphalt specimen A0, as an example, are shown in Figure 6.10. During the whole process of optimization, the range $[a, b]$ and the step length Δ decrease with the level of optimization; the accuracy, as shown by RMSE, of the optimized parameters (α and h/k) increases with the level of optimization. This verifies that the proposed ARS optimization method is effective at balancing the accuracy and efficiency.

The optimized parameters (α and h/k) and the predicted temperature under the optimized parameters for asphalt specimen A0 and concrete specimen C0 are shown in Figures 6.11 and 6.12, respectively.

6.6.3 Thermal Properties from the Optimized Parameters

According to the procedure developed previously, the thermal properties k and c can be calculated from the optimized α and h/k if h and ρ are

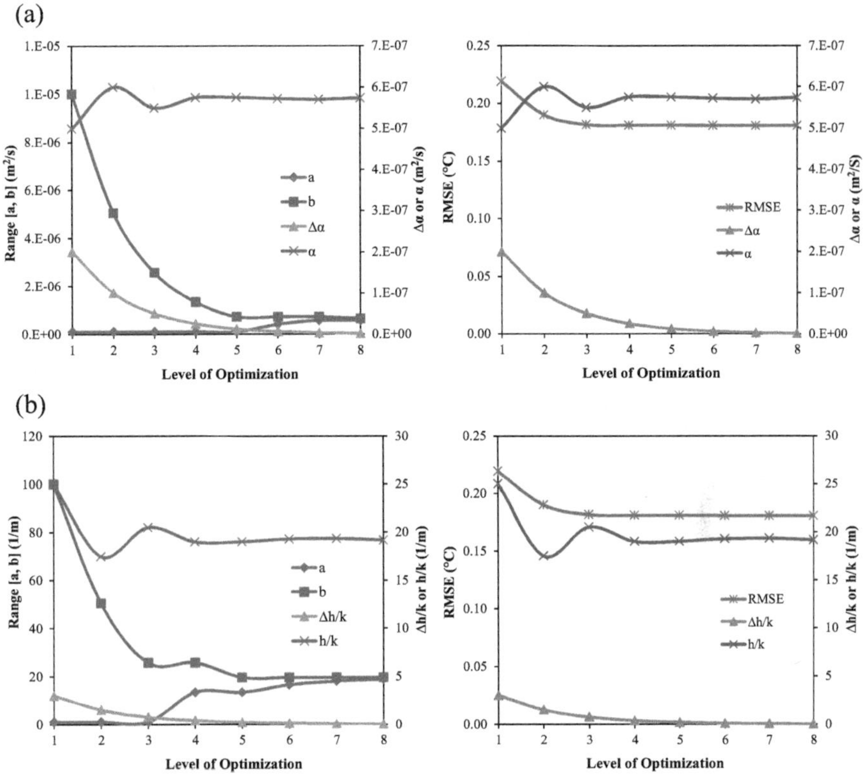

Figure 6.10 Adaptive range and step length, optimized parameters, and RMSE for various levels of optimization (A0). (a) Thermal diffusivity α, (b) ratio h/k.

known. The density ρ is known, as shown in Table 6.3. The convection coefficient h can be calculated from the airflow parameters (at 300 K) according to the model shown in Eq. (6.12). The convection coefficients h for the asphalt specimen A0 and the concrete specimen C0 are calculated and listed in Table 6.4. Using the known h and ρ, the thermal properties k and c can be calculated from the optimized α and h/k and are listed in Table 6.5. There are some differences in the optimized parameters α and h/k and thus the k and c between locations. The results at location 3 of the asphalt specimen and locations 2 and 4 for the concrete specimen C0 are quite different from those of other locations on the same specimen. The reason for that might be that these locations are closer to the specimen surface (with large absolute values of z and r, as shown in Table 6.3 and Figure 6.11). As noticed during the trial testing, the measured temperature

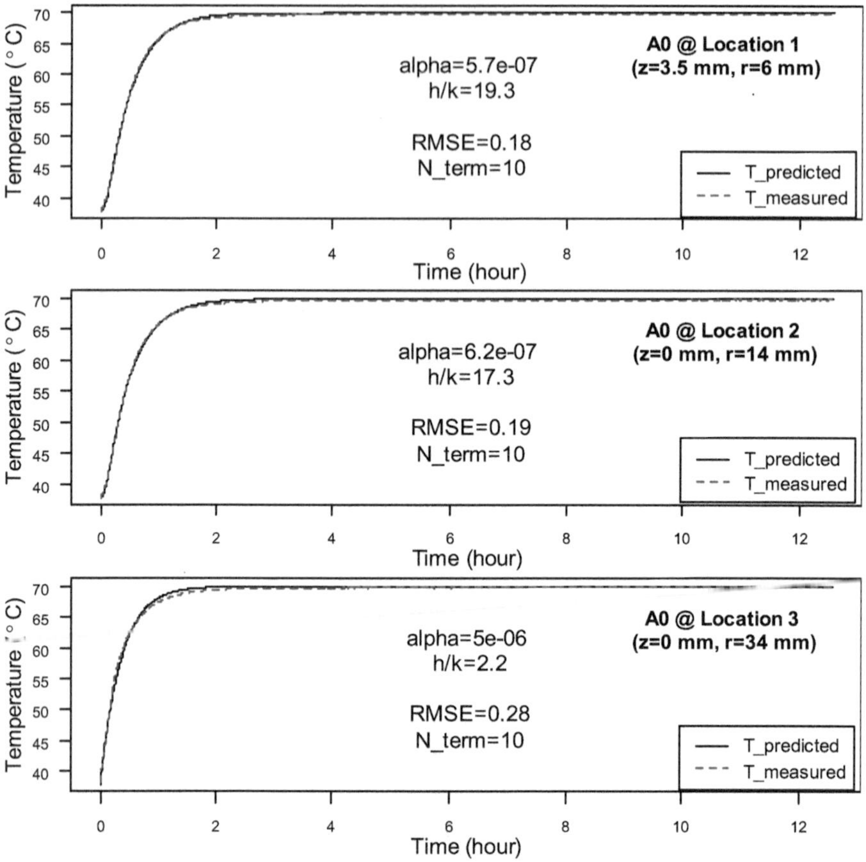

Figure 6.11 Predicted temperature with the optimized thermal properties compared with measured temperature: asphalt specimen A0 (units: α in m^2/s; h/k in 1/m; RMSE in °C).

profiles on the specimen surface and those from the location close to the surface are not very smooth and might cause error when used to back-calculate the thermal parameters. Therefore, to reduce the errors and obtain the back-calculated thermal parameters of high accuracy, the measured temperature profiles used for back-calculation should be as close to the center of a specimen as possible.

Based on this observation, the effective averages of thermal properties of the asphalt and concrete specimens (A0 and C0) were calculated and are shown in Table 6.5. The thermal conductivity k and heat capacity c of the concrete specimen are larger than those of the asphalt specimen. These values are comparable to those found in the literature (Table 6.1). This

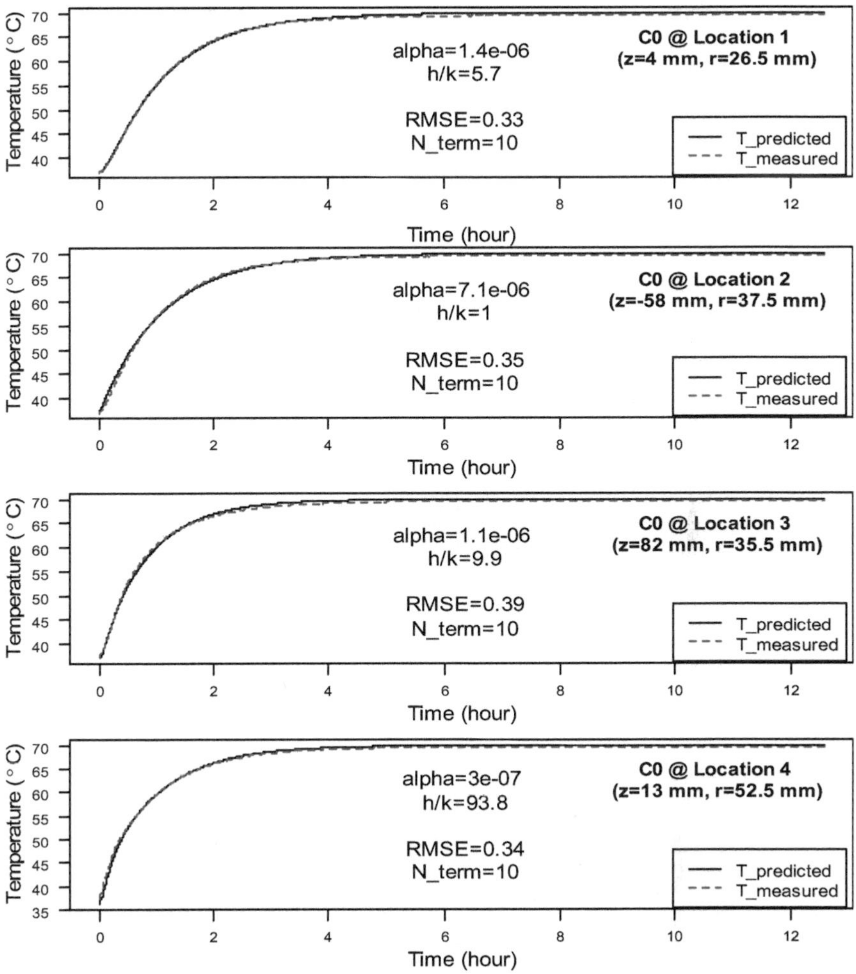

Figure 6.12 Predicted temperature with the optimized thermal properties compared with measured temperature: concrete specimen C0 (units: α in m^2/s; h/k in 1/m; RMSE in °C).

verifies partly that the method and procedure developed here are practical and valid for measuring the thermal properties of building materials.

6.6.4 Influence of the Length of Testing Time on the Optimized Parameters

The optimized parameters shown in Figures 6.11 and 6.12, and Table 6.5 were obtained according to the measured temperature profiles, which were completely developed curves for the transient heat transfer between the

Table 6.4 Parameters for convection coefficient *h*

Parameter	Unit	Value	Parameter	Unit	Value
Thermal conductivity k	W/(m °C)	2.63×10^{-2}	Kinematic viscosity v	m^2/s	1.59×10^{-5}
Heat capacity c	J/(kg °C)	1007	Density ρ	kg/m^3	1.1614
Thermal diffusivity α	m^2/s	2.25×10^{-5}	Prandtl number Pr	—	0.707
For asphalt specimen A0			For concrete specimen C0		
Airflow speed U	m/s	6	Airflow speed U	m/s	6
Specimen diameter D	m	0.102	Specimen diameter D	m	0.145
Re_D	—	3.85×10^4	Re_D	—	5.48×10^4
C	—	0.0266	C	—	0.0266
m	—	0.805	m	—	0.805
Nu	—	116.44	Nu	—	154.56
h	$W/(m^2\,°C)$	30.02	h	$W/(m^2\,°C)$	28.03

Note: $Re_D = UD/v$; $Nu = CRe_D{}^m Pr^{1/3}$; $h = kNu/D$; $Pr = v/\alpha$.

Table 6.5 Thermal properties calculated from the optimized parameters

Asphalt specimen A0						
Location	1	2	3	4	Avg[a]	St dev[a]
Thermal diffusivity α ($\times 10^{-7}$ m^2/s)	5.7	6.2	50.2	–	6.0	0.33
Ratio h/k (1/m)	19.3	17.3	2.2	–	18.3	1.4
Density ρ (kg/m^3)	2405					
Convection coefficient h ($W/(m^2\,°C)$)	30.02					
Thermal conductivity k($W/(m\,°C)$)	1.56	1.74	13.39	–	1.65	0.13
Heat capacity c ($J/(kg\,°C)$)	1133.6	1167.5	1108.2	–	1150.6	24.0
RMSE (°C)	0.18	0.19	0.28	–	0.19	0.01
Concrete specimen C0						
Location	1	2	3	4	Avg[b]	St dev[b]
Thermal diffusivity α ($\times 10^{-7}$ m^2/s)	14.2	70.9	11.0	3.0	12.6	2.2
Ratio h/k (1/m)	5.7	1.0	9.9	93.8	7.8	3.0
Density ρ (kg/m^3)	2481					
Convection coefficient h ($W/(m^2\,°C)$)	28.03					
Thermal conductivity k ($W/(m\,°C)$)	4.94	28.03	2.83	0.30	3.88	1.49
Heat capacity c ($J/(kg\,°C)$)	1399.9	1592.9	1033.4	405.9	1216.7	259.2
RMSE (°C)	0.33	0.35	0.39	0.34	0.36	0.04

[a]Average/standard deviation on locations 1 and 2.
[b]Average/standard deviation on locations 1 and 3.

specimens and the surrounding airflow. The total time was over 12 h, which is a long time. Two questions arise, which are whether the length of testing time influences the optimized results and what the shortest time needed to run the test for back-calculation of the thermal properties.

To answer these questions, taking the asphalt specimen as an example, the optimized parameters with the measured temperature profiles with different testing time lengths were back-calculated and plotted in Figure 6.13. From the plots, it is noticed that the back-calculated parameters do change with the length of testing time. When the testing time is over 4 h, the back-calculated parameters tend to be constant. Therefore,

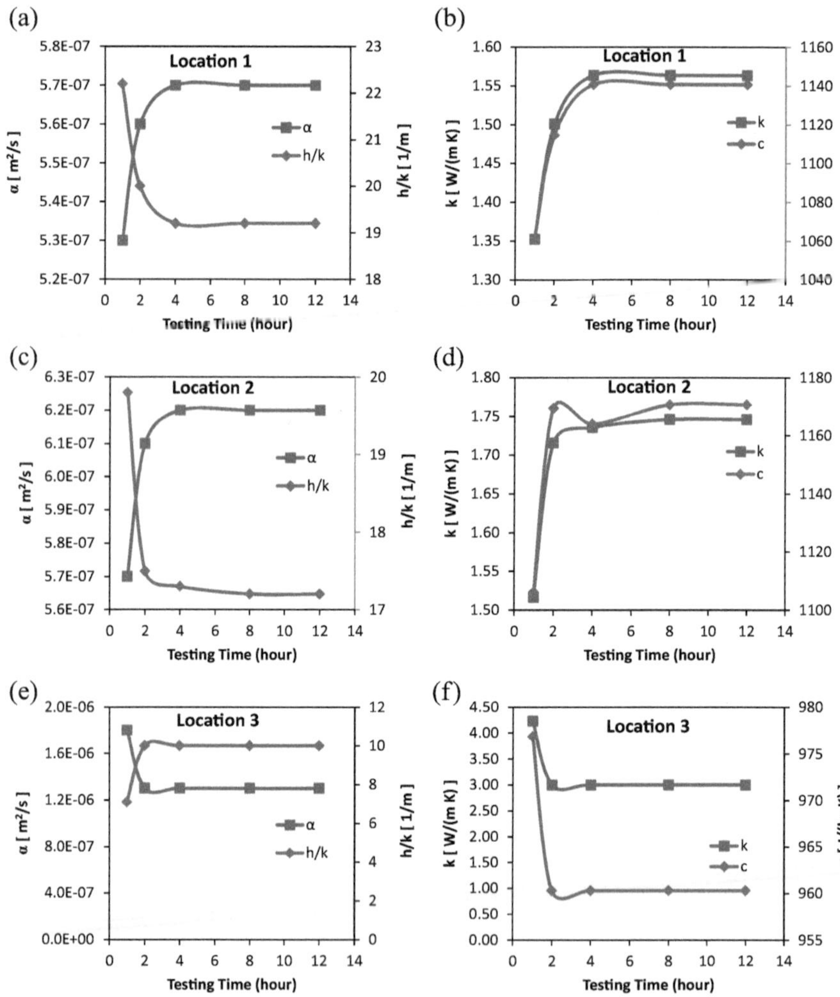

Figure 6.13 The influence of testing time on the optimized parameters (A0). (a) α and h/k at location 1, (b) k and c at location 1, (c) α and h/k at location 2, (d) k and c at location 2, (e) α and h/k at location 3, (f) k and c at location 3.

the testing time for back-calculation is recommended to be at least 4 h to balance the accuracy and cost in terms of both testing time and energy consumption. Otherwise, some error might occur in the back-calculated parameters.

6.7 THERMAL PROPERTIES OF SURFACE MATERIALS USED IN EXPERIMENTAL SECTIONS

The developed and validated procedure was used to measure the thermal properties of pavement materials, specifically some surface materials used for the experimental sections in this study. Cylinder specimens made of some of the materials (mainly concrete and asphalt, both dense graded and open graded) were tested at various locations (two specimen sizes for concrete: 4 × 6 and 6 × 12 in). Some concrete specimens used for testing and test setup are shown in Figure 6.14. Results are listed in Tables 6.6 and 6.7. These values will be used for modeling and simulation of thermal performance of pavements in Chapters 11 and 12.

From Table 6.7, it is noted that the dense-graded materials (concrete or asphalt) have a higher thermal conductivity and heat capacity than the open-graded materials. This might give a potential explanation for why the permeable pavement constructed with open-graded materials presents a higher daytime temperature and a lower nighttime temperature under the dry condition compared to the impermeable pavement constructed with dense-graded materials, as observed in Chapter 8. The permeable asphalt pavement sections (B2 and B3) presented a higher daytime surface temperature than the impermeable pavement (B1) under the dry condition. This might be because the lower thermal conductivity increases the thermal

Figure 6.14 Example concrete specimens used for testing and test setup.

Table 6.6 Thermal properties of pavement surface materials at various test locations

Sample[a]	T_0 (°C)	T_a (°C)	Location[b]	α (m²/s)	h/k (1/m)	ρ (kg/m³)	h (W/(m² °C))	k (W/(m °C))	c (J/(kg °C))
C1S4	32.9	60.0	1	7.85E-07	16.5	2203	30.0	1.82	1052
C1S4	32.9	60.0	2	8.27E-07	15.5	2203	30.0	1.93	1059
C1S4	32.9	60.0	3	8.47E-07	17.1	2203	30.0	1.75	938
C1B4	35.6	59.8	1	8.13E-07	15.5	2311	27.8	1.79	953
C1B4	35.6	59.8	2	8.20E-07	15.3	2311	27.8	1.81	955
C1B4	35.6	59.8	3	7.78E-07	14.7	2311	27.8	1.89	1051
C2S1	33.2	60.0	1	7.81E-07	24.6	1866	30.1	1.22	837
C2S1	33.2	60.0	2	7.45E-07	22.3	1866	30.1	1.35	971
C2S1	33.2	60.0	3	9.69E-07	21.3	1866	30.1	1.41	780
C2B4	35.2	59.8	1	7.25E-07	21.1	2065	27.8	1.32	882
C2B4	35.2	59.8	2	7.12E-07	19.3	2065	27.8	1.44	979
C2B4	35.2	59.8	3	7.57E-07	18.4	2065	27.8	1.51	966
C2B4	35.2	59.8	4	6.96E-07	20.0	2065	27.8	1.39	967
C3S1	42.9	54.9	1	7.64E-07	22.8	2051	30.1	1.32	842
C3S1	42.9	54.9	2	7.31E-07	24.9	2051	30.1	1.21	807
C3S1	42.9	54.9	3	7.72E-07	21.7	2051	30.1	1.39	878
C3B4	42.0	54.9	1	8.18E-07	22.1	2176	27.8	1.26	708
C3B4	42.0	54.9	2	8.29E-07	22.1	2176	27.8	1.26	698
C3B4	42.0	54.9	3	7.98E-07	21.3	2176	27.8	1.31	754
C3B4	42.0	54.9	4	8.21E-07	21.6	2176	27.8	1.29	722
B1S1	29.6	59.8	1	8.52E-07	18.2	2399	30.1	1.65	807
B1S1	29.6	59.8	2	8.33E-07	17.5	2399	30.1	1.72	861
B1S1	29.6	59.8	3	8.53E-07	16.5	2399	30.1	1.82	889
B2S1	29.0	59.8	1	7.16E-07	23.4	2269	30.2	1.29	794
B2S1	29.0	59.8	2	7.09E-07	24.5	2269	30.2	1.23	765
B2S1	29.0	59.8	3	7.19E-07	25.4	2269	30.2	1.19	729
B3S1	59.9	79.8	1	7.05E-07	24.2	2239	30.1	1.24	785
B3S1	59.9	79.8	2	7.95E-07	23.3	2239	30.1	1.29	724
B3S1	59.9	79.8	3	6.33E-07	25.8	2239	30.1	1.17	824

T_0 = initial specimen temperature; T_a = chamber air temperature; α = thermal diffusivity; ρ = density; h = convection coefficient; k = thermal conductivity; c = heat capacity.
[a]Sample.
[b]Location.

Table 6.7 Summary of average thermal properties of pavement surface materials

Section	Material	Density (kg/m^3)	Thermal diffusivity (m^2/s)	Thermal conductivity (W/(m °C))	Heat capacity (J/(kg °C))
C1	Concrete-D	2257 (59)	8.12E-07 (2.61E-08)	1.83 (0.07)	1001 (58)
C2	Concrete-O	1980 (106)	7.69E-07 (9.25E-08)	1.38 (0.09)	912 (80)
C3	Concrete-O	2134 (65)	7.95E-07 (3.74E-08)	1.29 (0.06)	761 (69)
B1	Asphalt-D	2399 (—)	8.46E-07 (1.13E-08)	1.73 (0.09)	852 (42)
B2	Asphalt-O	2269 (—)	7.15E-07 (5.13E-08)	1.24 (0.05)	763 (32)
B3	Asphalt-O	2239 (—)	7.11E-07 (8.15E-08)	1.23 (0.06)	798 (50)

D, dense graded; O, open graded. Standard deviation listed in parentheses.

resistance and thus the difficulty in conducting heat into deeper pavement layers, which keeps the heat around the surface. Moreover, because of their lower heat capacity compared to dense-graded materials, the open-graded materials will be heated up to a higher temperature with the same amount of energy absorbed from solar radiation or surroundings during daytime or hot periods. On the other side, owing to just their lower heat capacity, the open-graded materials have less thermal energy (or heat) stored in the solid body around the surface to lose into the cold ambient air during the nighttime or cold periods. Also, because of their lower thermal conductivity compared to the dense-graded materials, it is much more difficult for the open-graded materials to conduct heat to the surface from underlying layers to supply more energy for loss. These two aspects will significantly increase the possibility that the open-graded materials will produce a lower surface temperature when the same amount of heat is lost into the ambient air during the nighttime or cold periods. This theoretically confirms that the permeable pavement composed of open-graded materials can be a potential strategy to counter the nighttime heat island effect because of its lower surface temperature and less heat released into the ambient air during the night. However, as mentioned previously, attention should be given to its potentially higher surface temperature during the daytime under the dry condition.

In addition, it is noted that concrete materials generally show a slightly higher thermal conductivity and heat capacity than asphalt materials. This implies that when all the other factors and conditions (e.g., albedo, solar radiation, convection, etc.) are identical, the concrete pavements will still give a lower daytime surface temperature compared to the asphalt pavements. More details on the effects of thermal properties on pavement temperature will be discussed in Chapter 12.

6.8 SUMMARY AND CONCLUSIONS

This chapter was devoted to discussion of questions related to the multidimensional modeling and simulation of transient temperatures at any location on a beam or cylinder specimen of various sizes subject to convective heat transfer, and back-calculation of the thermal properties of specimens of various shapes and sizes from the measured transient temperatures profile of the specimen. The effects of thermal properties on the thermal behavior of thermal resistance pavements are discussed.

1. The model and tool developed can accurately predict transient temperature at any location on a beam or cylinder specimen of various sizes subject to convective heat transfer. The case studies presented in this chapter verify that they can also be used, if the thermal properties are known, to simulate the transient temperature and predict the time it takes to reach a specified target testing temperature at any location for specimens of various shapes and sizes, when the specimens are preheated or precooled in a forced convection oven or temperature chamber for mechanical or other temperature-related testing.
2. Based on the developed temperature simulation model, the procedure and tool for back-calculating thermal properties were developed and validated in part by case studies on both asphalt and concrete materials. The partly validated procedure and tool can be used to obtain the thermal properties of a specimen of pavement and building materials from its measured transient temperature profile, regardless of the shape and size of the specimen. With the fully validated procedure and tool, thermal properties of novel building materials (various innovative cool materials such as porous concrete and high-thermal-resistance materials) can be easily measured and then used for evaluating and modeling the thermal performance of a built environment composed of these materials.
3. The developed and validated procedure was employed to measure the thermal properties of pavement materials using cylinder specimens, particularly some surface materials used for the experimental sections in this study. It was found that the dense-graded (i.e., nonporous)

materials (concrete or asphalt) have a higher thermal conductivity and heat capacity than the open-graded (i.e., porous) materials. Concrete materials generally show a slightly higher thermal conductivity and heat capacity than asphalt materials.

The open-graded (i.e., porous) thermal-resistant materials (e.g., permeable pavement) could present a higher daytime surface temperature than the dense-graded (i.e., nonporous) materials (e.g., impermeable pavement) under the dry condition. This might be because the lower thermal conductivity increases the thermal resistance and thus the difficulty in conducting heat into deeper pavement layers, which keeps the heat around the surface. Moreover, because of their lower heat capacity compared to dense-graded materials, the open-graded materials will be heated up to a higher temperature under the same amount of energy absorbed from solar radiation or the surroundings during the daytime or hot periods. On the other side, owing just to their lower heat capacity, the open-graded materials have less thermal energy (or heat) stored in the solid body around the surface to lose into the cold ambient air during the nighttime or cold periods. Also, because of their lower thermal conductivity compared to the dense-graded materials, it is much more difficult for the open-graded materials to conduct heat to the surface from the underlying layers to supply more energy for loss. These two aspects will significantly increase the possibility that the open-graded materials will produce a lower surface temperature when the same amount of heat is lost into the ambient air during the nighttime or cold periods. This theoretically confirms that the permeable pavement composed of open-graded materials can be a potential strategy to counter the nighttime heat island effect because of its lower surface temperature and less heat released into the ambient air during the night. However, as mentioned previously, attention should be given to its potentially higher surface temperature during the daytime under dry conditions.

CHAPTER 7

Evaporation Rate and Evaporative Cooling Effect of Pavement Materials

Contents

7.1 INTRODUCTION

7.1.1 Background

Cool pavement strategies can be used to mitigate urban heat island effects and improve outdoor thermal environments in urban areas. Improved outdoor thermal environments in urban areas could potentially help reduce the negative impacts of heat islands such as increased energy consumption for air conditioning of buildings and vehicles and impaired air quality (ground-level ozone) [32,33,65,66,71,76,141,152–154]. In addition, as a strategy to reduce vehicle miles traveled by creating livable and walkable communities, improving the street thermal environment is attracting increasing attention from practitioners, academics,

Pavement Materials for Heat Island Mitigation
ISBN 978-0-12-803476-7
http://dx.doi.org/10.1016/B978-0-12-803476-7.00007-6

and competing industries [8–12]. Increased walking or cycling also provides an opportunity for improving human health and thus improving the quality of life [6,7].

With respect to the pavement type, the heat island might not just be a "black or white" issue (asphalt versus concrete, although it is known that albedo is a function of material microstructure and rugosity, not just color), but also might be an "impervious or pervious" issue. To mitigate local heat islands and reduce the associated impacts mentioned previously, some impervious surface coverage can be replaced by pervious coverage. This is also a requirement for limiting disruption of natural hydrology according to LEED (Leadership in Energy & Environmental Design) [155] criteria, which require various options for project sites with impervious area greater or less than 50%. Permeable pavement, as a low-impact development, can help minimize impervious surfaces and could potentially improve the quality of life in a community and reduce other environmental impacts, for example, reducing or at least slowing (where there is not full infiltration) storm-water runoff and associated water pollution, reducing storm-water management facilities, enhancing on-site infiltration for vegetation growth, and recharging underground water [56,76,152,156–160].

As one potential cool pavement type, permeable pavements have many environmental benefits beyond conventional impermeable pavements as mentioned above (note that asphalt, concrete, and integrated concrete paver pavements can be considered for applications as permeable pavements). The main potential cooling mechanisms for permeable pavements are evaporative cooling [161,162], heat resistance [153,161,162], and reflection and evaporative cooling, if using reflective permeable pavement [162,163].

Evaporation is energy transmitted away from the pavement surface by the latent heat of water vapor to achieve the phase change of water from liquid to gas. Water from moist soil or wet surface changes to vapor when heated by the sun or other heat sources. Water vapor then rises into the atmosphere, taking the solar energy with it, resulting in a cooling effect. The evaporation term also includes evapotranspiration, a more complicated process plants use to keep cool. During evapotranspiration, water is drawn from the soil by the roots of the plant and evaporates through the stoma on the plant's leaves. Both evaporation and evapotranspiration increase when there is more moisture available, when wind speeds are greater, and when the air is drier and warmer [2]. The evaporation latent heat loss q_{evap} (W/m^2) (i.e., theoretical maximum cooling effect) can be described as follows when water is completely exposed to air:

$$q_{evap} = L \cdot ER \tag{7.1}$$

where ER is the evaporation rate and L is the specific latent heat of water vaporization.

Evaporative cooling could reduce pavement temperature and consequent air temperature through latent heat absorbed during the phase change of water (from liquid to gas) when moisture exists in the pavement or in the underlying soil or is sprinkled on the pavement surface. Permeable pavements can provide these benefits. From the equation above, it is seen that the latent heat loss (q_{evap}) from pavement is linearly and positively correlated with the evaporation rate (ER) for water exposed to air. The effect of evaporative cooling of permeable pavement depends highly on the evaporation rate of the permeable pavement materials [161,162]. Therefore, it is of great interest to explore and better understand the evaporation rate for the various pavement materials used for permeable pavements.

7.1.2 Objective and Scope

The objectives of this study are to measure and compare the evaporation rates of various pavement materials under outdoor conditions and provide typical evaporation rate values that will be useful for the modeling and simulation of the cooling effects of evaporation from permeable pavements and to explore the first-order factors affecting the evaporation rates of permeable pavement materials.

7.2 MATERIALS AND METHODS

7.2.1 Description of Test Materials

The materials used for measurement of evaporation rate include water in six different pavement materials along with fully exposed water for comparison. These six pavement materials fall into two categories: permeable pavement surface-layer materials and base-layer materials or bedding-layer materials. All these six types of materials are open-graded materials used in experimental pavement sections constructed at the University of California Pavement Research Center (UCPRC) test facilities in Davis, California.

A total of nine experimental pavement sections (A1–A3, B1–B3, and C1–C3) were constructed for the cool pavement study at the UCPRC.

These nine test sections include three different pavement surfacing materials, namely integrated concrete pavers (surfacing type A), open-graded asphalt concrete (surfacing type B), and Portland cement concrete (surfacing type C). For each pavement surface type, one impermeable pavement design (design 1) and two permeable pavement designs (designs 2 and 3, can also be referred to as porous or pervious depending on the material; the word permeable is used for convenience in this chapter) were constructed. Six of the nine sections are permeable pavements. Each section is 4 m by 4 m in size (see previous chapters and Refs [161–164] for more details).

Three permeable pavement surface-layer materials were chosen for the evaporation testing in this chapter, which were permeable asphalt B3 and permeable concretes C2 and C3. The permeable asphalt B2 was open-graded asphalt concrete with a nominal maximum aggregate size (NMAS) of 9.5 mm and PG 64-10 asphalt binder. The main differences between permeable concretes C2 and C3 were the gradation and cement type. Section C2 had a finer gradation, with an NMAS of 4.75 mm and conventional gray cement, while C3 had a coarse gradation with an NMAS of 9.5 mm and a whiter cement. Owing to the whiter cement with much lighter color, the C3 concrete sample had an albedo of 0.26, which was larger than that of 0.18 for the C2 concrete sample (see Table 7.1).

The three base-layer materials or bedding-layer materials included gravel S1 and S2 and sand S3. The gravel S1 was used as open-graded base aggregate reservoir layers in the six experimental permeable pavement sections mentioned above. The size of S1 was 19 mm. The gravel S2 was an ASTM No. 8 aggregate (with an NMAS of 12.5 mm, finer than S1) and was used as a bedding-layer material for permeable paver sections A2 and A3. The gravel S3 was an ASTM C33 sand (with an NMAS of 9.5 mm, finer than S2) and was used as bedding-layer material for impermeable paver section A1. In addition, a fully exposed water sample, S0, was used for reference and comparison.

The gradations of these six materials are presented in Figure 7.1. Section C3 had the same NMAS as B3, which was 9.5 mm. However, C3 was more open graded with coarser aggregate compared to B3, as shown in Figure 7.1. The gravel S1 was the coarsest material, followed by S2, while the sand S3 was the finest material.

In addition to the gradation, the albedo, permeability, air void content, and density of these materials were measured and are listed in Table 7.1 for reference (see Refs [161,163–165] for details on measurements). The summary of each material along with material designs and other characteristics is listed in Table 7.1.

Table 7.1 Characteristics of permeable pavement materials used in this chapter

Material no.	Material type (NMAS)[a]	Thickness (mm)	Albedo [b]	Permeability[c] (cm/s)	Air void	Density (kg/m^3)
B2	Asphalt-O (9.5 mm) + aggregate-O (19 mm)	60 + 90	0.08	0.11	12%	2270
C2	Concrete-O (4.75 mm, No. 4)	150	0.18	0.21	15%	1980
C3	Concrete-O (9.5 mm)-white cement	150	0.26	0.29	17%	2050
S1	Aggregate-O (19 mm, ~ASTM No. 57)	150	0.18	2.5	46%	1650
S2	Aggregate-O (12.5 mm, ASTM No. 8)	150	0.20	1.2	25%	2100
S3	Sand (9.5 mm, ASTM C33)	150	0.30	5×10^{-3}	15%	2400
S0	Bare water	150	—[d]	—[d]	—[d]	1000

[a]O = open graded.
[b]Using dual pyranometer (see Ref. [163]).
[c]Using ASTM C1701 method (see Ref. [164]).
[d]— = Not applicable.

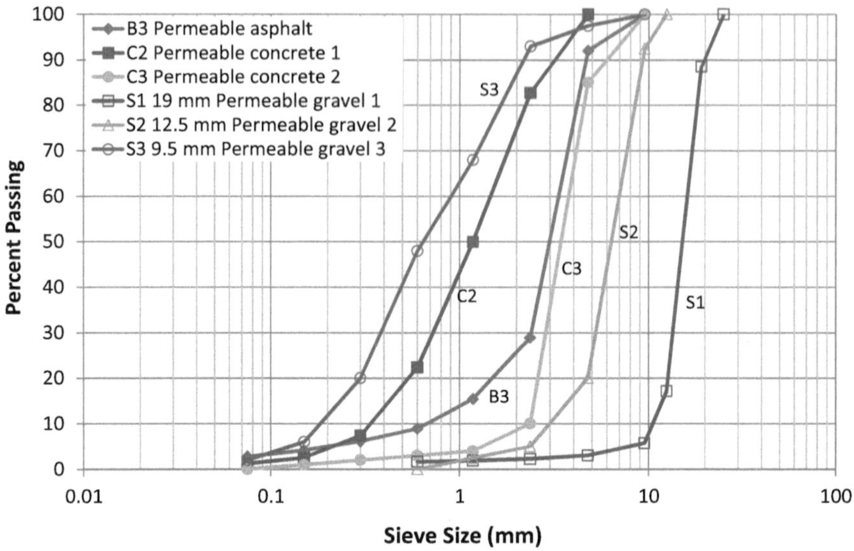

Figure 7.1 Gradations of materials tested in this chapter.

7.2.2 Experimental Plan

The samples of these materials were put into 100-mm diameter × 150-mm height plastic cylinder containers (Figure 7.2). The sample of permeable asphalt B3 was thin (60 mm), and the gravel S1 (19 mm NMAS) was used

Figure 7.2 Sample preparation (B3, C3, and C2 are dark in (d) because of wetting). (a) Samples and cylinder containers; (b) Gravel S1 used to fill up the cylinder under B3; (c) Samples in cylinder containers; (d) Sample in cylinders filled with water.

to fill up the cylinder (Figure 7.2(b)) under B3, simulating the full permeable pavement structure (surface layer + reservoir layer).

Each dry sample and cylinder container were then weighed together and recorded as m_1. The water was then added into each container slowly, ensuring it was filled up with water (Figure 7.3). The permeable pavement or aggregate base will not be full of water and only partially saturated for most of the time and regions. However, to simulate the permeable pavement just after a heavy rain event or extensive irrigation, the cylinders were first completely filled with water and the water level was allowed to drop through evaporation. This provides the opportunity to monitor well the change in evaporation rate over time and better investigate the effect of water level on evaporation rate. The overflow and surface water was dried using a towel. Then the total weight of the sample, container, and water was measured and recorded as m_{20}. After that, the samples in containers were moved outdoors and placed under the sun for evaporation (Figure 7.4). The total weight of each sample plus the container and remaining water was measured over time t and recorded as m_{2t}. The water weight left in the container at time t would be

$$m_{\mathrm{wt}} = m_{2t} - m_1 \tag{7.2}$$

where m_{wt} is the mass (kg) of water left in the container at time t,

m_{2t} is the total mass (kg) of a sample plus the container and remaining water at time t, and

m_1 is the total mass (kg) of a sample plus the container.

The weight loss (i.e., the water evaporated) over time t for each sample under outdoor conditions can be calculated as

$$\Delta m_{\mathrm{wt}} = m_{20} - m_{2t} \tag{7.3}$$

Figure 7.3 Adding water to fill up the cylinder containers.

Figure 7.4 Evaporation testing under outdoor conditions.

where Δm_{wt} is the mass (kg) of water evaporated from the container at time t,

m_{20} is the initial total mass (kg) of a sample plus the container and remaining water at time 0, and

m_{20} is the total mass (kg) of a sample plus the container and remaining water at time t.

The evaporation rate (ER, in kg/m^2/h or mm/h) during the time period t_1 through t_2 will be

$$ER = (m_{2t_1} - m_{2t_2})/[A(t_2 - t_1)] \tag{7.4}$$

where ER is evaporation rate, in kg/m^2/h or mm/h,

m_{2t_1} is the total mass (kg) of a sample plus the container and remaining water at time t_1 (h),

m_{2t_2} is the total mass (kg) of a sample plus the container and remaining water at time t_2 (h), and

A is the sample surface area, in m^2.

The latent heat flux q_{evap} (kJ/m^2/h) lost from evaporation (i.e., cooling effect) could be calculated as

$$q_{evap} = L \cdot ER \tag{7.5}$$

where q_{evap} is the theoretical maximum latent heat flux lost from evaporation (i.e., cooling effect), in kJ/m^2/h,

L is the specific latent heat for water vaporization, 2260 kJ/kg, and
ER is the evaporation rate of water, in kg/m^2/h or mm/h.

In addition, temperatures on the surfaces of the various materials were also measured using a thermometer. The weather data (including air temperature, relative humidity, solar radiation, wind speed, rain, and air pressure) were also monitored using a nearby weather station. The evaporation test was conducted on clear summer days in July 2012 (9 to 25 July) at Davis, California. The weight and the surface temperature for each sample were manually measured every hour. The weather data were automatically measured every half hour. The data in the first 3 days (July 9 to July 11) are presented in this chapter (same trends for other days).

7.2.3 Data Analysis and Presentation

The weather data, the surface temperature, the water weight left in the containers, the evaporation rate, and the latent heat loss of each sample were plotted using time series plots over the experimental period (9–11 July) to examine the changes in these fundamental variables over time. In addition, the average evaporation rates over 3 days (9–11 July) and 1 day (10 July) were statistically analyzed and plotted using box plots for each type of material, to compare the differences in evaporation rate between materials and provide average values of evaporation rate for each materials on clear summer days. The effects of permeability and air void content and the water level on the evaporation rate are also examined through data plotting.

7.3 RESULTS AND DISCUSSION

7.3.1 Weather Data and Surface Temperature Change over Time

The weather data during the experimental period of 9–11 July 2012 are presented in Figure 7.5. The weather was clear and sunny with high air temperatures of 36 –39 °C. The peak solar radiation intensity around noon was approximately 1000 W/m^2 for all 3 days. The wind speed was lower than 2 m/s for most of the time. The low daytime relative humidity was approximately 20% while the nighttime relative humidity was approximately 80%. The weather conditions during these 3 days were similar, except the air temperature on the third day (11 July) was slightly higher than on the other 2 days.

The surface temperatures of each sample are presented in Figure 7.6. The asphalt sample had the highest surface temperature during evaporation

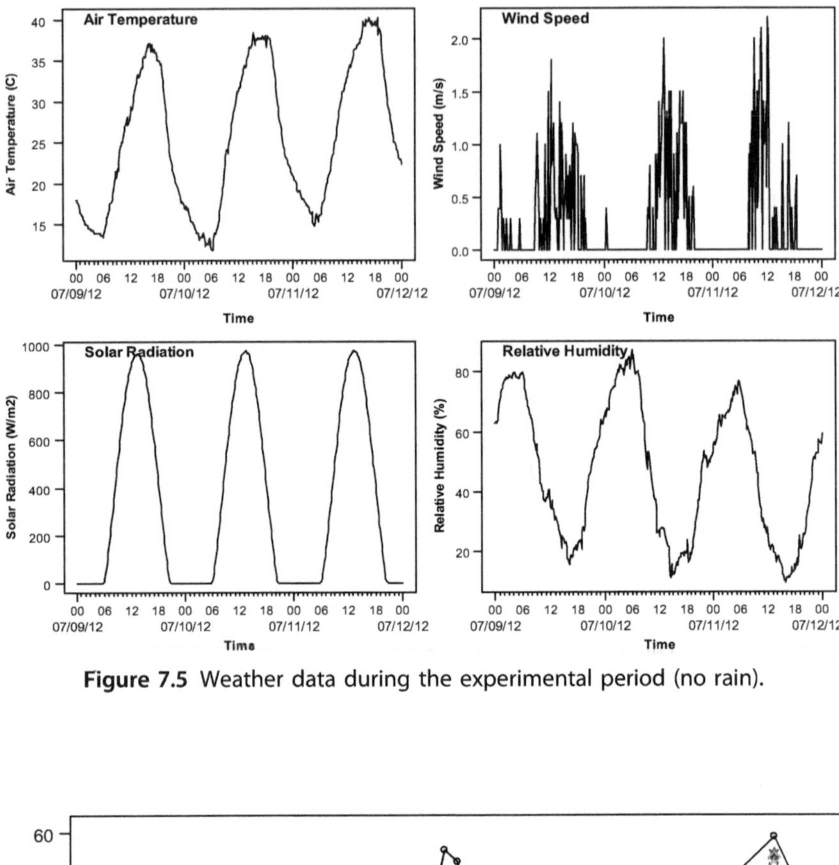

Figure 7.5 Weather data during the experimental period (no rain).

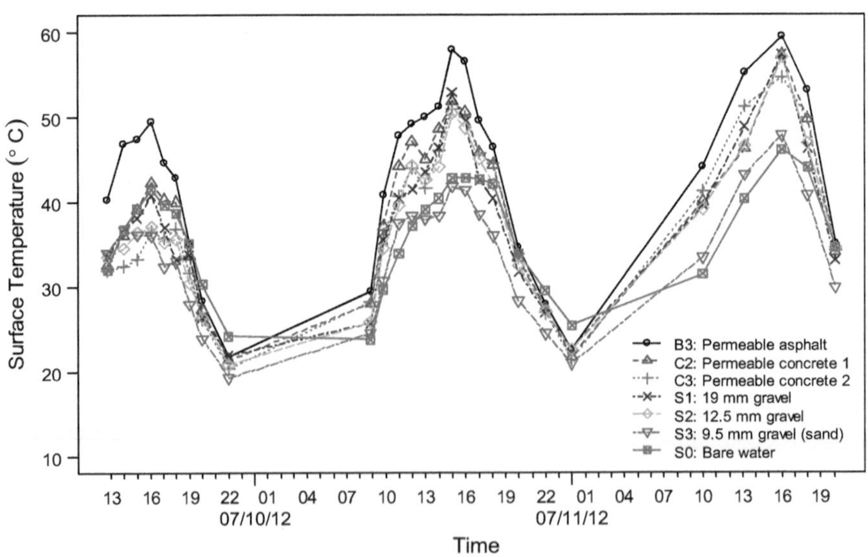

Figure 7.6 Surface temperature change over time.

testing because of its dark color and low albedo of 0.08 (see Table 7.1), which absorbs more solar energy compared to the other surfaces with higher albedos. The peak surface temperature was approximately 50 °C on the first experiment day (9 July) and reached 60 °C on the third day (11 July). The gravel sample S3 (sand) shows the lowest surface temperature owing to its higher albedo of 0.30 (see Table 7.1) and highest evaporation rate (see below for more details), except for the bare water, S0.

7.3.2 Water Weight Change over Time

The water weight change over time is presented in Figure 7.7. The bare water (S0) had the most water available and the largest change rate (slope of the curve). The gravel samples S1, S2, and S3 also had relatively larger change rates. The permeable concrete samples C3 and C2 and the permeable asphalt sample B3 had relatively smaller change rates. This implies that the permeable concrete samples C3 and C2 and the permeable asphalt sample B3 have lower evaporation rates compared to the gravels (S1–S3) or the bare water (S0). The reason for their low evaporation rates might be the low air void content and low permeability of these materials (see Table 7.1), which to some extent block the water evaporation from the materials. It should be noted that all samples still had more than 200 g of water available in the container even at the end of the third day (11 July).

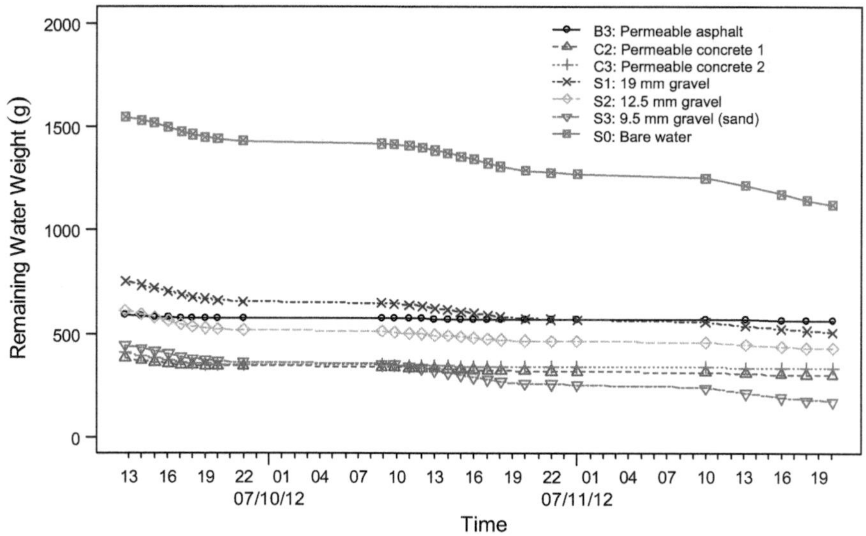

Figure 7.7 Water weight change over time.

7.3.3 Evaporation Rate and Cooling Effect (Latent Heat Flux) Change over Time

Figure 7.8 presents the evaporation rate and latent heat loss change over time for various materials. The evaporation rates of all materials show a

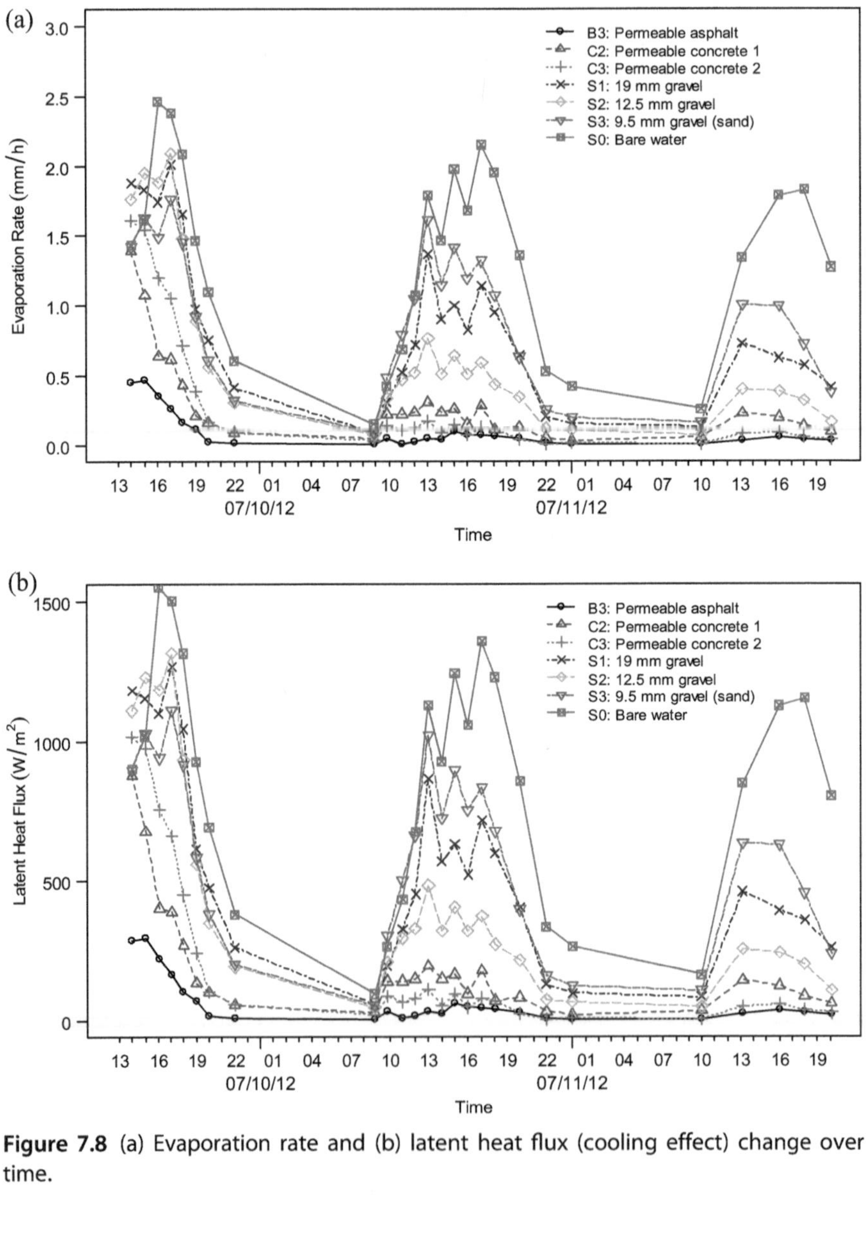

Figure 7.8 (a) Evaporation rate and (b) latent heat flux (cooling effect) change over time.

pattern similar to that of the air temperature, which is high in the daytime and low in the nighttime. The bare water shows the highest evaporation rate. The daily peak evaporation rate reaches 2.5 mm/h on the first day and decreases to 1.5 mm/h on the third day, although the air temperature is slightly higher on the third day (see Figure 7.5). Excluding the bare water, the gravel sample S3 (sand) generally has the highest evaporation rate apparently owing to its ability to move moisture to the surface for evaporation through capillary action; its peak evaporation rate drops from 1.8 mm/h on the first day to 1.0 mm/h on the third day. The surface materials of permeable concrete C2 and permeable asphalt B3 have the lowest evaporation rates, which are slightly lower than that of permeable concrete C3. Permeable concretes C3 and C2 and permeable asphalt B3 have much higher evaporation rates (0.5–1.5 mm/h) during the first experiment day when more water was available near the surface. Evaporation rates are lower (0.1–0.3 mm/h) on the second and third day, with less moisture available near the surface, although there still is adequate water available in the lower part of the containers and samples.

7.3.4 Average Evaporation Rates of Various Materials for Various Periods

The evaporation rates over time during the experimental period (9–11 July 2012) are calculated and plotted in Figure 7.9 using box plots for various materials, including over the 3 days (9–11 July), over the whole middle day (10 July), and in the morning, noon, afternoon, and night of the middle day. In addition, the descriptive statistics, including the number of measurements (N), mean, standard deviation (Std. Dev.), minimum (Min), first quartile (Q1), median, third quartile (Q3), and maximum (Max), for the evaporation rate for two main periods (3 days and 1 day) are presented in Table 7.2 for reference.

Figure 7.9(a) and (b) shows the evaporation rates over the 3 days (9–11 July) and over 1 day (10 July), respectively. Overall, the 3-day average evaporation rates are slightly higher than the 1-day average owing to the high availability of water close to surfaces and consequently the high evaporation rates on the first day (9 July, see Figure 7.7). This suggests that the availability of water close to surfaces is an important factor that influences the evaporation rate.

It is noted that the evaporation rate is high during the daytime, especially in the hot afternoon in summer (Figure 7.9(d) and (e)), and low during the nighttime and especially in the cold early morning (Figure 7.9(c) and (f)).

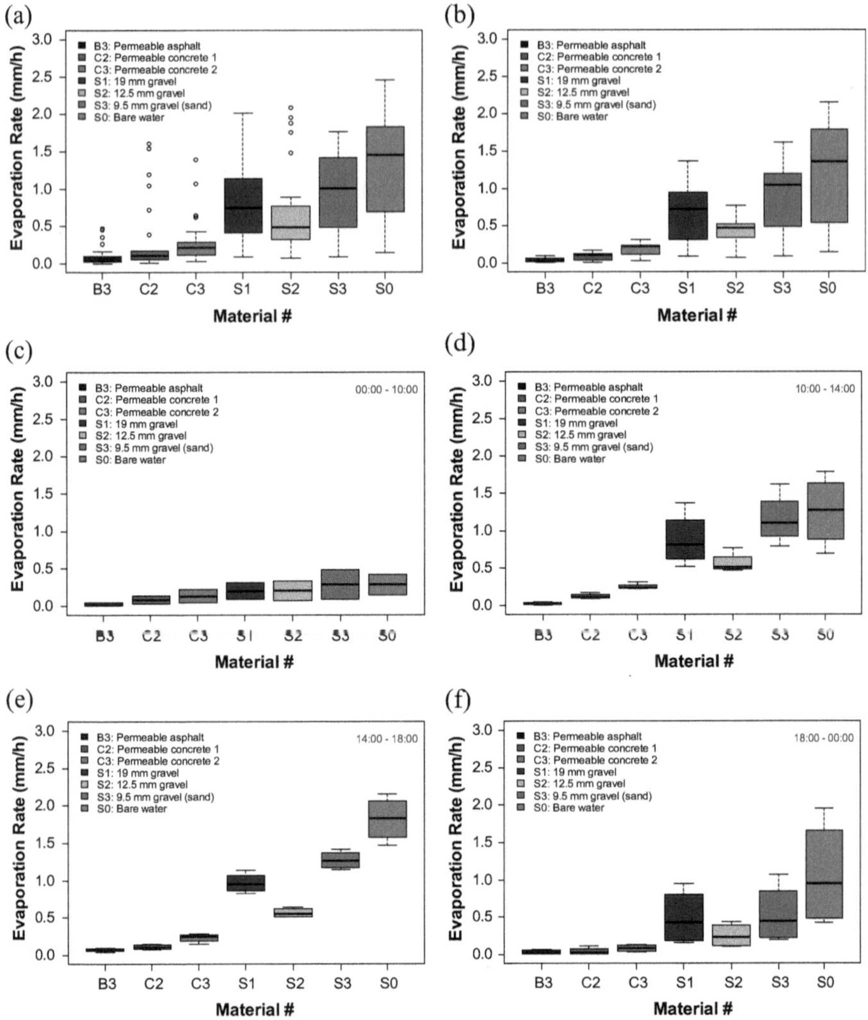

Figure 7.9 Average evaporation rates of various materials. (a) 3-day average (9–11 July). (b) 1-day average (10 July). (c) Morning (00:00–10:00 h, 10 July). (d) Noon (10:00–14:00 h, 10 July). (e) Afternoon (14:00–18:00 h, 10 July). (f) Night (18:00–0:00 h, 10 July). The center thick black horizontal line in each box of the box plots is the median value. The colored box indicates the first quartile (Q1) and the third quartile (Q3). The bars outside the box are the minimum and maximum values except for outliers. The circles are outliers defined as being more than 1.5 (Q3 – Q1) from Q1 or Q3.

As expected, the air temperature (also air humidity) also plays a vital role in the evaporation rate of the pavement materials. High air temperature and low humidity will help enhance the evaporation rate, which is just what is expected during hot periods.

Table 7.2 Descriptive statistics of the evaporation rate for various periods

Period	Sample ID	N	Mean	Std. dev.	Min	Q1	Median	Q3	Max
9–11 July	B3	26	0.10	0.13	0.00	0.02	**0.05**	0.09	0.47
	C2	26	0.31	0.48	0.00	0.06	**0.11**	0.17	1.61
	C3	26	0.29	0.32	0.03	0.12	**0.22**	0.28	1.39
	S0	26	1.36	0.67	0.15	0.78	**1.45**	1.82	2.46
	S1	26	0.87	0.57	0.09	0.44	**0.74**	1.10	2.01
	S2	26	0.68	0.61	0.07	0.32	**0.49**	0.73	2.08
	S3	26	0.93	0.51	0.09	0.51	**1.00**	1.39	1.76
10 July	B3	13	0.04	0.03	0.01	0.01	**0.05**	0.06	0.10
	C2	13	0.09	0.05	0.01	0.04	**0.11**	0.13	0.17
	C3	13	0.18	0.10	0.03	0.11	**0.22**	0.23	0.31
	S0	13	1.20	0.69	0.15	0.53	**1.36**	1.79	2.15
	S1	13	0.68	0.40	0.09	0.31	**0.72**	0.95	1.37
	S2	13	0.41	0.21	0.07	0.33	**0.47**	0.52	0.76
	S3	13	0.87	0.50	0.09	0.48	**1.05**	1.20	1.62

7.3.5 Effects of Permeability and Air Void Content on Evaporation Rate

To examine the effects of permeability and air void content on evaporation rate, the medians (highlighted in Table 7.2, which removes the impact of outliers shown in Figure 7.9(a)) of evaporation rate over the 3 days (9–11 July) were plotted against the permeability and air void content of the pavement materials (except the bare water), as shown in Figure 7.10.

Except for S3 (sand) and S0 (bare water), the evaporation rate rank is S1, S2, C3, C2, and B3. This is the same rank as the air void and permeability for these materials. Sample S1 (19 mm gravel) has larger air void content (46%, see Table 7.1) and higher permeability (0.29 cm/s, see Table 7.1) and the largest evaporation rate (0.8 mm/h, 3-day average) compared to S2, C3, C2, and B3 (Figure 7.9(a) and (b)). Among S1, S2, C3, C2, and B3, sample B3 (9.5 mm) has the smallest air void content (12%) and lowest permeability (0.11 cm/s) and the smallest evaporation rate (0.1 mm/h, 3-day average). This implies that the air void and permeability are positively correlated to the evaporation rate for these permeable surface layer materials and permeable gravel materials. A small air void and low permeability tend to block the water evaporation from the materials and seal the moisture inside the materials.

As shown in Table 7.1, the fine but well-graded sand S3 material (see Figure 7.1 for the gradation) has a relatively smaller air void (15%) and low permeability (5×10^{-3} cm/s) among the six materials (S1, S2, S3, C3, C2, and B3). However, it has the highest evaporation rate (1.0 mm/h, 3-day average) among all six materials and its evaporation rate is just lower than but close to that of the bare water S0 (1.5 mm/h, 3-day average), as shown in Figure 7.9(a) and (b). The reason for this might be that the

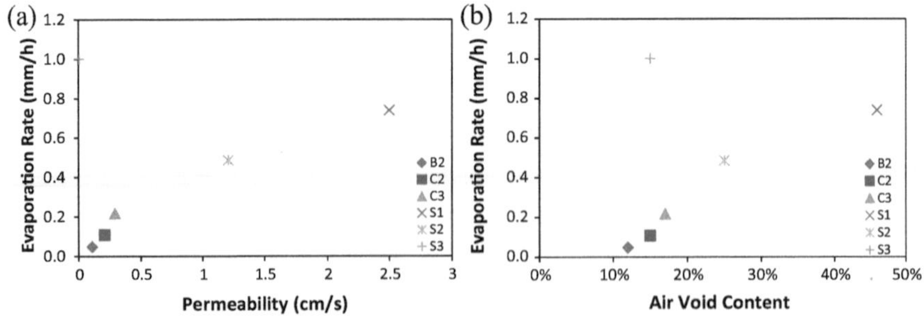

Figure 7.10 Effects of (a) permeability and (b) air void content on evaporation rate.

fine sand material has a small pore size and large surface-area-to-volume ratio and consequently a high capillary effect compared to the materials with a large pore size (e.g., the permeable concrete C2 or gravel S1). The high capillary effect of the fine materials will easily move the water from the bottom of the material to the top or surface, enhancing the water availability near the surface of the material and the consequent evaporation rate (see the discussion above). This suggests that improving the pore size and enhancing the capillary effect of materials could help increase the evaporation rate. However, reducing the pore size and increasing the capillary effect might reduce the permeability. Therefore, an optimal design of material with optimized pore size and adequate permeability is desired.

All these findings imply again that high water availability near the surface or moisture exposure to the atmosphere is critical for the evaporation rate of pavement materials, as well as high air temperature and low humidity. An increase in the air void and the permeability is one way to improve the water availability near the surface or the moisture exposure to the atmosphere for the pavement materials. Keeping the surface wet by enhancing the capillary effect or sprinkling water on the surface is another way that could increase the evaporation rate and consequently produces a better evaporative cooling effect. The capillary effect depends on the air void content and structure of the surface materials and the size of the air void (i.e., pore size). This needs more experimental and theoretical research to optimally design the materials, which is out of the scope of this chapter and will be the next step of the study.

7.3.6 Effect of Water Level Depth on Evaporation Rate

As the water evaporates, the water level depth from the material surface increases. To explicitly examine the effect of water level depth on evaporation rate of permeable pavement materials, the relationship between evaporation rate and water level depth (estimated from remaining water weight) is plotted in Figure 7.11 for the three permeable surface materials (B3, C2, and C3, in (a)) and the three gravel materials (S1, S2, and S3, in (b)). It is clearly shown that the evaporation rate of permeable pavement materials decreases as the water level depth from the material surface increases. For the three permeable surface materials (Figure 7.11(a)), the evaporation rate decreases significantly when the water level depth is over 20 mm. The evaporation rate decreases significantly when the water level depth is over 100 mm for the three gravel

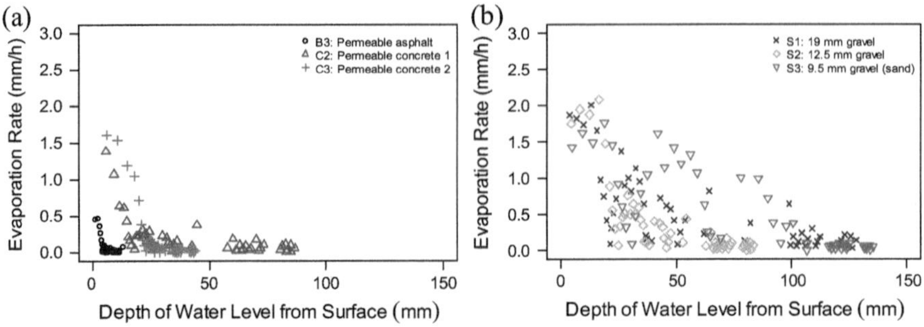

Figure 7.11 Effect of water level depth on evaporation rate: (a) permeable surface materials and (b) gravel materials.

materials (Figure 7.11(b)). This confirms again that the availability of water or moisture close to the surface is very important for the evaporation rate.

7.4 SUMMARY AND CONCLUSIONS

The evaporation rate is an important factor that influences the evaporative cooling effect of permeable pavements. It is determined by a complex system of factors, such as air temperature, relative humidity, water temperature, moisture content, air void content, air void size, and connecting structure. To take an initial look at this complex system, a simple experimental method was used to measure the evaporation rate and explore the main factors affecting various pavement materials under hot summer outdoor conditions at a test site in Davis, California. The results can provide some initial values of evaporation rate for the modeling and simulation of the cooling effects of evaporation from pavement. Some main findings from this experimental study include:

1. The peak evaporation rate of bare water is about 2.0–2.5 mm/h during hot days, with 3-day average of 1.5 mm/h, which is higher than that of water in any pavement materials tested.
2. The 3-day average evaporation rate of gravel materials (S1, S2, and S3) is in the range of 0.5–1.0 mm/h. Fine sand with small pore size had a higher evaporation rate compared to coarse gravel.
3. The permeable asphalt and concrete pavement surface materials have 3-day average evaporation rates in the range of 0.1–0.3 mm/h, which is lower than the gravel materials because of their smaller air void content and lower permeability.

4. Evaporation rate of pavement materials is high during the daytime, especially in the hot afternoon in summer, and low during the nighttime and especially in the cold early morning. High air temperature and low humidity help enhance the evaporation rate, which is just what is expected under hot conditions.
5. The air void and permeability are positively correlated to the evaporation rate for these permeable surface layer materials and permeable gravel materials except for sand. A small air void and low permeability tend to block the water evaporation from the materials and seal the moisture inside the materials.
6. Improving the pore size and enhancing the capillary effect of materials could help increase the evaporation rate.

All these findings imply that high water availability near the surface or large moisture exposure to the atmosphere are critical for the evaporation rate of pavement materials, as well as high air temperature and low humidity. Based on the findings, it can be concluded that increasing the air void and the permeability is one way to improve the moisture exposure to the atmosphere and enhance the evaporation for the pavement materials. Keeping the surface wet by enhancing the capillary effect or sprinkling water on the surface is another way that could increase the availability of water or moisture close to surfaces and the evaporation rate and consequently produce a better evaporative cooling effect. The capillary effect depends on the air void content and connecting structure in the surface materials and the size of the air void (i.e., pore size). More experimental (e.g., with larger sample size and sample number) and theoretical studies are recommended to evaluate and optimally design the evaporative cooling effect of pavement materials with adequate permeability.

CHAPTER 8

Thermal Performance of Various Pavement Materials

Contents

Pavement Materials for Heat Island Mitigation
ISBN 978-0-12-803476-7
http://dx.doi.org/10.1016/B978-0-12-803476-7.00008-8

8.1 OBJECTIVES

The objectives of this chapter are to: (1) instrument experimental sections of various pavement types, (2) observe the thermal behavior of the pavement types in different seasons and under various moisture conditions, (3) compare the thermal performances of the pavement types, and (4) examine the factors affecting the thermal performance of the pavements.

8.2 METHODOLOGY

8.2.1 Experimental Sections

The nine 4-m by 4-m (13-ft by 13-ft) experimental sections (see Section 4.3 in Chapter 4), which were constructed during the summer of 2011 at the University of California Pavement Research Center test facilities in Davis, California, were used to empirically examine the thermal behaviors of various pavement types in different seasons and under various moisture conditions. The experimental sections include three different pavement surfacing materials, namely interlocking concrete paver (surfacing type A), asphalt concrete (surfacing type B), and concrete (surfacing type C). For each pavement surfacing type, one impermeable pavement design (design 1) and two permeable pavement designs (designs 2 and 3) were prepared. Both of the permeable interlocking concrete paver pavements had the same cross-section, the difference is in the solar reflectivity of the pavers. Both of the porous asphalt sections had the same surface material, the difference is in the thicknesses of the surface layers. The two pervious concrete sections had different thicknesses and concrete surface material mix designs, meaning that the aggregate gradations, cement contents, and other ingredient proportions were different. The pervious concrete section C2 was darker than C3, which resulted in different albedos. More details on the material characteristics, which were measured previously, are summarized in Table 8.1.

8.2.2 Instrumentation for Experimental Sections

The pavement sections were instrumented for monitoring the temperature, albedo, weather conditions, and water table. The instruments for the test sections used for the experiments presented in this chapter and some other chapters are summarized in Table 8.2.

Eight type T thermocouple sensors were embedded into the pavement layers and near-surface air for monitoring the temperature of both pavements and near-surface air on each test section. Examples of the locations of the

Table 8.1 Summary of material characteristics for test sections

Section no.,/ Layer	Material	Hydraulic conductivity[a] (cm/s)	Albedo	Thermal conductivity (W/(m °C))	Heat capacity (J/(kg °C))	Density (kg/m^3)
A-1	Paver-I	—	0.28	—	—	—
A-2	Paver-P	0.62	0.25	—	—	—
A-3	Paver-P	0.50	0.28	—	—	—
B-1	Asphalt-D	—	0.09	1.73	852	2399
B-2	Asphalt-O	0.11	0.08	1.24	763	2269
B-3	Asphalt-O	0.06	0.08	1.23	798	2239
C-1	Concrete-D	—	0.29	1.83	1001	2257
C-2	Concrete-O	0.21	0.18	1.38	912	1980
C-3	Concrete-O	0.29	0.26	1.29	761	2050
Base	Aggregate-D	—	0.20	—	—	2250
Base	Aggregate-O	—	0.18	—	—	1650
Subgrade	Soil-U	1.08×10^{-3}	0.22	—	—	—
Subgrade	Soil-C	4.95×10^{-4}	0.22	—	—	—

I, impermeable; P, permeable; D, dense graded; O, open graded; U, uncompacted; C, compacted.
[a]ASTM C1701 method used.

Table 8.2 Instruments used for test sections

Item	Type/Made	Quantity	Note
Thermocouple wire	8TX20PP[a]	220 m (722 ft)	For pavement, near-surface air, and wall
Data logger	CR10X[b]	3 each	For temperature, albedo, and weather station
	CR1000[b]	2 each	For wall temperature and thermal properties
Multiplex	AM25T[b]	5 each	For pavement, near-surface air, and wall
Albedometer	240-8140[c]	2 each	For pavement and other land covers
Weather station	GRWS100[b]	2 each	For test sections and other on-site pavements
Water table well	Tube	6 each	One for each permeable section

[a]Omega Engineering, Inc.
[b]Campbell Scientific, Inc.
[c]NovaLynx Corporation.

thermocouple sensors for sections B1–B3 are shown in Figure 8.1. Six of eight thermocouple sensors were embedded into each pavement section to monitor the pavement temperature (two for the pavement surface and four for the in-depth pavement layers at 1.27, 3.81, 6.35, and 25.4 cm (0.5, 1.5, 2.5, and 10 in) below the surface) and two for the near-surface air (at 5.1 and 12.7 cm (2 and 5 in) above the surface), as shown in Figure 8.1 and Figure 8.2(a).

The temperature data were recorded and collected with a CR-10X data logger with a time interval of 10 min (Figure 8.2(b) and (c)). In addition to the temperature data of pavements and near-surface air on the test sections, the on-site weather data, including air temperature and humidity, solar radiation, wind speed and direction, rainfall, and air pressure, were monitored as well with the nearby weather station instruments and recorded using the CR-10X data logger with a time interval of 30 min, as shown in Figure 8.2(d).

8.2.3 Overall Experiment Plan

The overall experiment plan is presented in Table 8.3, showing the two main experiments:

1. thermal performance of various pavements (including asphalt, concrete, and paver; permeable and impermeable) in different seasons;
2. thermal performance of various permeable pavements compared with impermeable pavements under dry and wet conditions in the summer.

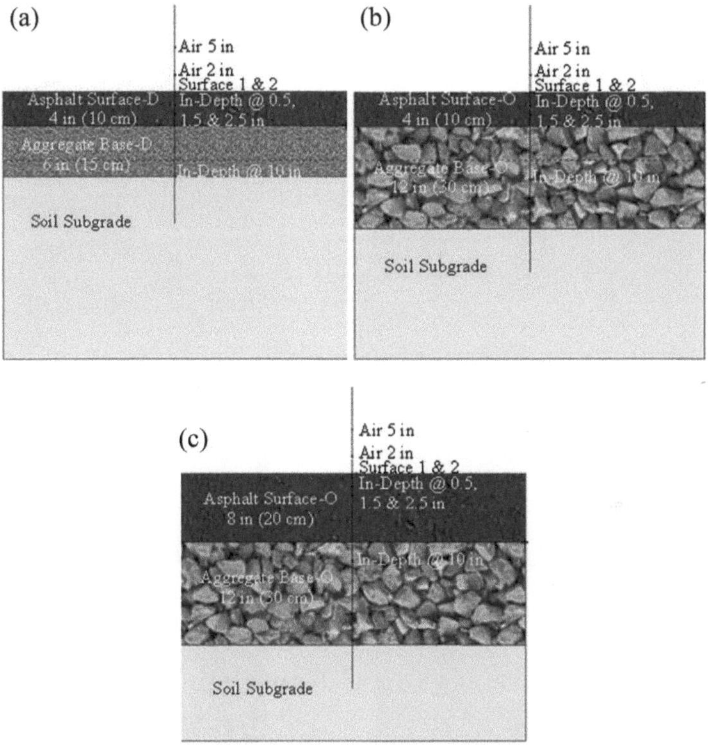

Figure 8.1 Cross-sections and sensor locations for the test sections. (a) Section B1. (b) Section B2. (c) Section B3. *Note: D, dense-graded; O, open-graded. 1 in = 25.4 mm.*

8.3 THERMAL PERFORMANCE OF VARIOUS PAVEMENTS IN DIFFERENT SEASONS

This section presents the experimental results and discussion regarding the thermal performance of the pavements (including asphalt, concrete, and paver; permeable and impermeable) in different seasons, including:

1. overview of temperature profiles at various locations over a hot 3-day period;
2. diurnal variation of surface temperatures of the various pavements;
3. times of maximum and minimum air temperature, solar radiation, and pavement surface temperature;
4. seasonal variation of surface temperatures of the pavements;
5. in-depth pavement temperatures ($\geq$25.4 cm (10 in) deep);
6. near-surface air temperatures of the various pavements;

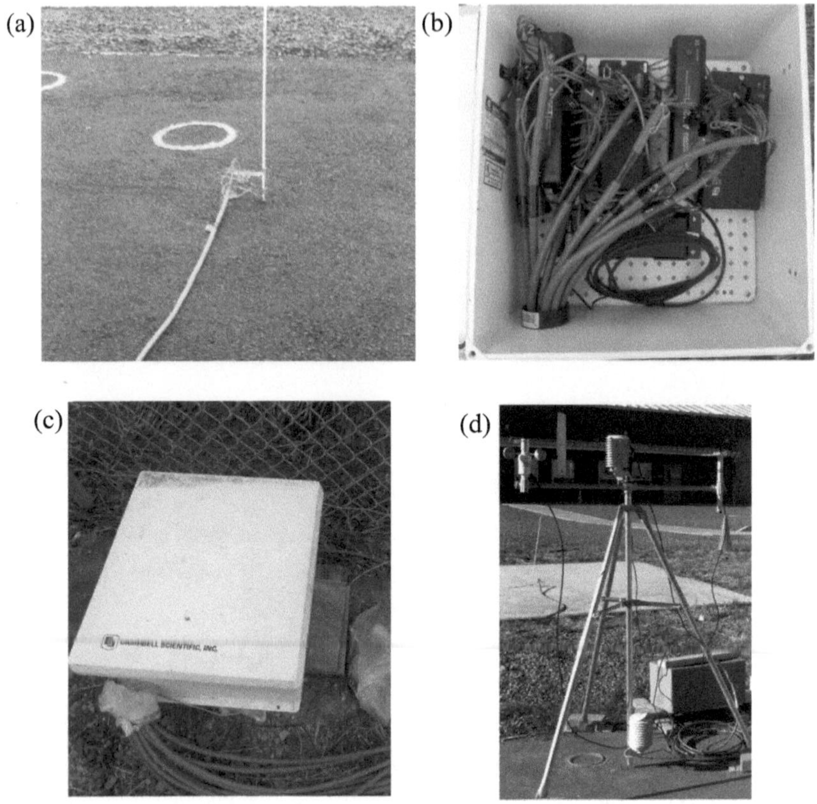

Figure 8.2 Instrumentation and data collection system. (a) Thermocouple sensors. (b) Data collection system (inside). (c) Data collection system (outside). (d) Weather station.

7. initial comparison of the thermal performances of permeable and impermeable pavements;
8. calculated heat exchange between pavement and near-surface air.

8.3.1 Overview of Temperature Profiles at Various Locations over a Hot 3-Day Period

This section presents an overview of the temperature profiles at various locations on the pavements in hot summer, to briefly examine the thermal behavior of the pavements (asphalt, concrete, and paver).

The 3-day temperature profiles over time in September 2011 at various locations are plotted in Figure 8.3 for one permeable paver pavement (A3), one permeable asphalt pavement (B3), and one permeable concrete

Table 8.3 Overall experiment plan for test sections

Experiment	Monitoring variable	Monitoring section	Date	Condition
Thermal performance of pavements in different seasons	Temperatures of in-depth pavement, surface, and near-surface air	All sections (A1–A3, B1–B3, and C1–C3)	15 September 2011–31 October 2012	As is (natural) for different seasons
Thermal performance of permeable pavements under dry and wet conditions in summer (compared with impermeable pavements)	Temperatures of in-depth pavement, surface, and near-surface air; thermal images	All sections (A1–A3, B1–B3, and C1–C3)	Irrigation on a hot day for each summer of 2011 and 2012 (21 September 2011 and 10 July 2012)	Dry and wet (irrigation on six permeable sections: A2, A3, B2, B3, C2, and C3)

pavement (C3). The properties of these three pavements are summarized in Table 8.1. The ambient air temperature (Air Temp. in Figure 8.3), which was measured from a nearby weather station at around 2 m above the ground, is plotted as well for comparison. It is noted that the pavement temperature profiles change over time following the variation pattern of the ambient air temperature. The highest pavement temperatures happened at the surface around noon and were about 15–30 °C higher than the highest ambient air temperature. The lowest pavement temperatures happened around the surface in the early morning and were about 3–8 °C higher than the lowest ambient air temperature. The pavement surface had the largest fluctuation in temperature compared to the other in-depth locations. As the depth increased, the temperature fluctuation went down. When the depth was 61 cm (24 in) below the surface, the temperature (C3_24 in in Figure 8.3(c)) was almost constant at 30 °C over the whole 3-day period.

The near-surface air temperatures (5.1 and 12.7 cm (2 and 5 in) above the pavement surface) show a variation pattern similar to that of the ambient air temperature. Owing to heating by the pavement surface, the near-surface air temperatures were higher than the ambient air temperatures

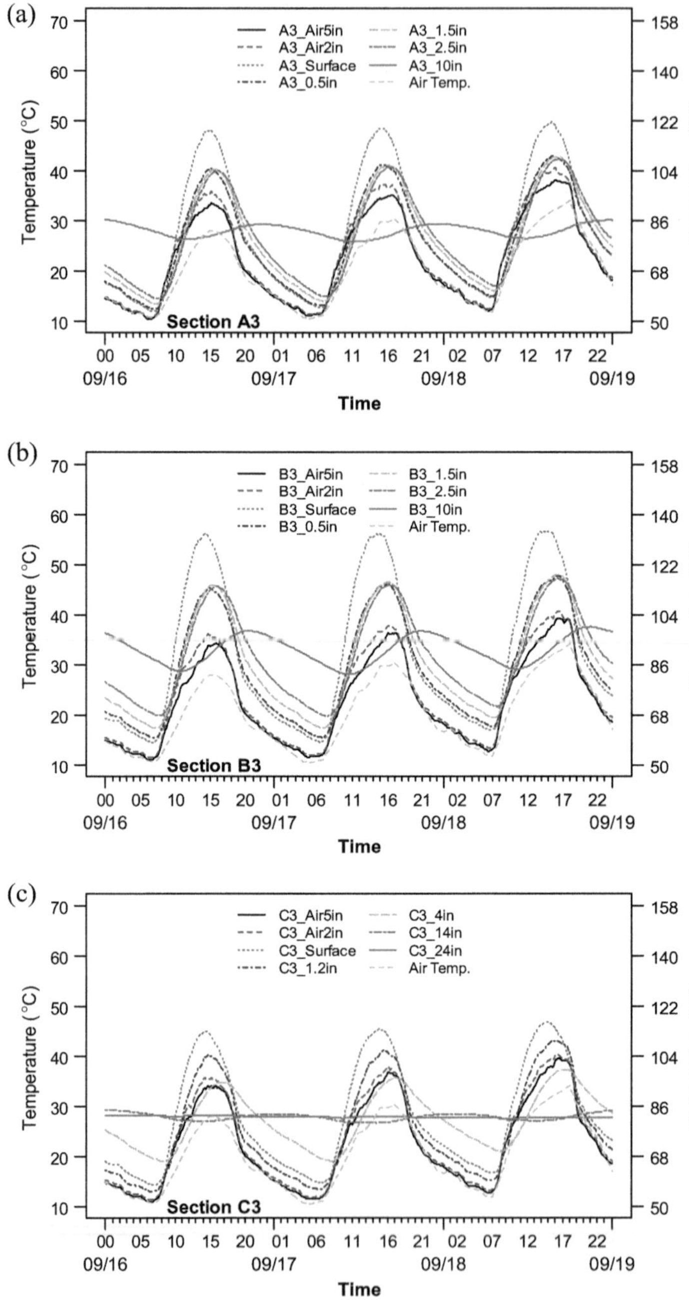

Figure 8.3 Temperature profiles of various permeable pavements at various locations in summer 2011. (a) Paver pavement (A3). (b) Asphalt pavement (B3). (c) Concrete pavement (C3). *Air temperature (Air Temp.) was measured from a nearby weather station at 2 m above the ground.* Dates are given as month/day.

at 2 m above the ground. The daytime temperature differences between the near-surface air at 12.7 cm (5 in) above the surface and the ambient air were around 5–10 °C, and the largest differences happened around 4:00–5:00 PM, when the ambient air temperature reached its peak. During the nighttime, the temperature differences were not that significant, typically between about 0 and 2 °C at 12.7 cm (5 in) above the surface. The temperature differences were larger closer to the pavement surface (the near-surface air temperature at 5.1 cm (2 in) above the surface was higher than that at 12.7 cm (5 in) above).

8.3.2 Diurnal Variation of Surface Temperatures of Various Pavements

In the previous section, it was shown that the extreme pavement temperatures happen at the pavement surface. The critical pavement surface temperatures have a great influence on the near-surface air temperature and consequently affect human thermal comfort and air quality. This section is dedicated to further examining the diurnal variation of pavement surface temperature in different seasons for various pavements (concrete pavement section C1, paver pavement section A1, and asphalt pavement section B1).

The surface temperatures of the three different pavements on one clear sunny day of each season were monitored and are plotted in Figure 8.4. The weather data (ambient air temperature, wind speed, and solar radiation), which were taken from a nearby weather station at around 2 m above the ground, are also presented in Figure 8.4 for reference.

It is clearly shown that there are differences in temperature for different pavements. The darker asphalt pavement (B1) has a higher surface temperature than the concrete (C1) and paver (A1) pavements with light color. The temperatures of the paver pavement (A1) are slightly higher than those of the concrete (C1) owing to the slightly lower albedo of the paver (0.28 vs 0.29). The surface temperature of the asphalt pavement (B1) reached up to almost 70 °C (158 °F) in summer (Figure 8.4(c)), compared to 50 °C (122 °F) for the concrete pavement (C1). The differences in peak temperature between asphalt (B1) and concrete (C1) pavements were about 10–20 °C, depending on the weather conditions and season. During nighttime, the surface temperatures of the three pavements were very close to one another, but still higher than the ambient air temperature by 2–10 °C. The difference in surface temperatures between asphalt and concrete pavements was determined mainly by the color of pavement

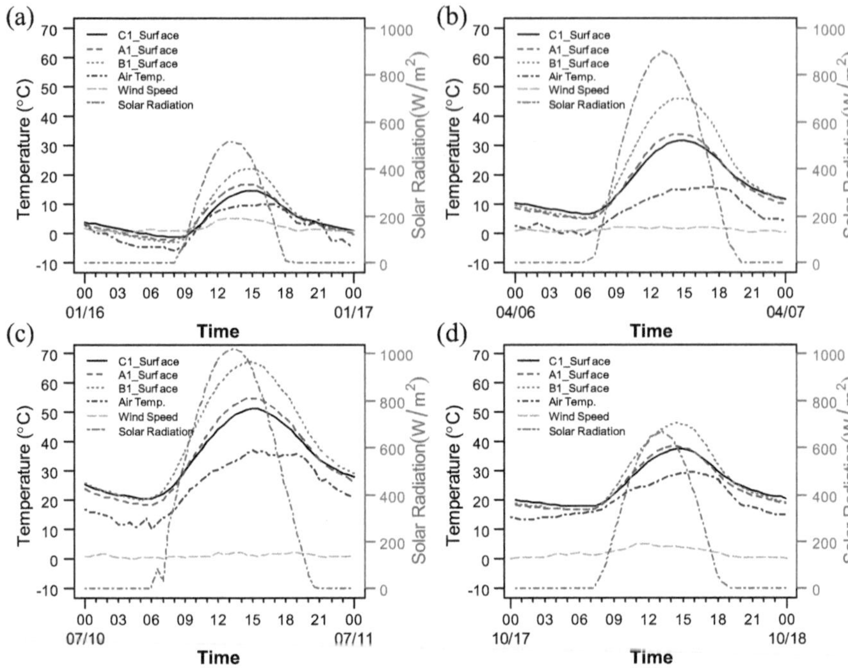

Figure 8.4 Diurnal variation of surface temperature and weather data on 1 day of each season. (a) Winter. (b) Spring. (c) Summer. (d) Fall. *Weather data (air temperature, wind speed, and solar radiation) were measured from a nearby weather station at around 2 m above the ground.* Dates are given as month/day.

surface (i.e., the solar reflectivity or albedo, ~0.1 for asphalt, 0.18–0.29 for concrete, 0.25–0.28 for paver, see Table 8.1). The higher the albedo of the pavement, the more solar radiation is reflected and the less is absorbed by the surface, which will produce a lower surface temperature. The asphalt pavement had a darker color and consequently a lower albedo and a higher surface temperature, compared to the concrete pavement. The difference in surface temperatures between the concrete and the asphalt was higher (~20 vs ~10 °C) during summer, with a high solar radiation (peak intensity of ~1000 W/m^2 in Figure 8.4(c)), than in winter, with a low solar radiation (peak intensity of ~500 W/m^2 in Figure 8.4(a)). This implies that increasing the albedo is an effective strategy to reduce the surface temperature, especially for the climates and seasons with high solar radiation. The observation discussed above is further confirmed by the diurnal variation in surface temperatures in 3 days of each season for all nine experimental pavements as shown in Figure 8.5.

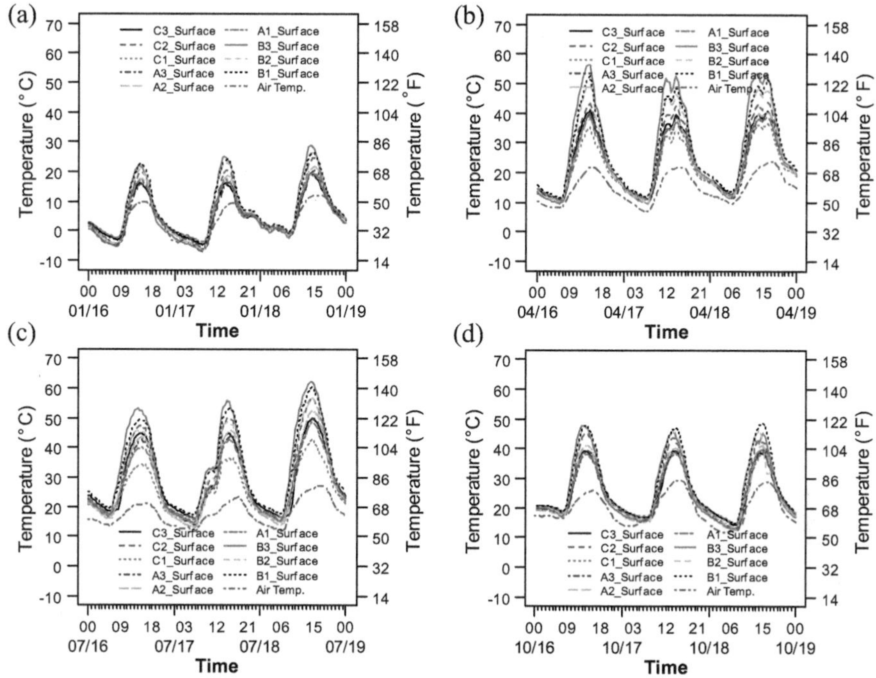

Figure 8.5 Diurnal variation in surface temperatures in 3 days of each season. (a) Winter. (b) Spring. (c) Summer. (d) Fall. *Air temperature (Air Temp.) was measured from a nearby weather station at around 2 m above the ground.* Dates are given as month/day.

8.3.3 Times of Maximum and Minimum Air Temperature, Solar Radiation, and Pavement Surface Temperature

The times when the maximum and minimum air temperatures, solar radiation, and pavement surface temperatures happen in 1 day are important because they are important for human thermal comfort analysis related to the use of outdoor space and have a potential influence on the peak power for building cooling and heating. This section presents an examination on this issue of times of maximum and minimum for various pavements (concrete pavement section C1, paver pavement section A1, and asphalt pavement section B1) in different seasons.

Figure 8.6 presents the times of maximum air temperature, solar radiation, and pavement surface temperature on one sunny day of each season. The peak solar radiation intensity happens around 13:00 h. The maximum air temperature happens at 16:00–17:00 h. The peak pavement surface temperature occurs at around 15:00 h. The times when the minimum air and pavement surface temperatures happen are very close to each other, at

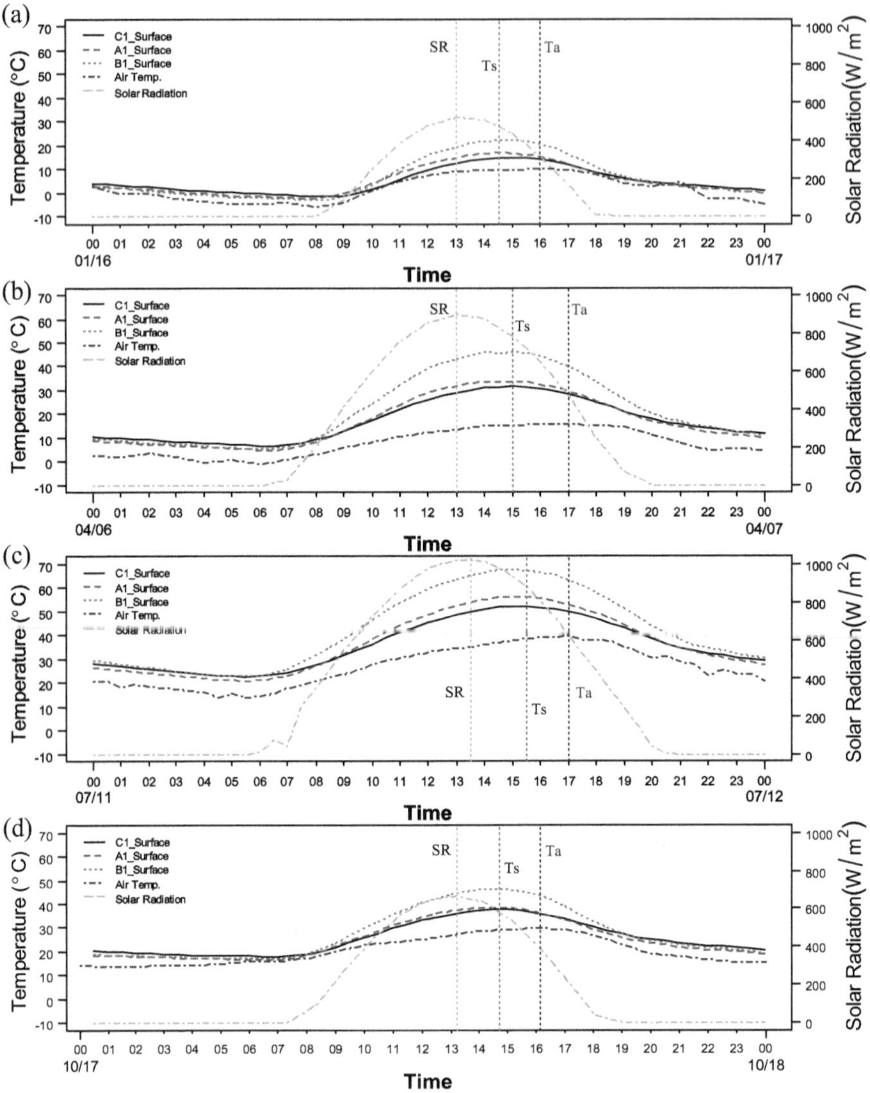

Figure 8.6 Times of maximum air temperature (Ta), solar radiation (SR), and pavement surface temperature (Ts) in 1 day. Dates are given as month/day. (a) Winter. (b) Spring. (c) Summer. (d) Fall.

around 06:00 h in the early morning. The lowest solar radiation in a day is zero and happens during night after sunset.

Statistical times of maximum and minimum air temperature, solar radiation, and pavement surface temperature on each day of 1 year (August 2011–July 2012) are presented using histograms in Figure 8.7. The frequency

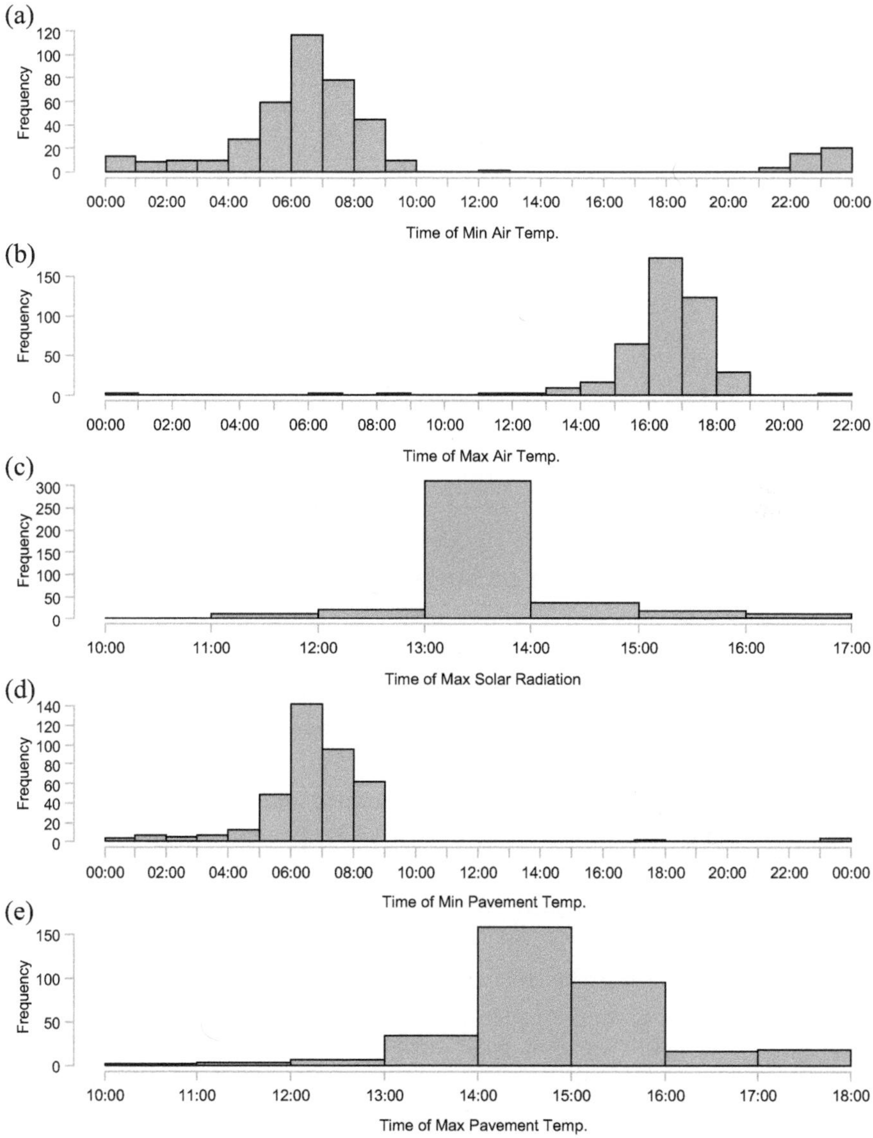

Figure 8.7 Statistical times of maximum and minimum air temperature, solar radiation, and pavement surface temperature in 1 year. (a) Min air temperature. (b) Max air temperature. (c) Max solar radiation. (d) Min pavement surface temperature. (e) Max pavement surface temperature.

distributions of the times are also shown in Figure 8.7. The median times of maximum and minimum air temperatures, solar radiation, and pavement surface temperatures are listed in Table 8.4. This table shows that the solar radiation peaks first at 13:00 h, and the air temperature reaches its peak at

Table 8.4 Median times of maximum and minimum air temperatures, solar radiation, and pavement surface temperatures

Item	Time of min	Time of max
Air temperature	06:00	16:00
Solar radiation	—[a]	13:00
Pavement surface temperature	06:30	14:30

[a]All are zero and minimum during nighttime.

16:00 h. The pavement surface temperature reaches its peak at 14:30 h, which is 1.5 h later than the solar radiation and 1.5 h earlier than the air temperature. The lowest air temperature happens at 06:00 h, which is half an hour earlier than that of pavement surface temperature (06:30 h). This information can be used for setting boundary conditions for modeling and simulation of pavement temperature, which will be addressed later in this dissertation.

8.3.4 Seasonal Variation of Surface Temperatures of Various Pavements

This section examines the seasonal change in surface temperatures in 1 year for various pavements (concrete (C1), paver (A1), and asphalt (B1) pavements), which is of significance to comprehensively evaluate the thermal performance of the pavements.

The daily maximum and minimum surface temperatures of concrete (C1), paver (A1), and asphalt (B1) impermeable pavements over 1 year are extracted and plotted in Figure 8.8, along with daily maximum and minimum ambient air temperatures for reference. As observed previously, the changes in daily maximum and minimum pavement surface temperatures follow the same pattern as the ambient air temperature. The pavement surface temperatures are generally higher than the ambient air temperatures over the year, even for the daily minimum temperature during winter (Figure 8.8(b)). The daily maximum surface temperature of asphalt pavement (B1) was higher than that of the concrete (C1) and paver (A1) pavements and the ambient air temperature as well. The temperature differences between them were higher during summer than winter owing to the difference in solar radiation of the different seasons. During summer, the daily maximum surface temperature of asphalt pavement (B1) was around 20 °C higher than that of the concrete (C1) (owing to the lower albedo of asphalt pavement compared to concrete pavement, 0.09 versus 0.29 as shown in Table 8.1) and around 16 °C higher than that of paver (A1)

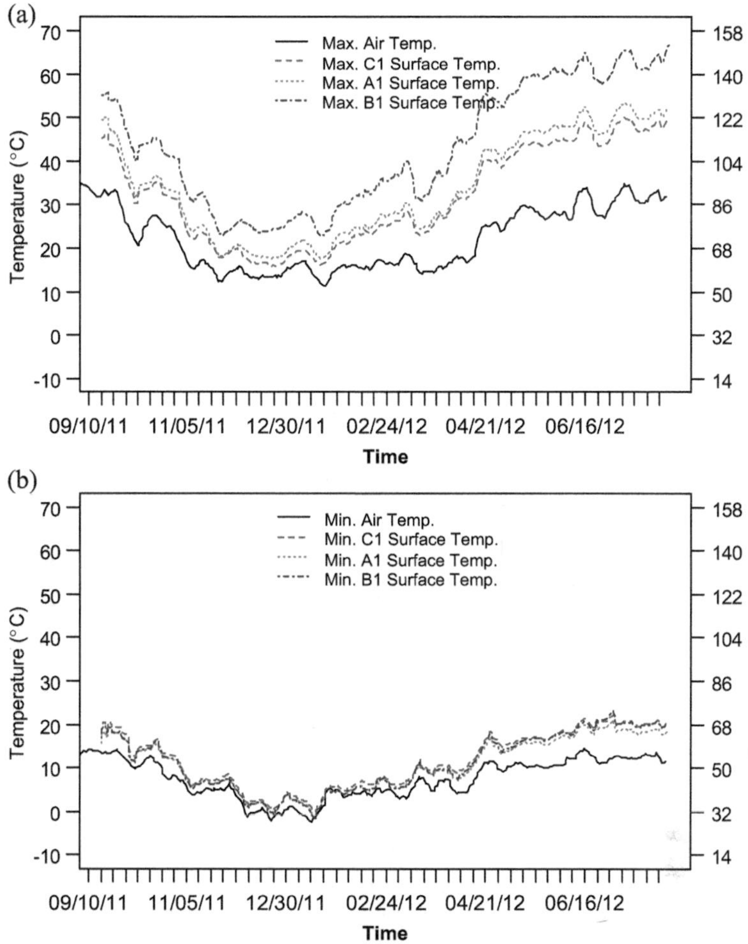

Figure 8.8 Daily maximum and minimum air temperatures and pavement surface temperatures over 1 year. *Concrete (C1), paver (A1), and asphalt (B1) pavements.* Dates are given as month/day/year. (a) Max temperature. (b) Min temperature.

pavement (higher albedo of A1 0.28 as shown in Table 8.1) and about 30 °C higher than the ambient air temperature. During winter, the daily maximum surface temperature of asphalt pavement (B1) was around 10 °C higher than that of the concrete (C1) and around 8 °C higher than that of the paver (A1) pavements and about 15 °C higher than the ambient air temperature. There were no significant differences in the nighttime daily minimum surface temperatures between the three pavements for both summer and winter, and they were about 10 and 5 °C higher than the ambient air temperature for summer and winter, respectively.

8.3.5 In-depth Pavement Temperatures (≥25.4 cm (10 in) Deep)

This section examines the profiles and seasonal changes in temperature of the deeper layers of the impermeable and pervious concrete pavements (C1–C3), providing some understanding of the thermal behavior of in-depth pavements beyond the pavement surface.

The in-depth temperatures of impermeable and pervious concrete pavements (C1–C3) at various depths over 3 days of each season are presented in Figure 8.9. The temperatures at 24 in (C3_24 in) are generally constant for each season and are around 10, 20, 30, and 25 °C for winter, spring, summer, and fall, respectively. More detailed in-depth temperatures can be found from the year-round plot (Figure 8.9(e)). This information can be used for temperature boundary conditions for modeling and simulation of pavement temperature, which will be addressed later in this dissertation.

8.3.6 Near-Surface Air Temperatures of Various Pavements

This section examines the behavior of near-surface air temperatures of various pavements (concrete pavement section C1, paver pavement section A1, and asphalt pavement section B1) in different seasons, providing here a brief understanding of the thermal behavior of the near-surface air above pavements and thermal interactions between near-surface air and pavement.

The near-surface air temperatures of three different pavements (2 in above the pavement surface) are presented in Figure 8.10 for 1 day of each season. The near-surface air temperatures are generally higher than the ambient air temperature measured at 2 m high, especially during the daytime. High near-surface air temperatures heated up by pavement surfaces will potentially have an influence on the air quality (especially the formation of ground-level ozone) and thermal comfort (especially for children, whose body mass is closer to the surface). More details about the near-surface air temperatures and thermal interactions between them and pavement surfaces will be discussed in Chapter 9.

8.3.7 Initial Comparison of Thermal Performance of Permeable and Impermeable Pavements

This section comparatively examines the thermal behavior of permeable and impermeable pavements, under the same weather conditions, due to the differences in thermal properties (i.e., thermal conductivity and heat capacity) as shown in Table 8.1. To reduce the influence of differences in albedo, the permeable asphalt pavement (B3) and impermeable asphalt

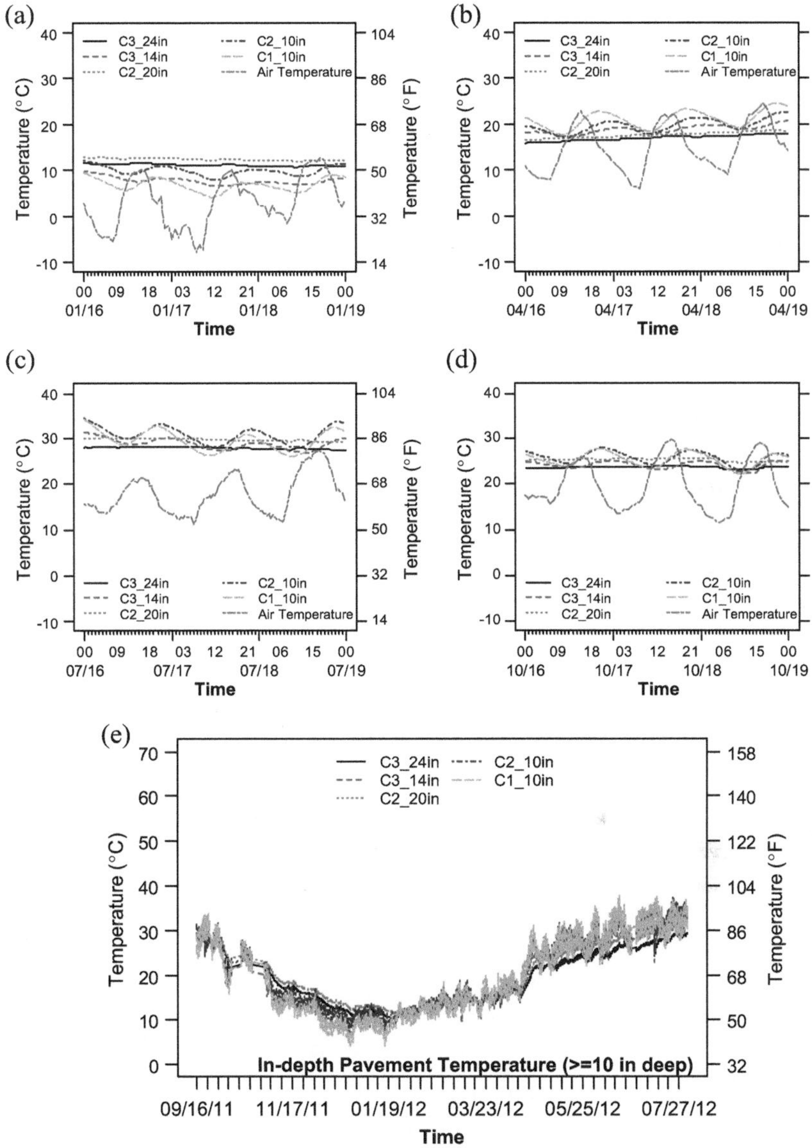

Figure 8.9 Variation of in-depth pavement temperatures (≥25.4 cm (10 in) deep). Dates are given as month/day or month/day/year. (a) Winter. (b) Spring. (c) Summer. (d) Fall. (e) Year-round.

pavement (B1), which have similar albedos (0.08 vs 0.09 as shown in Table 8.1), were selected for comparison of thermal performance.

The surface temperatures of the permeable pavement (B3) and impermeable pavement (B1) are plotted in Figure 8.11 for 1 day of each season, along with the ambient air temperature for reference. It is observed that the

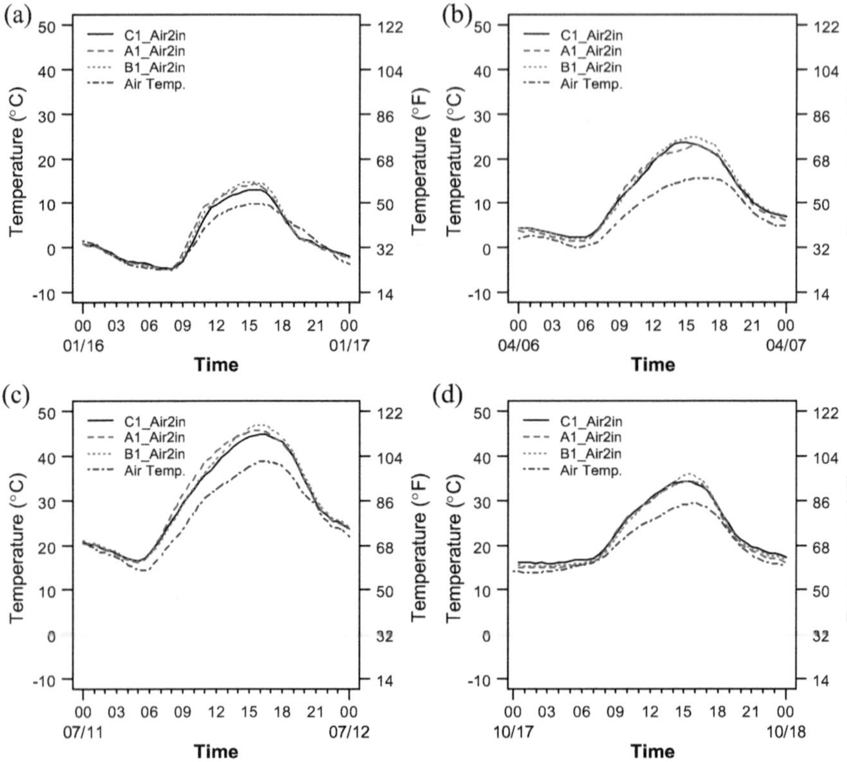

Figure 8.10 Near-surface air temperatures of various pavements (2 in (5 cm) above the pavement surface). Dates are given as month/day. (a) Winter. (b) Spring. (c) Summer. (d) Fall.

permeable pavement (B3) shows a higher daytime peak temperature than the impermeable pavement (B1) by about 5 °C during the dry seasons of spring (Figure 8.11(b)) and summer (Figure 8.11(c)). However, during the wet seasons (see Figure 8.12 for rainfall data) of winter (Figure 8.11(a)) and fall (Figure 8.11(d)), the permeable pavement (B3) shows a slightly lower daytime peak temperature. In addition, the permeable pavement (B3) shows a lower temperature during nighttime than the impermeable pavement (B1). This implies that the permeable pavement can help to mitigate the nighttime heat island effect (usually the heat island intensity, i.e., the temperature difference, is larger during nighttime than daytime), compared to the impermeable pavement, although it produces a higher daytime temperature under dry conditions.

The open-graded materials (in permeable pavement B3) have a lower thermal conductivity and heat capacity than the dense-graded materials

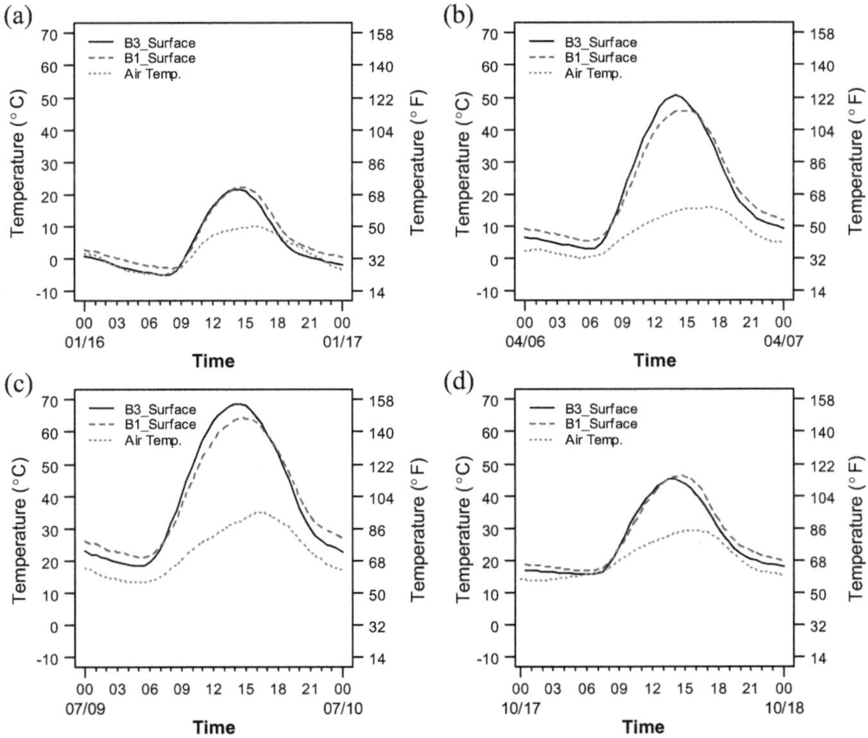

Figure 8.11 Comparison of thermal performance of permeable (B3) and impermeable (B1) pavements. Dates are given as month/day. (a) Winter. (b) Spring. (c) Summer. (d) Fall.

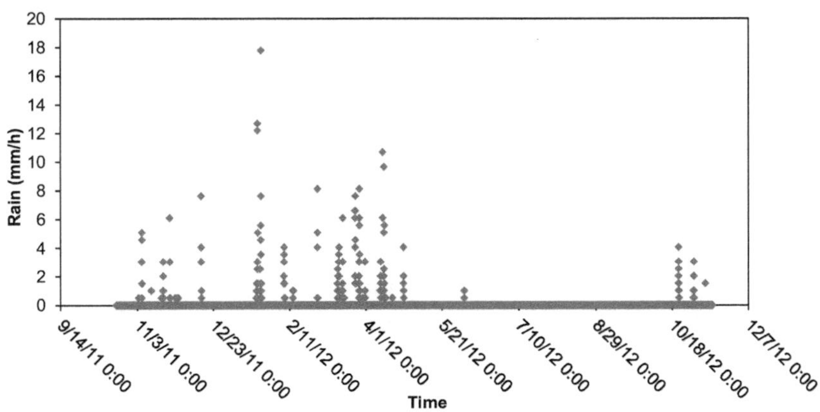

Figure 8.12 Rainfall data in 1 year (October 2011 to November 2012). Dates are given as month/day/year. 0:00 indicates midnight.

(in impermeable pavement B1) (as discussed in Chapter 6 and shown in Table 8.1). The permeable asphalt pavement section (B3) presented a higher daytime surface temperature than the impermeable pavement (B1) under the dry condition. This might be because the lower thermal conductivity increases the thermal resistance and thus the difficulty in conducting heat into the deeper pavement layers, which levels and keeps the heat near the surface. Moreover, because of its lower heat capacity compared to dense-graded material, the open-graded material will be heated up to a higher temperature under the same amount of energy absorbed from solar radiation or the surroundings during the daytime or hot periods. On the other side, just owing to its lower heat capacity, the open-graded material has less thermal energy (or heat) stored in the solid body around the surface to lose into the cold ambient air during the nighttime or cold periods. Also, because of its lower thermal conductivity compared to the dense-graded material, it is much more difficult for the open-graded material to conduct heat to the surface from the underlying layers to supply more energy to lose into the cold surface and near-surface air. These two aspects will significantly increase the possibility of the open-graded material producing a lower surface temperature when the same amount of heat is lost into the ambient air during the nighttime or cold periods. This theoretically confirms that the permeable pavement composed of open-graded materials can be a potential strategy to counter the nighttime heat island effect because of its lower surface temperature and the smaller amount of heat released into the ambient air during the night. However, as mentioned previously, attention should be given to its potentially higher surface temperature during the daytime under the dry condition.

To make an initial examination of the thermal performance and cooling effect of the permeable pavement under wet conditions during summer, water was irrigated into permeable pavement section B3 from 15:00 h (the time with highest surface temperature) to 23:00 h on 10 July 2012. The water flow rate was about 0.35 m^3/h during irrigation. The pavement section was irrigated until it was filled up with water. The surface temperatures of the permeable pavement (B3) are presented in Figure 8.13 for the period of 9–12 July 2012, which includes 1 day before the irrigation and 2 days after. The surface temperatures of the impermeable pavement (B1) are also plotted for reference as well as the ambient air temperature during the period.

Under the dry condition before irrigation (9 July 2012), the permeable pavement (B3) produced a higher daytime surface temperature than the

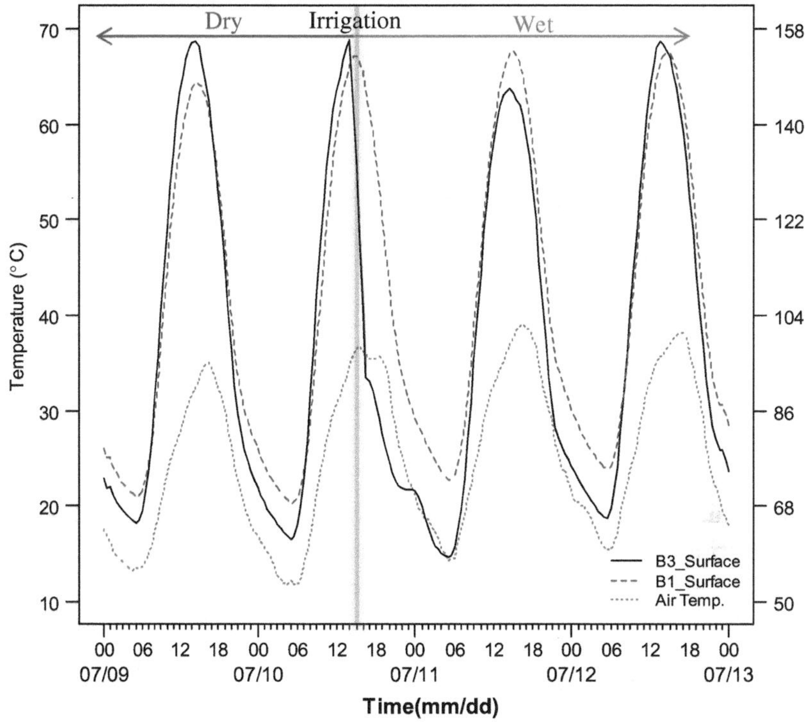

Figure 8.13 Thermal performance of permeable pavement (B3) with and without irrigation.

impermeable pavement (B1), by about 5 °C. However, under the wet condition after irrigation (11 July 2012), the permeable pavement (B3) showed a lower daytime surface temperature than the impermeable pavement (B1), by about 5 °C as well, although the peak ambient air temperature under the wet condition (9 July 2012) was about 3 °C higher than that under the dry condition (9 July 2012). During the irrigation, especially immediately after the irrigation started (16:00 h on 10 July 2012), the surface temperature of the permeable pavement (B3) was lowered by over 30 °C. It is also noted that the cooling effect of irrigation into the permeable pavement vanishes over time as the moisture surface level moves down inside the pavement owing to evaporation and infiltration into the subgrade. This is verified by the decreased cooling effect 2 days after irrigation (12 July 2012).

These observations imply that irrigation can help to lower the daytime pavement surface temperature of permeable pavements and consequently mitigate the heat island effect and improve thermal comfort. The cooling effect depends on the availability of moisture around the pavement surface and will vanish over time as the moisture decreases. Therefore, water can be irrigated into the permeable pavement during the late afternoons or evenings in summer. Especially when the weather report forecasts a very hot day coming, this strategy can be conducted during the night before to mitigate the heat wave coming the next day and improve thermal comfort.

8.3.8 Calculated Heat Exchange between Pavement and Near-Surface Air

This section comparatively examines the thermal interaction (or heat exchange) between the pavement and the near-surface air and other surroundings such as walls and buildings, which is another important aspect to examine in addition to pavement temperatures. The heat released by the pavement to its surroundings is the focus of this study.

The heat exchange processes include reflected short-wave solar radiation, emitted long-wave radiation, and convective heat. The sum of the reflected short-wave solar radiation and emitted long-wave radiation from pavement surface is called radiosity. It is the total radiation (sensible heat) from the pavement surface that might hit and be absorbed by the surroundings such as building surfaces or human bodies. The convective heat is the energy exchanged through convection between the pavement and the near-surface air.

Based on the theory of heat transfer and radiation (see Chapter 11, Section 11.3, for more details), the heat flux of these heat exchanges of pavements were calculated and are presented in Figure 8.14 for one sunny day in summer. The convective heat of the asphalt pavement (B1) is higher than that of the other pavements because of its high surface temperature. This means more convective heat will be released into the near-surface air by the asphalt pavement than the concrete pavement (C1 and C2) and the interlocking concrete paver pavement (A1). In addition, more long-wave radiation is emitted by the asphalt pavement with higher surface temperature. However, the concrete pavement with high albedo reflects more short-wave solar radiation than the asphalt pavement. The sum of the reflected short-wave solar radiation and the emitted long-wave radiation, i.e., radiosity, is higher for the concrete pavement (C1 and C2) than the

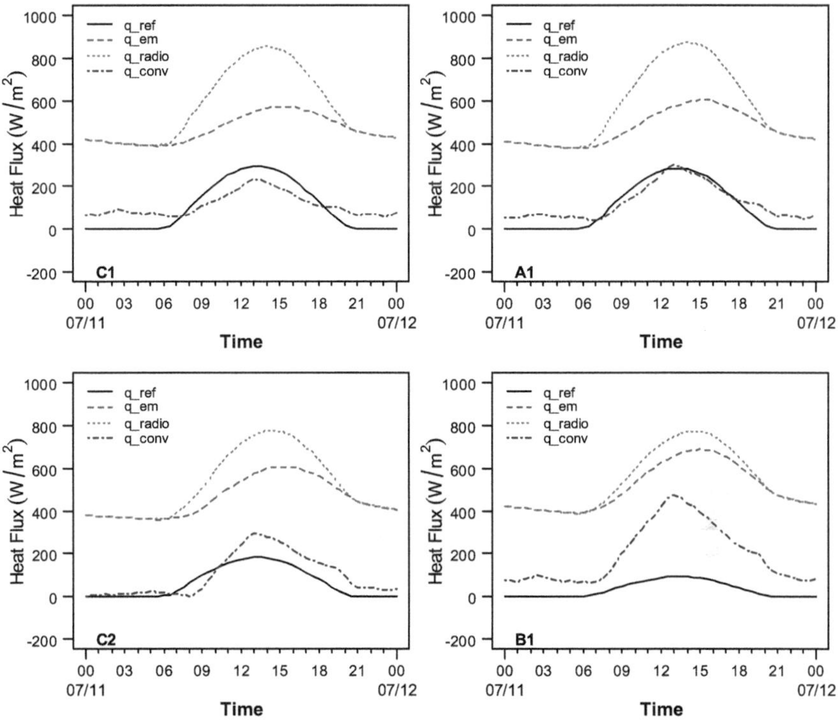

Figure 8.14 Heat flux from pavement surfaces. *q_ref is reflected short-wave solar radiation; q_em is emitted long-wave radiation; q_radio is radiosity and equal to q_ref + q_em; q_conv is convective heat. C1 is light concrete pavement; C2 is dark concrete pavement; A1 is paver pavement; B1 is asphalt pavement.* Dates are given as month/day.

asphalt pavement (B1). This implies that, although the concrete pavement produces a lower surface temperature and lower emitted long-wave radiation and convective heat owing to the low surface temperature led by high albedo, it will increase the reflected short-wave solar radiation and might increase the radiosity, the total sensible energy released by the surface. The increased radiosity of concrete pavement might hit and be absorbed by its surroundings, such as building surfaces and human bodies, leading to increased building energy use for cooling and reduced human thermal comfort. Therefore, optimal context-sensitive design of the pavement albedo is of great significance to ensure a net benefit in terms of energy use or human thermal comfort or both.

8.4 THERMAL BEHAVIOR AND COOLING EFFECT OF PERMEABLE PAVEMENTS UNDER DRY AND WET CONDITIONS

This section is dedicated to the experimental investigation of the cooling effect and the uncertainty about permeable pavements. All the permeable pavements, including porous asphalt, pervious concrete, and permeable paver, might have similar thermal behaviors and cooling effects compared to the corresponding impermeable pavements. Since the albedos were close for the permeable and impermeable asphalt sections in this experimental study (0.008 vs 0.09 as shown in Table 8.1), they were directly comparable for examining the cooling effects of permeable compared to impermeable pavement. Therefore, this section focuses on the three asphalt sections (B1–B3). The same results can be inferred for concrete pavement and interlocking concrete paver.

The experimental results on the thermal behavior and cooling effects of permeable asphalt pavements under both dry and wet conditions compared to the conventional impermeable asphalt pavements are presented here, and the factors affecting the thermal behavior of permeable asphalt pavements are quantified through field measurement on experimental sections.

8.4.1 Wetting/Irrigation Experimental Procedure

Three asphalt sections (B1–B3) were used for this experimental investigation. B1 was impermeable asphalt pavement used as a control. Both of the porous asphalt sections (B2 and B3) had the same surface material, the difference was in the thickness of the surface layers (see Figure 8.1 for cross-sections and sensor locations). The temperature data of pavements under dry conditions were recorded from 15 to 21 September 2011. In the late afternoon of 21 September 2011, as a pavement thermal management strategy, water was irrigated into the two permeable test sections to put the pavements under the wet condition. Approximately 3.19 and 3.29 m^3 of water was irrigated into sections B2 and B3, respectively. The water levels were then free to drop under the natural processes of infiltration and evaporation. The temperature data for these two permeable pavements under the wet condition were continuously recorded from 21 to 30 September 2011.

8.4.2 Statistical Analysis

First of all, time-series plots were used as an illustrative method to analyze and interpolate the data of the temperature profiles. This method is a convenient way of graphically depicting the variation of temperature over time for each location and each test section. In addition, a descriptive statistics analysis was performed for each test section and the mean values of temperature differences and cooling degree hours and heating degree hours were used to quantitatively compare the thermal behavior and cooling effects of the permeable asphalt pavements. Finally, discontinuity-based ordinary least squares linear regression analysis was performed to quantitatively evaluate the cooling effects of the permeable pavements under both dry and wet conditions.

8.4.3 Results and Discussion

8.4.3.1 Temperature Profiles on Pavements

The temperature profiles of eight locations on each test section (two for near-surface air at 2 and 5 in (5 and 12.5 cm) above the surface, two for the pavement surface, and four for in-depth pavement layers at 0.5, 1.5, 2.5, and 10 in (1.25, 3.75, 6.25, and 25 cm) below the surface) are shown in Figure 8.15. To more easily compare the temperature profiles, the temperature differences between permeable pavements (section B2 and B3) and conventional impermeable pavements (section B1) were plotted over time in Figure 8.16. Generally, under the dry condition, the daytime high temperatures of the permeable pavements were higher than those of the impermeable pavement; the nighttime low temperatures of the permeable pavements were slightly lower than those of the impermeable pavement. This indicates that the permeable pavement under dry conditions will produce a worse thermal behavior with higher diurnal fluctuation of temperature and gradient compared to impermeable pavement. However, under the wet condition, the permeable pavement will produce a better thermal behavior with lower diurnal fluctuation of temperature and gradient, which means it will produce a lower daytime high temperature and a higher nighttime low temperature. The cooling effect of permeable pavement depends on the moisture available and the evaporation rate on the pavement. As discussed in Chapter 7, the water table after irrigation will drop over time owing to infiltration and evaporation (see Figure 8.21) and the moisture available near the pavement surface will be reduced, and consequently the evaporation rate will decrease (see Figure 7.6). Therefore, the cooling effect of irrigation of the permeable pavement diminishes over time as the moisture surface level moves down inside the pavement because of evaporation and infiltration into the subgrade. This implies that irrigation can help to reduce the daytime pavement surface temperature of

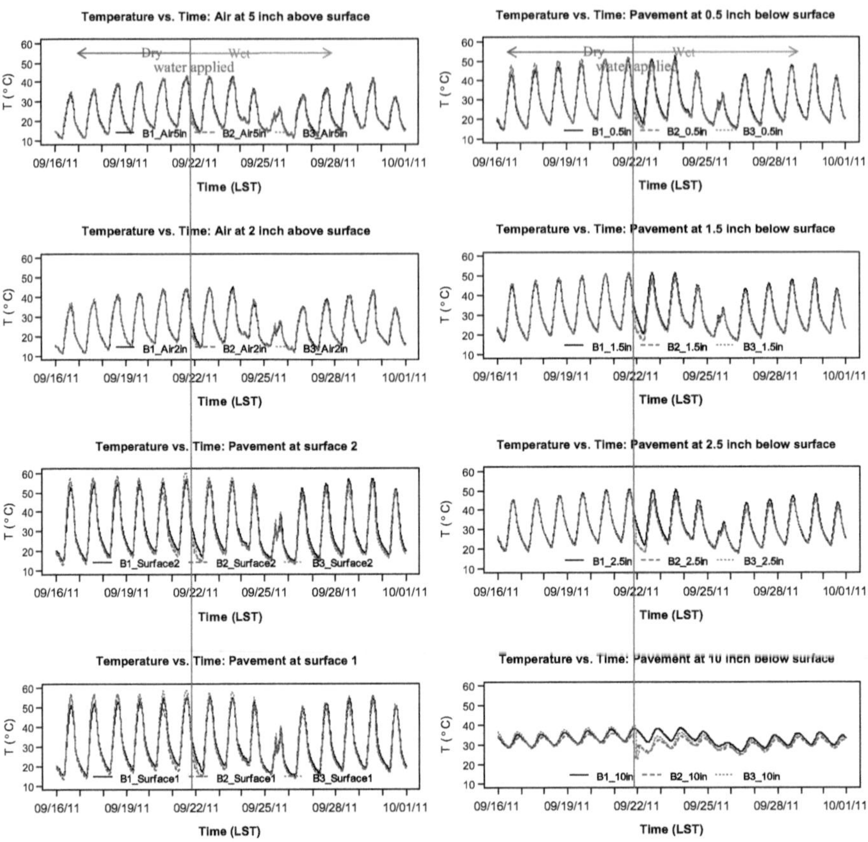

Figure 8.15 Temperature profiles at eight locations on each test section, B1–B3, versus local standard time (LST). Dates are given as month/day/year.

permeable pavements and consequently mitigate the heat island effect and improve thermal comfort. The cooling effect depends on the availability of moisture around the pavement surface and will vanish over time as the amount of moisture decreases. One approach is that water can be irrigated into the permeable pavement during the late afternoons or evenings in summer. This is especially beneficial when the weather report forecasts a very hot day coming; this strategy can be conducted during the night before to mitigate the heat wave coming the next day and improve thermal comfort.

Statistical temperature differences (overall cooling effect during the test period) between the permeable pavements (B3 and B2) and the impermeable pavement (B1) under dry and wet conditions are plotted in

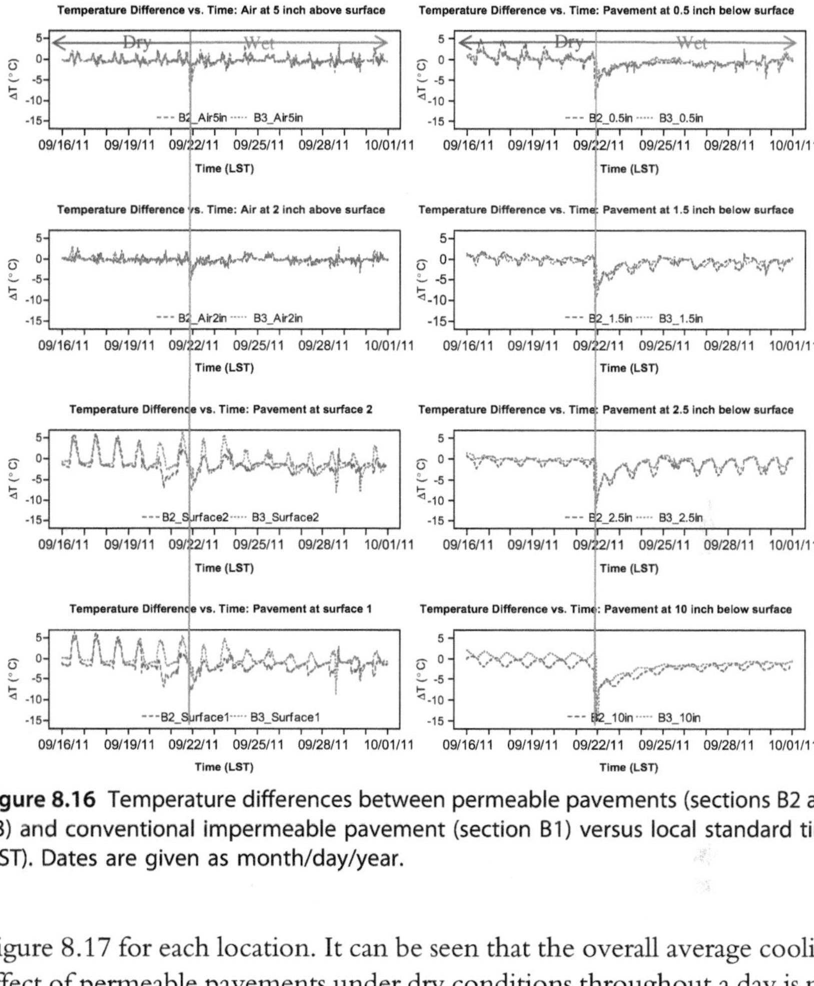

Figure 8.16 Temperature differences between permeable pavements (sections B2 and B3) and conventional impermeable pavement (section B1) versus local standard time (LST). Dates are given as month/day/year.

Figure 8.17 for each location. It can be seen that the overall average cooling effect of permeable pavements under dry conditions throughout a day is not as significant as that under the wet condition (even hotter around noon as observed previously). This also confirms the findings from the study on porous concrete pavements conducted by Kevern et al. [61].

8.4.4 Degree Hours for Quantitative Temperature Comparison over a Period

Degree hours (or degree days) are essentially a simplified representation of outside air-temperature data. They are widely used in the energy industry for calculations relating to the effect of outside air temperature on building energy consumption [166]. Degree hours include cooling degree hours and heating degree hours.

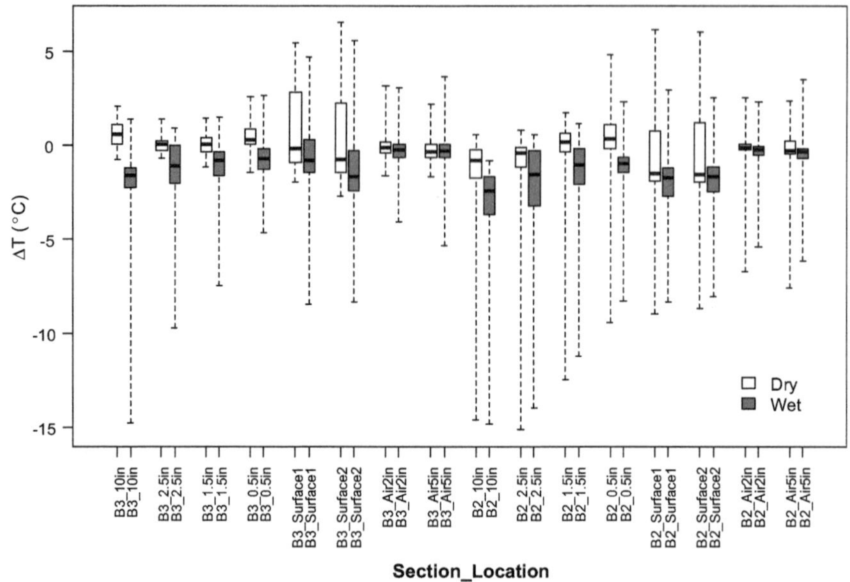

Figure 8.17 Statistical temperature difference (overall cooling effect through a day) of permeable pavements (B2 and B3) compared to conventional impermeable pavement (B1) under dry and wet conditions over the whole test period. Negative difference means permeable pavement is cooler.

"Cooling degree hours," or "CDH," are a measure of how much (in degrees), and for how long (in hours), outside air temperature is *higher* than a specific "base temperature" (or "balance point"). They are used for calculations relating to the energy consumption required to *cool* buildings.

"Heating degree hours," or "HDH," are a measure of how much (in degrees), and for how long (in hours), outside air temperature is *lower* than a specific base temperature. They are used for calculations relating to the energy consumption required to *heat* buildings.

In this case, our focus is on the comparison of pavement temperature profiles over a period, which would affect the near-surface air temperature profile determining the energy consumption required to cool and heat buildings and the human comfort. Therefore, cooling and heating degree hours are employed here as indirect indicators of the energy consumption required for cooling and heating.

If the average temperature for the hour falls below the heating base temperature of 18 °C (lower limit for human comfort, see Table 13.2), the degree hour is recorded as a heating degree hour. A cooling degree hour is recorded if the average air temperature for the hour rises above the cooling base temperature of 26 °C (upper limit for human comfort, see Table 13.2).

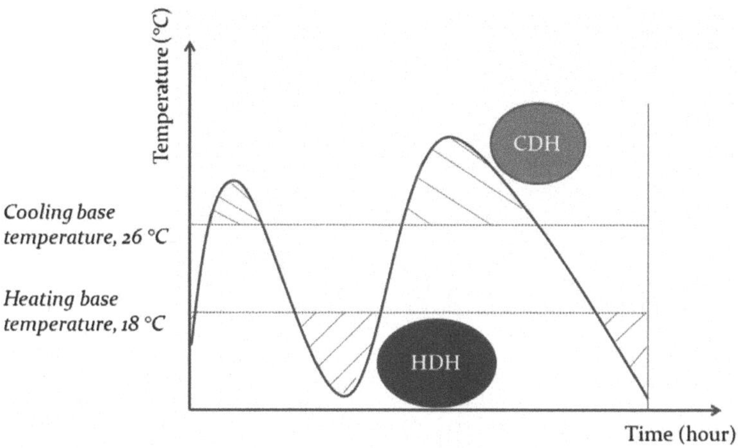

Figure 8.18 Cooling degree hours (CDH) and heating degree hours (HDH) over a period.

Over a period, the total cooling degree hours and heating degree hours are, respectively, the blue and red area shown in Figure 8.18.

The total cooling degree hours and heating degree hours of sections B1, B2, and B3 over a period of 7 days (converted to the same period of 7 days for both dry and wet conditions to allow direct comparison for the hot weather, and this is an example of the type of analysis that can be performed) are listed in Table 8.5.

The major findings from the results include:

1. Compared to the impermeable pavement B1, the permeable ones of B3 and B2 produce lower CDH and slightly higher HDH under wet conditions owing to evaporative cooling, although the CDH and HDH generally are slightly higher under dry conditions (owing to lower thermal conductivity and heat capacity).
2. For the two permeable pavement sections B3 and B2, both CDH and HDH are generally lower under wet conditions than under dry conditions as the effect of water evaporation contributes to more moderate temperatures of pavement surface and near-surface air.
3. Compared to B3, with a thick asphalt layer (8 in), the thin section B2 (4 in) has slightly lower CDH and HDH under both dry and wet conditions as less heat is stored in and later released from thin pavement.
4. These findings imply that thin permeable pavement can produce more moderate temperatures of pavement surface and near-surface air and thus reduce the thermal stress on human bodies and potentially lower the cooling and heating load of buildings and vehicles.

Table 8.5 Thermal load of cooling and heating degree hours (a) CDH anc HDH of various sections in 7 days and (b) differences in CDH and HDH from B1 in 7 days

(a)

Section	Location		Dry				Wet			
			CDH	n^a	HDH	n^a	CDH	n^a	HDH	n^a
B3 (thick asphalt permeable)	In-depth	B3_10 in	1316	168	0	0	799	168	0	0
		B3_2.5 in	1372	168	0	0	967	168	0	0
		B3_1.5 in	1297	163	2	5	923	163	3	5
		B3_0.5 in	1225	155	19	13	883	153	19	15
	Surface	B3_Surface1	1652	153	25	15	1239	144	42	24
		B3_Surface2	1650	149	31	19	1228	139	59	29
	Near-surface air	B3_Air2 in	716	124	137	44	541	120	110	48
		B3_Air5 in	598	119	161	49	466	115	126	53
B2 (thin asphalt permeable)	In-depth	B2_10 in	1057	168	0	0	658	168	0	0
		B2_2.5 in	1295	168	0	0	894	167	0	0
		B2_1.5 in	1308	162	4	6	888	163	4	5
		B2_0.5 in	1271	153	28	15	851	151	22	17
	Surface	B2_Surface1	1514	142	53	26	1072	141	51	27
		B2_Surface2	1574	141	54	27	1126	141	49	27
	Near-surface air	B2_Air2 in	735	123	140	45	531	121	107	47
		B2_Air5 in	612	119	159	49	440	117	123	51
B1 (dense control)	In-depth	B1_10 in	1221	168	0	0	1147	168	0	0
		B1_2.5 in	1375	168	0	0	1152	168	0	0
		B1_1.5 in	1283	162	4	6	1053	165	2	3
		B1_0.5 in	1154	154	26	14	952	159	9	9
	Surface	B1_Surface1	1474	155	21	13	1243	154	18	14
		B1_Surface2	1526	156	20	12	1283	154	19	14
	Near-surface air	B1_Air2 in	720	124	134	44	551	124	95	44
		B1_Air5 in	609	121	148	47	470	121	106	47

(b)

Section	Location		Dry				Wet			
			CDH	n[a]	HDH	n[a]	CDH	n[a]	HDH	n[a]
B3 (thick asphalt permeable)	In-depth	B3_10 in	95	0	0	0	−348	0	0	0
		B3_2.5 in	−4	0	0	0	−185	0	0	0
		B3_1.5 in	14	1	−2	−1	−130	−3	1	2
		B3_0.5 in	71	1	−7	−1	−69	−6	10	6
	Surface	B3_Surface1	178	−3	3	3	−4	−11	24	11
		B3_Surface2	124	−7	11	7	−55	−14	40	14
	Near-surface air	B3_Air2 in	−4	0	3	0	−10	−4	15	4
		B3_Air5 in	−11	−2	13	2	−3	−6	21	6
B2 (thin asphalt permeable)	In-depth	B2_10 in	−164	0	0	0	−488	0	0	0
		B2_2.5 in	−81	0	0	0	−258	−1	0	0
		B2_1.5 in	25	0	0	0	−166	−3	2	3
		B2_0.5 in	117	−1	2	1	−101	−8	14	8
	Surface	B2_Surface1	40	−13	32	13	−171	−14	33	14
		B2_Surface2	48	−15	34	15	−157	−13	31	13
	Near-surface air	B2_Air2 in	15	−1	7	1	−20	−3	13	3
		B2_Air5 in	4	−2	12	2	−29	−4	18	4

[a]Number of hours.

8.4.5 Influence of Weather during the Wet Period

From the results above, it is apparent that the temperatures and degree hours of permeable pavements during the wet period are somewhat lower than those in the dry period. The temperatures would be expected to be affected by the weather conditions (such as air temperature, solar radiation, wind speed). A change in weather conditions might be the reason for the differences in temperatures and degree hours of permeable pavements during the wet and dry periods (respectively after and before 21:00 h on 21 September 2011). Actually, it is noted from Table 8.5 that there are indeed some differences in temperatures and degree hours between the wet and the dry period for impermeable pavement (B1), even though no water irrigation was applied to B1 at all. This indicates that the weather might change during the dry and wet periods and influence the thermal behavior of the pavements. Therefore, further exploration is needed of the temperature data together with the weather data for both dry and wet periods, to isolate and quantify the cooling effect of wetting from the effect of weather.

8.4.5.1 Weather Data

First, the weather data from the on-site weather station were checked, including air temperature (at 2 m height), solar radiation (at 3 m height), wind speed, rainfall (at 2 m height), relative humidity (at 2 m height), and air pressure (at 2 m height), shown in Figure 8.19. It is noted that, except for the rainfall, which was always zero, the weather, in fact, did change over the dry and wet periods (before and after 21 September 2011).

8.4.5.2 Quantifying the Cooling Effect of Wetting

To quantify the cooling effect of wetting, we define a dummy variable wet for the wet condition as follows,

$$\text{Wet} = \begin{cases} 0, & \text{if dry (before 21.00 } 9/21/2011) \\ 1, & \text{if wet (after 21.00 } 9/21/2011) \end{cases} \tag{8.1}$$

The empirical framework for the temperature is defined as Eq. (8.2),

$$T \sim \text{Wet} + \text{Weather} \tag{8.2}$$

where T is the temperature for each location and the covariate vector Weather includes air temperature, solar radiation, wind speed, rainfall, relative humidity, and air pressure.

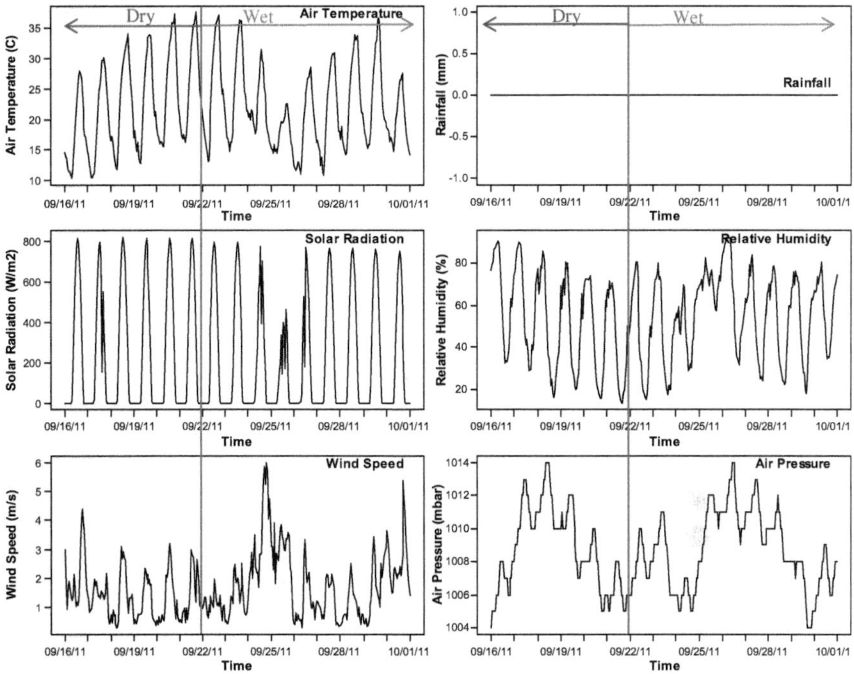

Figure 8.19 Weather data during the dry and wet periods (before and after 21:00 h, 21 September 2011). Dates are given as month/day/year.

According to the empirical framework above, the discontinuity-based ordinary least squares (DB-OLS) [167] method was used for the linear regression on the temperature during the whole period of dry and wet conditions (16 to 30 September 2011). The results of DB-OLS regression are listed in Table 8.6. The estimates of the coefficients of wet, air temperature, solar radiation, and wind speed, are all statistically significant at the level of 0.01 for most locations, and relative humidity is statistically significant at the level of 0.1 for most locations. The estimate of the coefficient of wet indicates the temperature difference between wet (wet = 1) and dry (wet = 0) conditions under the same weather, which is just the cooling effect of wetting (wet = 1) alone. From the results in Table 8.6, it is noted that just the overall cooling effect of wetting alone (wet = 1) for near-surface air (up to 5 in (12.5 cm) above the surface) is about 0.2–0.45 °C; for the surface it is about 1.2–1.6 °C, and for the in-depth layers it is about 1.5–3.4 °C.

Table 8.6 Cooling effect of irrigation from discontinuity-based ordinary least squares regression

	Estimate	Std. error	t value	Pr (>|t|)	Estimate	Std. error	t value	Pr (>|t|)	Estimate	Std. error	t value	Pr (>|t|)	Estimate	Std. error	t value	Pr (>|t|)
B3	Air 5 in				Air 2 in				Surface 1				Surface 2			
(Intercept)	−10.1416	0.722	−14.1	0	−7.6848	0.678	−11.3	0	3.8886	1.582	2.46	0.014	4.648	1.5047	3.09	0.002
Wet	−0.2387	0.055	−4.31	0	−0.4484	0.052	−8.61	0	−1.6225	0.122	−13.35	0	−1.6826	0.1156	−14.55	0
Air temperature	1.2939	0.018	71.39	0	1.2366	0.017	72.63	0	1.0907	0.04	27.46	0	1.0411	0.0378	27.55	0
Solar radiation	0.0057	1E−04	46.32	0	0.0082	1E−04	70.45	0	0.0212	3E−04	77.91	0	0.0237	0.0003	91.46	0
Wind speed	0.2028	0.026	7.74	0	0.192	0.025	7.8	0	0.5122	0.057	8.92	0	0.3298	0.0546	6.03	0
Relative humidity	0.0816	0.006	13.51	0	0.0681	0.006	12	0	−0.0199	0.013	−1.5	0.1338	−0.0232	0.0126	−1.84	0.066
B3	0.5 in				1.5 in				2.5 in				10 in			
(Intercept)	3.4154	1.262	2.71	0.0069	7.59	1.485	5.11	0	14.9016	1.675	8.9	0	34.5072	1.0316	33.45	0
Wet	−1.5003	0.097	−15.5	0	−1.8304	0.114	−16	0	−2.3555	0.129	−18.31	0	−3.3875	0.0793	−42.74	0
Air temperature	1.1597	0.032	36.58	0	1.1213	0.037	30.07	0	0.9955	0.042	23.67	0	0.1919	0.0259	7.41	0
Solar radiation	0.0065	2E−04	29.79	0	0.0016	3E−04	6.28	0	−0.0051	3E−04	−17.81	0	−0.0111	0.0002	−62.49	0
Wind speed	0.4215	0.046	9.19	0	0.5892	0.054	10.92	0	0.7367	0.061	12.11	0	0.3449	0.0375	9.21	0
Relative humidity	−0.0126	0.011	−1.19	0.2335	−0.0361	0.012	−2.9	0.0037	−0.071	0.014	−5.07	0	−0.0535	0.0086	−6.19	0
B2	Air 5 in				Air 2 in				Surface 1				Surface 2			
(Intercept)	−6.8975	0.767	−8.99	0	−5.8288	0.799	−7.29	0	10.9821	1.923	5.71	0	12.3779	1.9254	6.43	0
Wet	−0.3616	0.059	−6.14	0	−0.4472	0.061	−7.28	0	−1.3305	0.148	−9.01	0	−1.204	0.1479	−8.14	0
Air temperature	1.2044	0.019	62.53	0	1.2205	0.02	60.8	0	0.8992	0.048	18.62	0	0.8644	0.0484	17.87	0
Solar radiation	0.006	1E−04	45.19	0	0.0071	1E−04	51.37	0	0.0186	3E−04	56.4	0	0.0204	0.0003	61.66	0
Wind speed	0.136	0.028	4.88	0	0.1254	0.029	4.32	0	0.3234	0.07	4.63	0	0.2499	0.0699	3.57	0.0004
Relative humidity	0.0609	0.006	9.48	0	0.0465	0.007	6.96	0	−0.085	0.016	−5.28	0	−0.0963	0.0161	−5.97	0
B2	0.5 in				1.5 in				2.5 in				10 in			
(Intercept)	6.5495	1.549	4.23	0	11.2298	1.729	6.49	0	16.4804	1.786	9.23	0	34.8946	0.9441	36.96	0
Wet	−1.7391	0.119	−14.6	0	−1.8891	0.133	−14.2	0	−1.9875	0.137	−14.49	0	−2.3362	0.0725	−32.21	0
Air temperature	1.0998	0.039	28.26	0	1.055	0.043	24.29	0	0.9544	0.045	21.28	0	0.055	0.0237	2.32	0.0205
Solar radiation	0.0065	3E−04	24.46	0	−0.0004	3E−04	−1.18	0.2383	−0.0063	3E−04	−20.49	0	−0.0073	0.0002	−45.07	0
Wind speed	0.2869	0.056	5.1	0	0.4997	0.063	7.96	0	0.5963	0.065	9.2	0	0.1538	0.0343	4.49	0
Relative humidity	−0.0429	0.013	−3.31	0.0009	−0.0673	0.015	−4.65	0	−0.0906	0.015	−6.06	0	−0.0464	0.0079	−5.87	0

Note: Rainfall and air pressure are not listed because of statistical insignificance.

8.5 THERMAL IMAGES OF EXPERIMENTAL PAVEMENT SECTIONS

To confirm the thermal behaviors of the various pavements and cooling effects of permeable pavements under both dry and wet conditions, another irrigation experiment was conducted in July 2012. In California, the summer season is the dry season with almost no rainfall. From 15:00 h on 10 July 2012 to ~01:00 h on 11 July 2012, water was irrigated into the permeable test sections. Approximately 3.5 m^3 water (~0.35 m^3/h × 10 h = ~3.5 m^3; 3.5 m^3/(4 × 4 m) = 0.22 m^3 of water per square meter of pavement or 48.3 gal/yd^2) was irrigated into each of the six permeable sections (C2, C3, B2, B3, A2, and A3). Thermal images were taken at various times on 9–11 July 2012 for all the experimental sections along with optical images (Figure 8.20). The water tables were monitored using small monitoring wells for the six permeable experimental sections after irrigation. The weather data were also monitored using a nearby weather station.

From 15:00 h on 10 July 2012, water was irrigated into the permeable test sections. The irrigation was continued until each permeable pavement section was filled up with water at around 01:00 h on 11 July 2012. After irrigation stopped, the water tables would drop over time owing to the nature processes of infiltration and evaporation. The water tables and water temperatures were monitored using small wells for each permeable section and are presented in Figure 8.21.

The weather data during the experiment period were also monitored and are shown in Figure 8.22. It is noted that the air temperature was increasing during the 3 days. The solar radiation was fairly constant for the period. The wind speed was increasing slightly and the relative humidity was decreasing slightly.

Figure 8.20 Thermal camera used in the study.

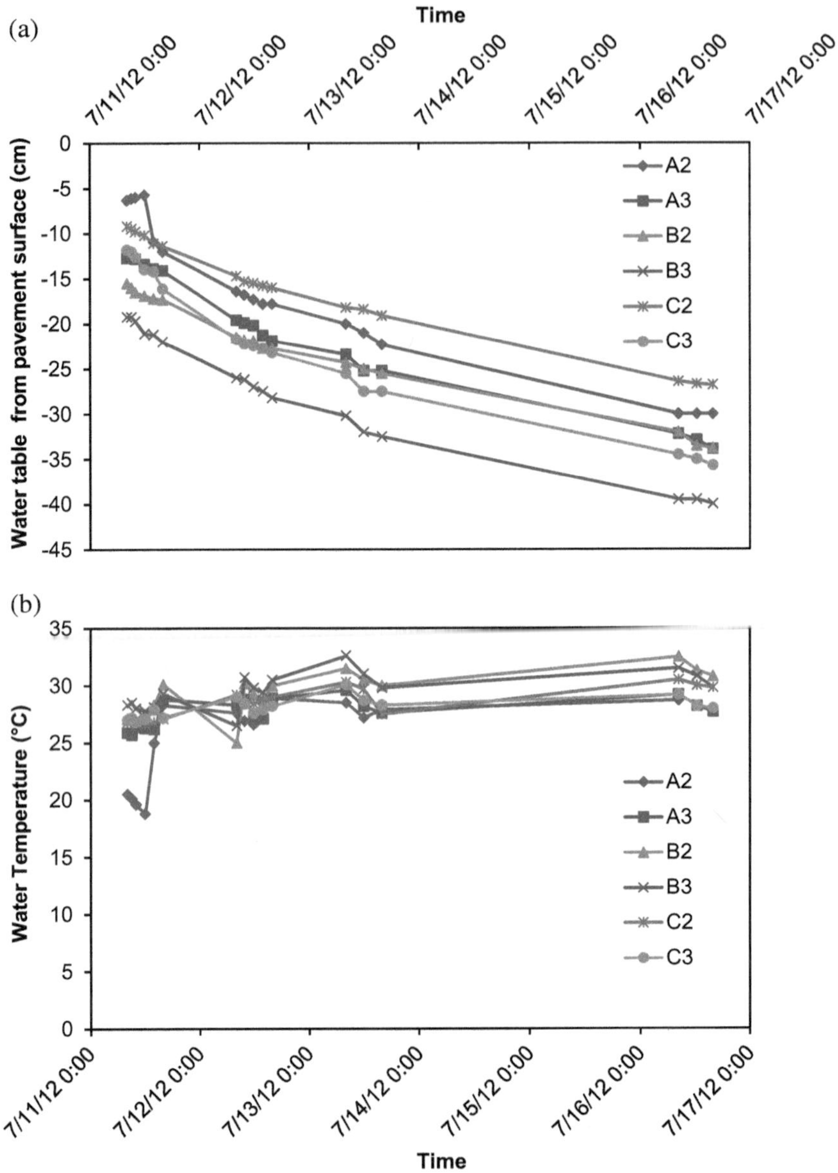

Figure 8.21 Water tables and water temperatures in the monitoring wells for six permeable sections. (a) Water tables. (b) Water temperatures. Dates are given as month/day/year. 0:00 indicates midnight.

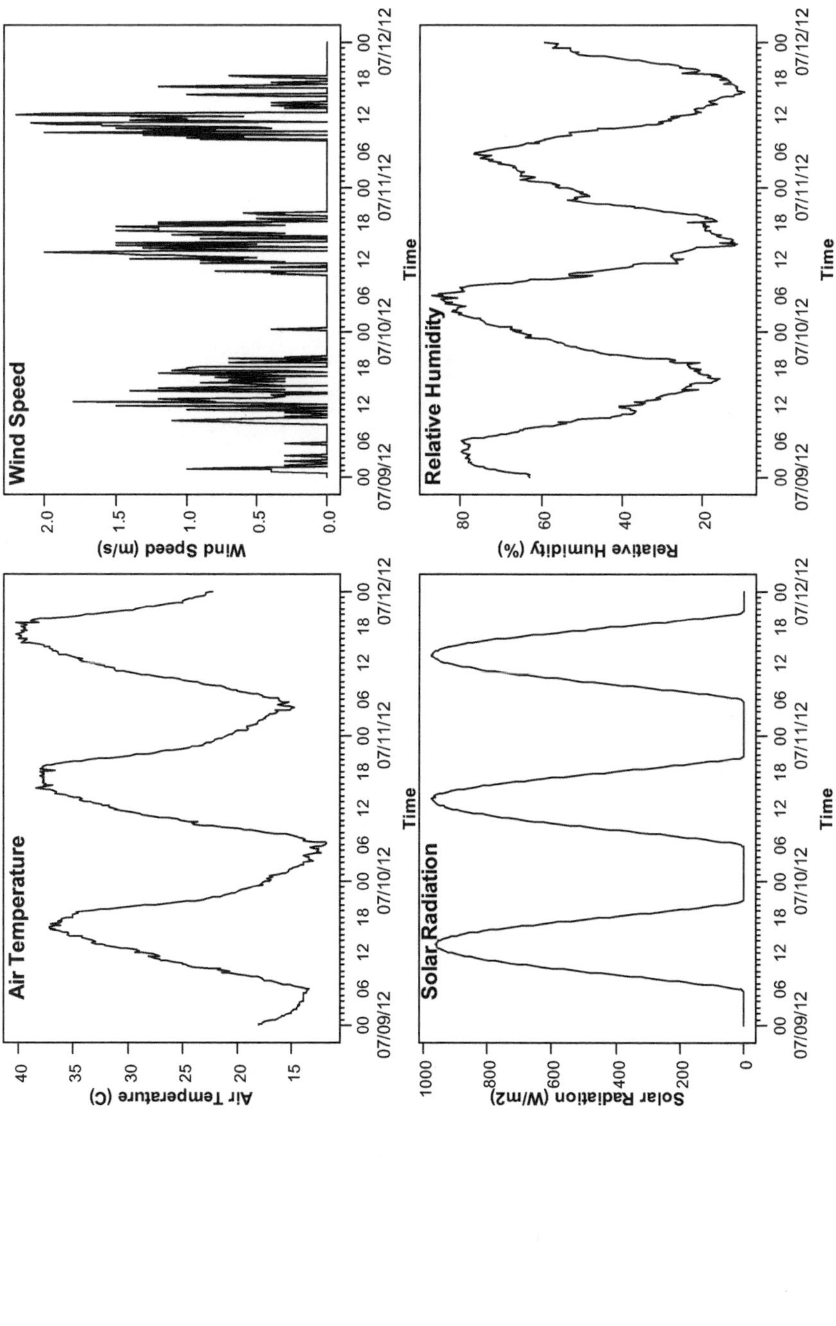

Figure 8.22 Weather data during the experiment period (no rain). Dates are given as month/day/year.

The thermal images at 16:00 h on 9 and 10 July 2012 are presented in Figures 8.23 and 8.24, respectively, along with the optical images. First, it is seen that the highest temperatures were given by the asphalt pavements (~70 °C for B2 and B3 in Figure 8.23) and were 10–25 °C higher than

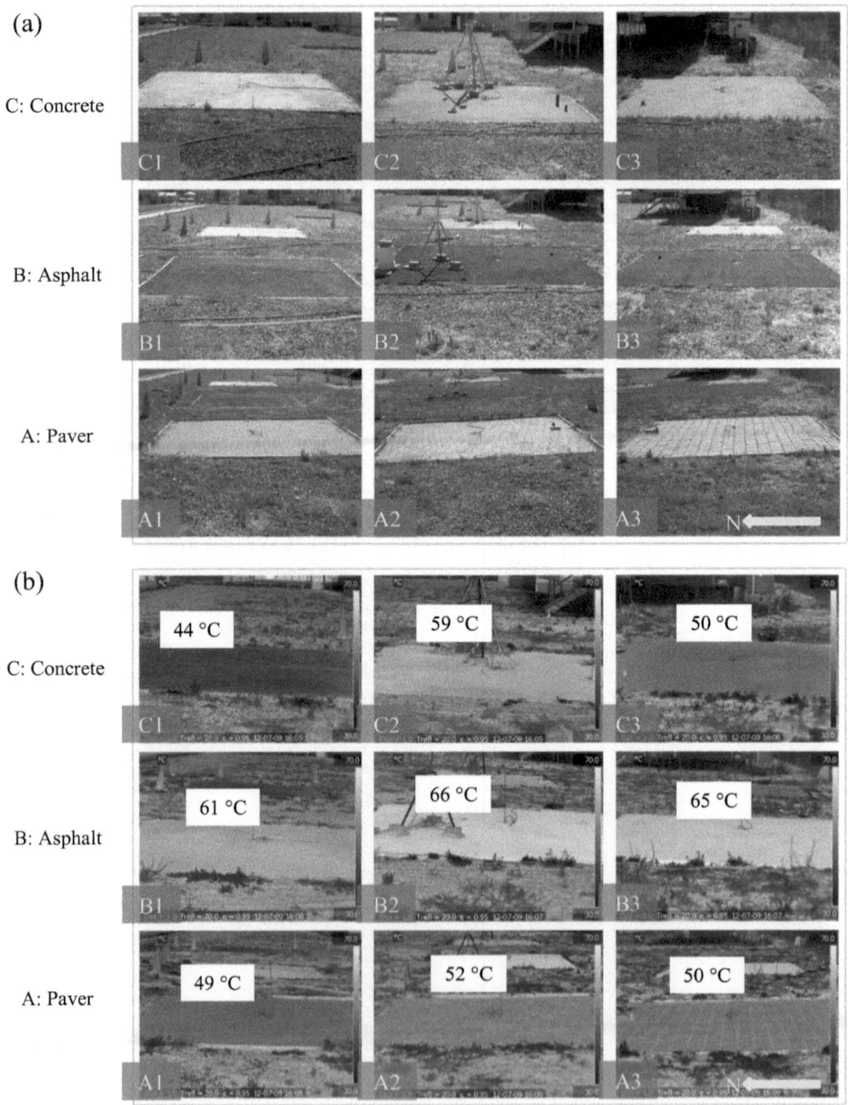

Figure 8.23 Optical and thermal images of surface temperature of experimental sections under dry condition on 9 July 2012. (a) Optical images. (b) Thermal images under dry condition (16:00 h on 9 July 2012). Lighter is hotter.

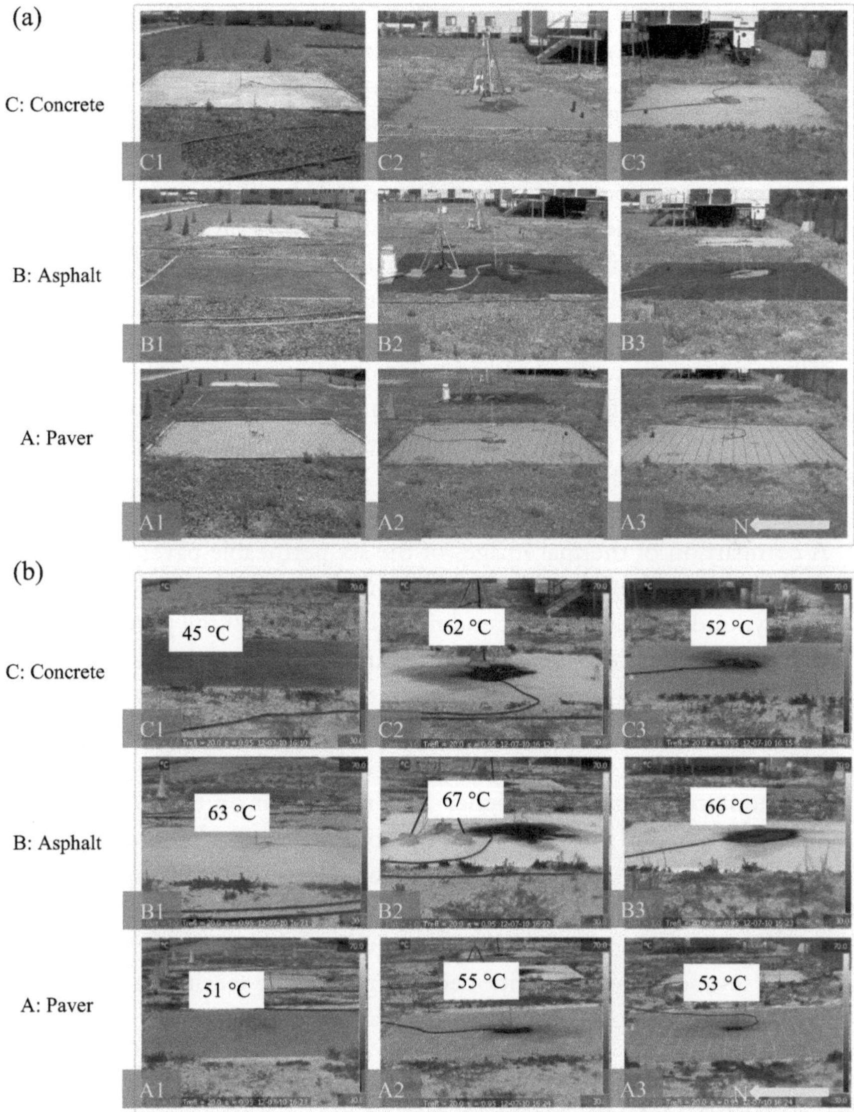

Figure 8.24 Optical and thermal images of surface temperature of experimental sections during watering on 10 July 2012. (a) Optical images. (b) Thermal images of pavements during watering (16:00 h on 10 July 2012). Lighter is hotter.

those of the concrete pavements (~45 °C for C1 and ~60 °C for C2 in Figure 8.23). The paver pavements were also cooler than the asphalt pavements by ~20 °C. It is also noted from Figure 8.23 that the permeable pavements B2 and B3 under dry conditions produced higher surface temperatures than the impermeable pavement B1 by ~10 °C, although the albedos were nearly equal. The same thermal behavior was shown by the concrete and paver (note the differences in albedo as shown in Table 8.1). The reason for the differences in thermal behavior is the differences in thermal properties (i.e., thermal conductivity and heat capacity) as discussed previously in Section 8.3.7.

When water was applied into the permeable pavements (Figure 8.24), the temperatures of the pavement portions with water were as low as ~30 °C, which was much lower than the pavement portion without water. This implies that watering can be an effective way to lower the pavement temperature in summer, provided sufficient water is available.

A comparison of thermal images of the six permeable pavements under dry, watering, and wet conditions is presented in Figure 8.25. As observed previously, under the watering condition (10 July 2012) the pavements had much lower temperatures than under the dry condition (9 July 2012). Even on the third day (11 July 2012) without watering but with higher air temperature (see Figure 8.22), the pavements still showed lower temperatures at 4:00 PM (25 h after watering) (by 2–7 °C compared to the second day, without considering the higher air temperature on the third day (11 July 2012); see Figure 8.25) owing to evaporative cooling of some moisture existing in the pavements. This implies that watering with cool water can effectively lower the pavement surface temperature, and evaporation of some moisture existing in the pavements also can help produce a low pavement temperature. As mentioned previously, the specific cooling effect depends on the evaporation rate on the pavement.

8.6 SUMMARY AND CONCLUSIONS

Through the design, construction, and instrumentation of nine experimental sections of various pavement types, this chapter presented the seasonal thermal behavior and cooling effects of the various pavement types, with a focus on permeable pavements under both dry and wet conditions, and investigated factors affecting the thermal behavior and cooling effects of permeable pavements.

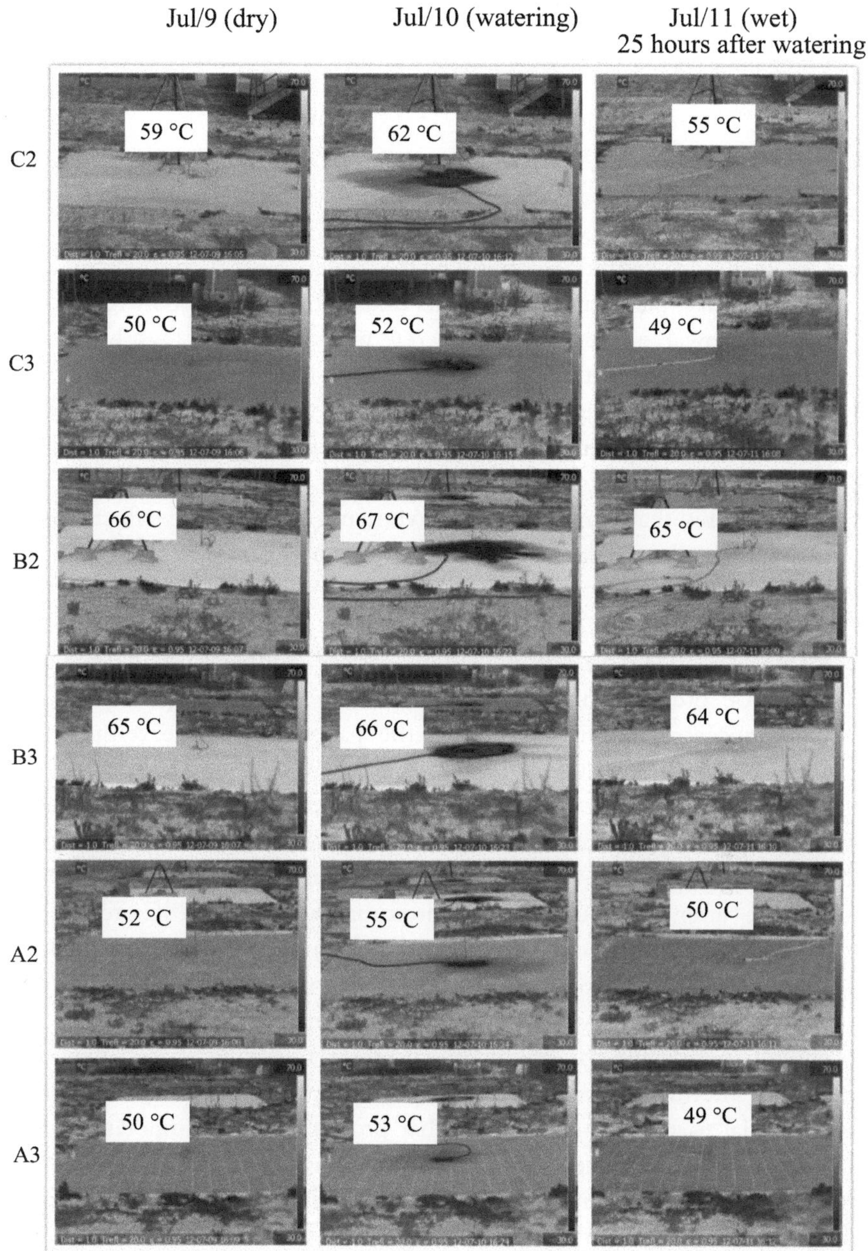

Figure 8.25 Comparison of thermal images of surface temperature of permeable pavements under various conditions (16:00 h on 9 through 11 July 2012). Lighter is hotter.

The major conclusions drawn from the study presented in this chapter include

1. Concrete and paver pavements (albedos of 0.18–0.29) in this study showed lower surface temperatures than asphalt pavements (albedos of 0.08–0.09) by 10–25 °C during hot summer at Davis, California; asphalt pavement with high albedo, through reflective coating or other treatments, could also produce a low surface temperature.
2. Under the dry condition, owing to the lower thermal conductivity and heat capacity, permeable pavements (including pervious concrete pavement, permeable interlocking concrete paver, and porous asphalt pavement) are hotter in the daytime but cool faster and consequently get colder during the night and help mitigate the nighttime heat island effect, compared to impermeable pavements.
3. Permeable pavements under wet conditions could have lower surface temperatures than impermeable pavements; the cooling effect highly depends on the availability of moisture near the surface layer and the evaporation rate.
4. The peak cooling effect of watering for the test sections was approximately 15–35 °C on the pavement surface temperature in the early afternoon during summer owing to the cool water and evaporation; the evaporative cooling effect on the pavement surface temperature at 4:00 PM on the third day (25 h after watering) was still 2–7 °C compared to that on the second day, without considering the higher air temperature on the third day.
5. The overall average cooling effect of wetting alone (wet = 1) over 1 week after irrigation was approximately 0.2–0.45 °C for near-surface air on permeable pavements; for the surface it was approximately 1.2–1.6 °C and approximately 1.5–3.4 °C for the in-depth layers.

Based on the findings from this study, compared with impermeable pavements, permeable pavements (including pervious concrete pavement, permeable interlocking concrete paver, and porous asphalt pavement) have the potential to be cool pavements that produce lower temperatures and help to mitigate local heat island effects. However, attention should be given to permeable pavements under dry conditions, which might produce a higher daytime temperature. Watering or irrigation and evaporation can help to reduce the daytime pavement surface temperature of permeable pavements and consequently mitigate the heat island effect and improve thermal comfort. The cooling effect depends on the availability of moisture around the pavement surface and will vanish over time as the water level decreases. As a pavement thermal management strategy, water collected from rain (where there is rain in summer, not in California) or irrigation can be irrigated into the permeable pavement during the late afternoons or evenings in summer. This is especially beneficial when the weather report forecasts a very hot day coming; this strategy can be conducted during the previous night to mitigate the heat wave coming the next day and improve thermal comfort.

CHAPTER 9

Thermal Interaction between Pavement and Near-Surface Air

Contents

Pavement surfaces could exchange heat with the near-surface air through reflected short-wave solar radiation, emitted long-wave radiation, and convection from the pavement surface. These proce3sses of thermal interaction will influence the temperature profile of the near-surface air and consequent human thermal comfort and air quality (e.g., ground-level ozone). Therefore, it is of great importance to investigate the thermal interaction between pavement and near-surface air.

9.1 OBJECTIVES

The objectives of the study in this chapter are to investigate the thermal interactions between pavement and near-surface air for various pavement

Pavement Materials for Heat Island Mitigation
ISBN 978-0-12-803476-7
http://dx.doi.org/10.1016/B978-0-12-803476-7.00009-X

types, measure the temperature profile of near-surface air, and explore the factors affecting the profile.

9.2 MATERIALS AND METHODOLOGY

9.2.1 Pavement Sections for Measurement

Four pavement sections were chosen to conduct the experiments for measuring the temperature profile of near-surface air above the pavement surface. Two of them were asphalt pavements and the other two were concrete pavements. They were of different shapes and sizes. The pavement sections are summarized in Table 9.1, along with the albedos (solar reflectivities) measured by an albedometer.

9.2.2 Measurement Method and Equipment

Type T thermocouple wires were used to measure the temperature on the pavement surfaces and of near-surface air at various heights above the pavement surface. Thermocouple sensor locations were 0, 2, 5, 10, 20, and 40 in above the pavement surface (Figure 9.1). The data were recorded at time intervals of 10 min using a CR1000 data logger. Local weather data were also monitored using an on-site portable weather station.

9.2.3 Correction for Near-Surface Air Temperature Measured by Thermocouple Sensors

To obtain the temperature profile of the near-surface air, conveniently conducting multiple measurements of near-surface air temperatures at various heights above the ground is critical. Using thermocouple wire

Table 9.1 Summary of pavement sections for measurement

Section no.	Type	Size	Albedo	Measurement date (month/day/year)
PA1	Asphalt	~30 m diameter	0.06	7/2/2012–7/17/2012
PC1	Concrete	15 m wide × 45 m long	0.26	7/18/2012–7/27/2012
PA2 (B1)[a]	Asphalt	4 m wide × 4 m long	0.08	7/28/2012–8/20/2012
PC2 (C1)[a]	Concrete	4 m wide × 4 m long	0.28	7/28/2012–8/20/2012

[a]PA2 and PC2 are the experimental test sections B1 and C1 as described previously, respectively.

Figure 9.1 Example experimental setups for temperature profile measurements. (a) Asphalt pavement (PA1). (b) Concrete pavement (PC1).

sensors is a simple, convenient, and cost-effective choice to perform these multiple measurements compared to the air temperature sensors with radiation shield commonly used in weather stations.

However, unlike the air temperature sensors with radiation shield, the thermocouple sensors open in the near-surface air would be influenced by the environmental conditions surrounding the sensors, such as solar radiation, mainly, and possibly wind speed and ambient air temperature. The near-surface air temperature measured from the thermocouple sensors is not the real air temperature, but represents a combined effect of these environmental factors (several efforts were made to create a radiation shield small enough for the soft thermocouple wires using light materials such as aluminum foil; however, no good effects were obtained). Usually a higher (by 1–4 °C) daytime air temperature would be given by the open thermocouple sensors because of the heating effect of solar radiation. Therefore, the near-surface air temperature measured from the thermocouple sensors must be corrected for these environmental factors to reveal the real air temperature, in particular for the daytime air temperature.

To achieve this goal, separate experiments were performed to compare the air temperature measured with an open thermocouple sensor to that measured from an air temperature sensor with radiation shield at the same location and height. A formula for the correction factor was developed against such simultaneous environmental conditions as solar radiation, wind speed, and ambient air temperature obtained from a nearby weather station. The correlation for the correction factor is

$$CF_T = (-1.152) + 0.003 * SR + (-0.105 * WS) + 0.042 * T_a \tag{9.1}$$

where CF_T is the correction factor for near-surface air temperature measured from an open thermocouple sensor, in °C; SR is the solar radiation measured at ~2 m height, in W/m^2, which is the most significant factor for this correction; WS is the wind speed measured at ~2 m height, in m/s; and T_a is the ambient air temperature measured at ~2 m height, in °C. The raw near-surface air temperature, T_{raw}, measured from the open thermocouple sensor, must be less the correction factor CF_T to obtain the corrected real value T_{cor}, i.e.,

$$T_{\text{cor}} = T_{\text{raw}} - CF_T \tag{9.2}$$

An example spatial profile of near-surface air temperature before and after correction is presented in Figure 9.2. On a clear sunny day with peak solar radiation intensity of ~1000 W/m^2 (10 July 2012), the daytime

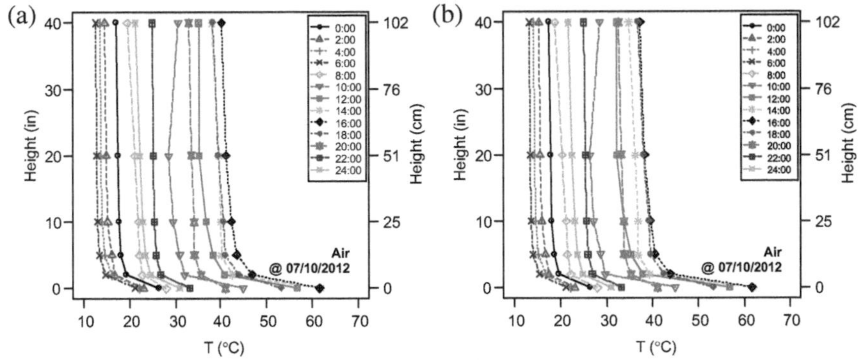

Figure 9.2 Example spatial profile of near-surface air temperature before and after correction (PA1). (a) Before temperature correction. (b) After temperature correction.

temperatures, specifically the peak value at around 16:00 h, are reduced by ~3 °C after correction. However, the nighttime temperatures do not change much because there is no solar radiation during the night.

9.3 RESULTS AND DISCUSSION

9.3.1 Example Results on Asphalt Pavement PA1 and Concrete Pavement PC1

The near-surface air temperatures on asphalt pavement PA1 and concrete pavement PC1 were monitored at various times from 2 through 17 July 2012 and from 18 through 27 July 2012, respectively. The weather conditions did not change much during this summer period, especially for the sunny days. This makes it comparable to examine the effects of asphalt and concrete pavement on near-surface air temperatures for some clear days. Some results for PA1 and PC1 are presented in Figures 9.4 and 9.5, as well as the example weather conditions during the experiment periods in Figure 9.3.

It is noted that the temporal profiles of near-surface air have patterns similar to those of ambient air temperature for both asphalt and concrete pavements. For the spatial profiles, the near-surface air temperatures gradually decrease as the distance from the pavement surface increases, with greater slope (change rate) in the first 25 cm above the pavement surface. In the range close to the pavement surface, the near-surface air temperatures are much higher than the ambient air temperature (at ~2 m height), especially for the black asphalt pavement, with higher surface temperature

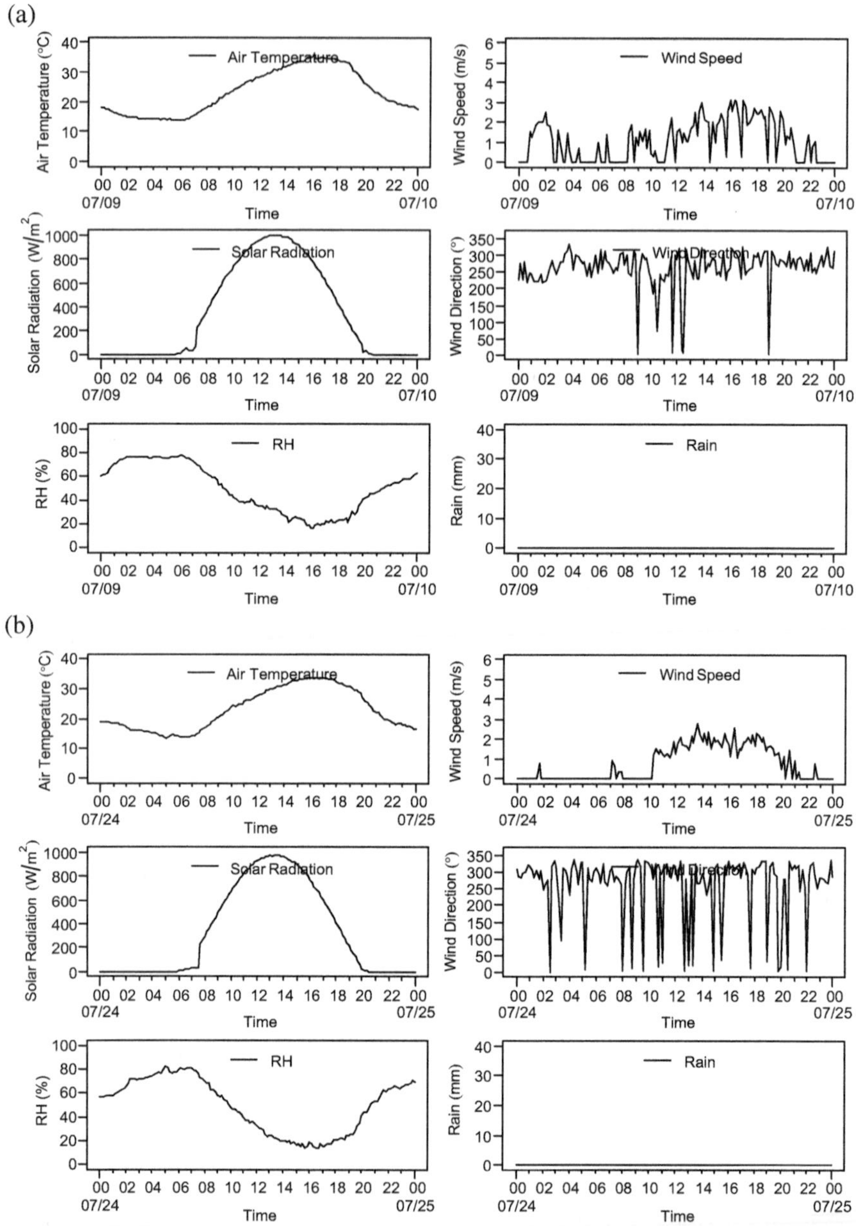

Figure 9.3 Example weather conditions during the experimental period for PA1 and PC1. (a) PA1. (b) PC1. Dates are given as month/day.

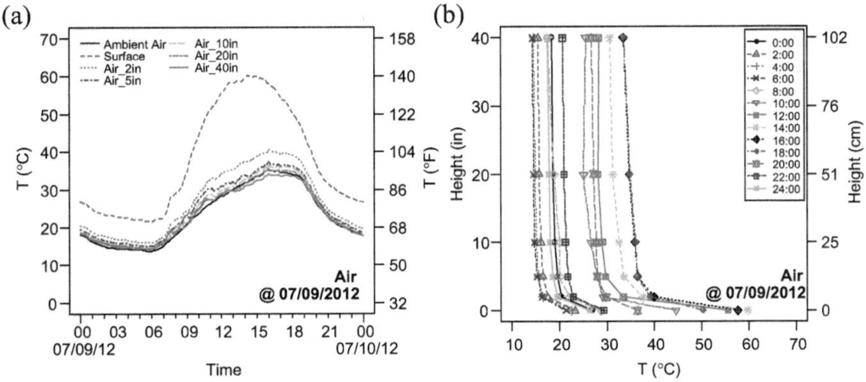

Figure 9.4 Example profiles of near-surface air temperatures on asphalt pavement PA1. (a) Temporal profile. (b) Spatial profile. *Ambient Air is the ambient air temperature from a sensor on a portable weather station at 67 in (1.7 m). Surface and Air_xin are the air temperatures at 0 in and x in above the pavement surface, respectively.* Dates are given as month/day/year.

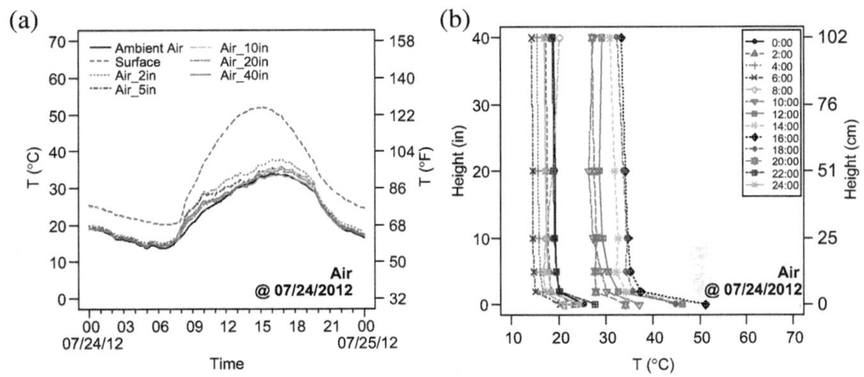

Figure 9.5 Example profiles of near-surface air temperatures on concrete pavement PC1. (a) Temporal profile. (b) Spatial profile. Dates are given as month/day/year.

compared to high reflectance concrete pavement. The high near-surface air temperatures would reduce the human thermal comfort, and they are more critical for babies and children, who are shorter and closer to the surface. In addition, the formation of ground-level ozone would be facilitated by these high near-surface air temperatures when smog (volatile organic compounds and nitrogen oxides) is emitted from vehicle exhaust pipes that are close to the pavement surface on roads or parking lots.

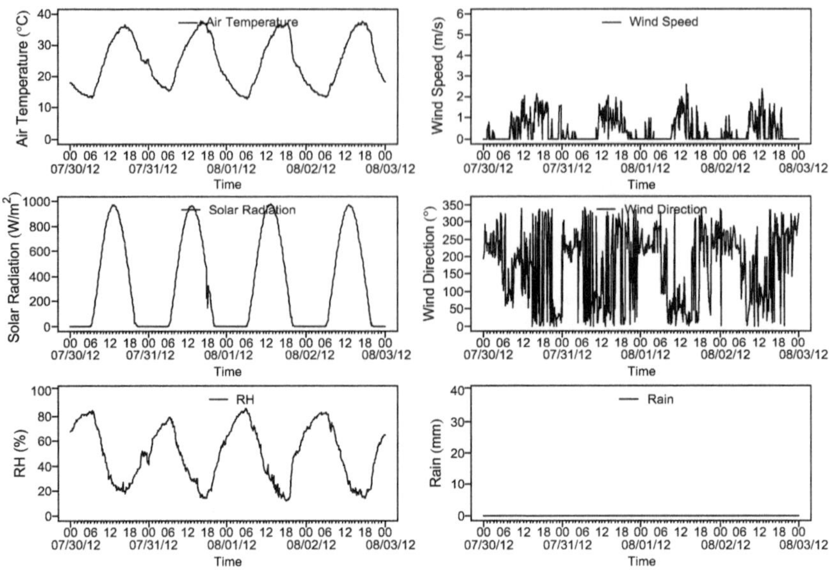

Figure 9.6 Example weather conditions during the experiment period for PA2 and PC2. Dates are given as month/day/year.

9.3.2 Temporal and Spatial Variation of Near-Surface Air Temperatures on Asphalt Pavement PA2 and Concrete Pavement PC2

The near-surface air temperatures on asphalt pavement PA2 and concrete pavement PC2 were monitored at the same time from 28 July 2012 through 20 August 2012. This made it easy to directly compare the effects of asphalt and concrete pavement on near-surface air temperatures. Some results are presented in Figures 9.7 and 9.8, as well as the example weather conditions during the experimental period (Figure 9.6).

In addition to the observations similar to those for PA1 and PC1 presented above, the daytime near-surface air temperatures on asphalt pavement (B1 in Figure 9.8) were much higher than those of the concrete pavement (C1 in Figure 9.8), although they were under the same weather conditions. The higher pavement surface temperature would heat up the near-surface air and produce a higher temperature and a steeper temperature gradient for near-surface air. However, the temperature profiles for near-surface air are quite similar for asphalt and concrete pavements during the nighttime, and the nighttime surface temperatures are close. This implies that pavement surface temperatures play a relatively important role in the near-surface air temperature profiles, which are critical for human thermal comfort and air quality.

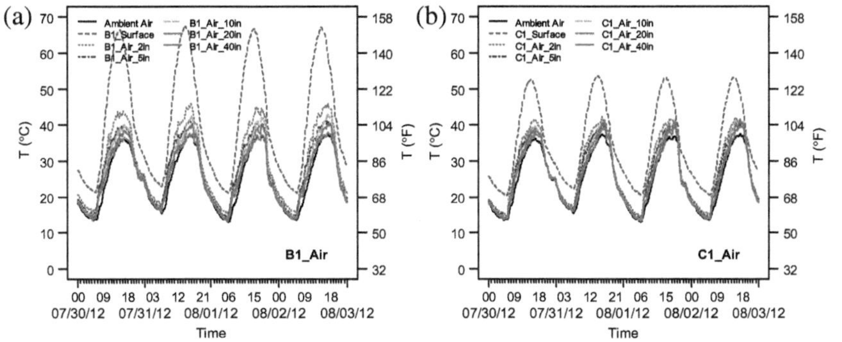

Figure 9.7 Example temporal profiles of near-surface air on asphalt pavement PA2 and concrete pavement PC2. (a) Asphalt pavement PA2 (B1). (b) Concrete pavement PC2 (C1). *Ambient Air is the ambient air temperature from a sensor on a portable weather station at 67 in (1.7 m). Surface and Air_xin are air temperatures at 0 in and x in above the pavement surface, respectively*. Dates are given as month/day/year.

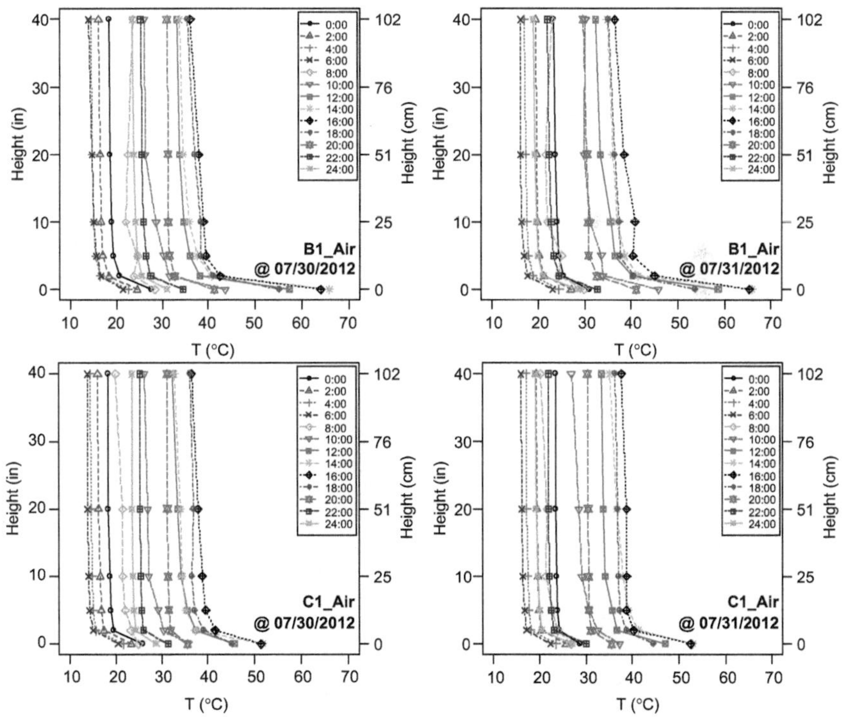

Figure 9.8 Example spatial profiles of near-surface air temperatures on asphalt pavement PA2 (B1) and concrete pavement PC2 (C1) on the same 2 days. Dates are given as month/day/year.

9.3.3 Influence of Wind Speed on Near-Surface Air Temperature Profile

Wind speed (air flow velocity) might also significantly influence the near-surface air temperature and the shape of the profiles of near-surface air temperature. A separate experiment was performed on asphalt pavement PA1 to examine the effects of wind speed.

Like other pavement sections, asphalt pavement PA1 was also open to the wind in a quite large and open space with few barriers to the wind. To break the wind around the temperature sensors, a box was installed around them at midnight of 10 July 2012 (i.e., 00:00 h, 11 July 2012) as shown in Figure 9.9. The temperature profiles of the near-surface air were measured and compared for both before and after installing the windbreak. The results are presented in Figures 9.11 and 9.12, along with the weather conditions during the experiment period in Figure 9.10. Owing to the lack of a handy windmeter, the specific wind speed in the box was not measured.

It is noted from Figure 9.11 that the temporal profiles of near-surface air were changed owing to the installation of windbreak. Just after the installation of the windbreak (00:00 h, 11 July 2012), the surface temperature and near-surface air temperatures increased to some extent because of the reduced wind speed inside the windbreak. During the daytime, the shapes of the temperature profiles, especially the surface temperature, were also influenced by the shading caused by the windbreak.

Figure 9.12 presents the spatial profiles of near-surface air before and after installing the windbreak. The daytime temperature profiles of the near-surface air became steeper and the temperatures around 50 cm (20 in)

Figure 9.9 Experimental setup for windbreak.

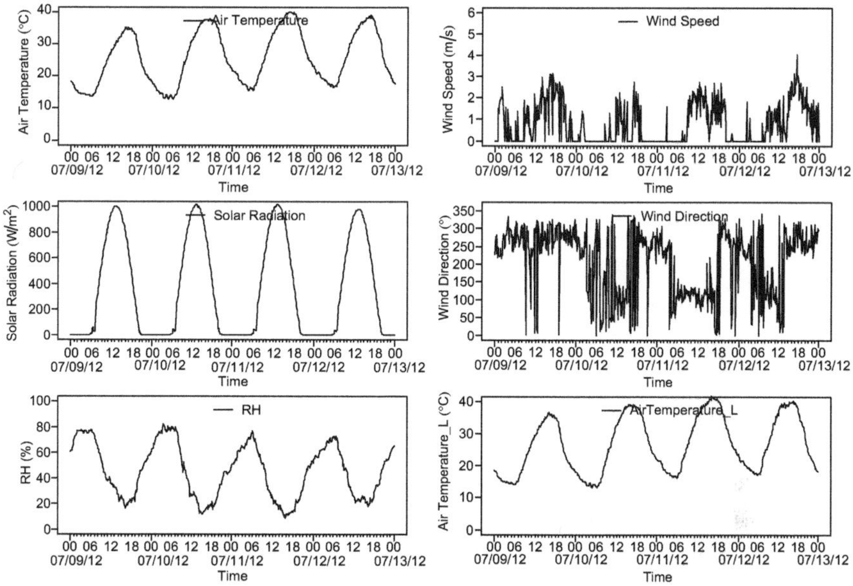

Figure 9.10 Weather conditions during the experiment period for PA1. Dates are given as month/day/year.

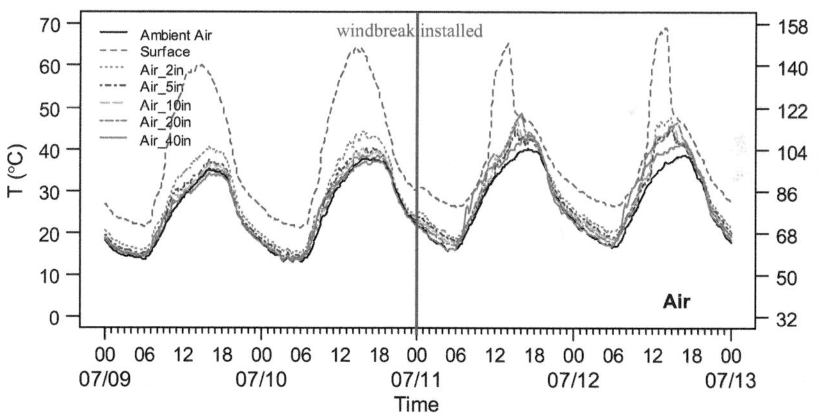

Figure 9.11 Temperature temporal profiles of near-surface air before and after installing the windbreak. Dates are given as month/day.

high increased compared to those without the windbreak. The daytime temperature profiles of near-surface air did not change significantly.

These findings imply that the wind speed will influence the shape of the temperature profiles of near-surface air. Lower wind speed will make the

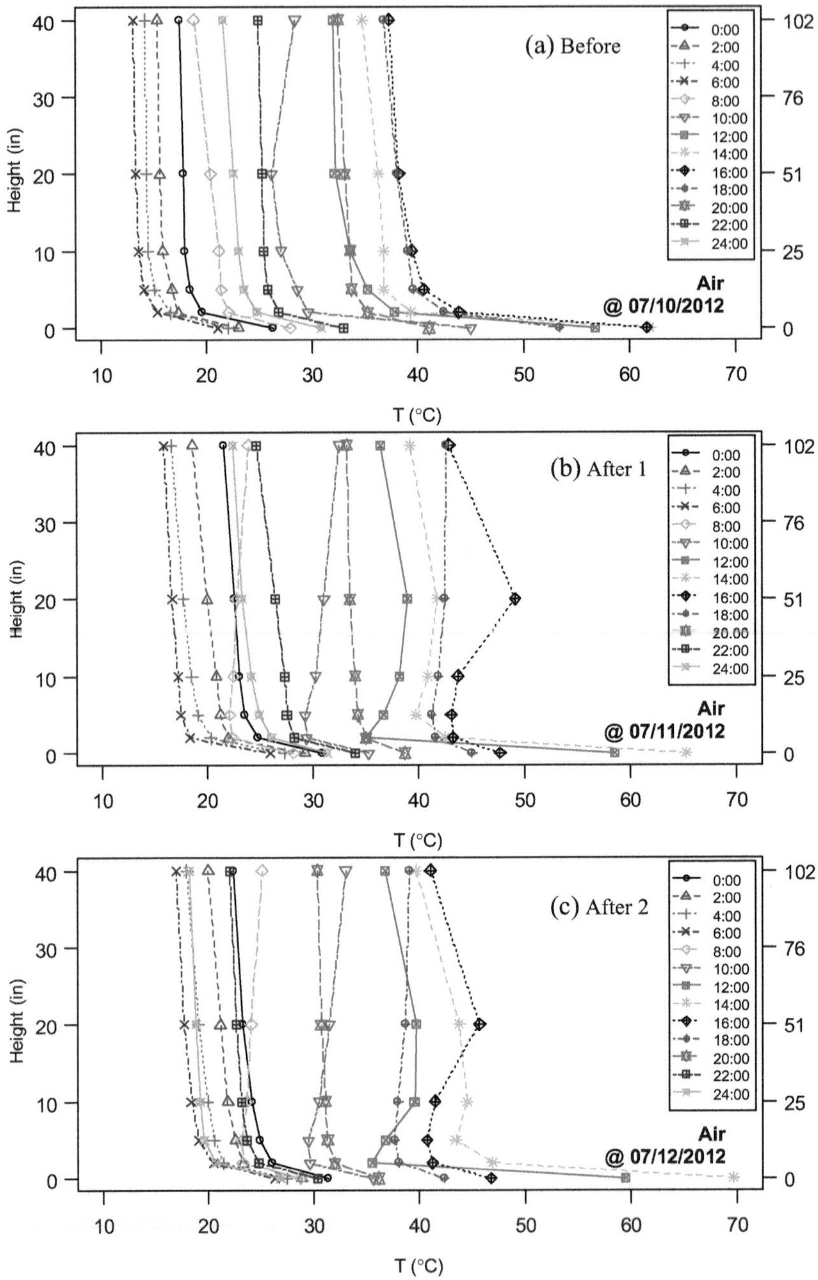

Figure 9.12 Temperature spatial profiles of near-surface air before and after installing the windbreak. Dates are given as month/day/year.

spatial profiles steeper owing to less heat dissipated by wind or airflow. The near-surface air temperatures at higher distances from the surface will be influenced by the pavement heat under lower wind speed. This further implies that the heat effects of pavements will be more critical in the environments with lower wind speed, such as highly developed dense urban areas.

9.4 MODELING OF NEAR-SURFACE AIR TEMPERATURE PROFILE

It is revealed from the findings above that during hot periods the near-surface air temperatures diminish gradually as the height above the surface increases, and the slope of temperature versus height decreases. The shape of the spatial profiles is influenced by the surface temperature, ambient air temperature, and wind speed.

This section will develop a model to describe the spatial profiles of near-surface air temperatures, which characterize the effects of surface temperature, ambient air temperature, and wind speed.

The assumptions for developing the model include:

1. The spatial variation of near-surface air temperature is considered to be in the range of up to 1 m from the surface.
2. The influences of ambient air temperature and wind speed are characterized by the ambient air temperature and wind speed at ~2 m height, which are usually measured by a nearby weather station.
3. The size effect of the paved area is not explicitly taken into account here.

9.4.1 Developing the Dimensionless Parameters for the Model

To simplify the model, the normalized dimensionless parameters are developed for the model as follows,

$$T_n = \frac{T_z - T^+}{T_s - T^+} \tag{9.3}$$

$$z_n = \frac{Z}{Z^+} \tag{9.4}$$

where

T_n is normalized dimensionless temperature, $[-1, 1]$;
z_n is normalized height above the surface, $[0, 1]$;
Z is the height above the pavement surface, in m, $[0, Z^+]$;

T_z is the near-surface air temperature at height Z, in °C;
Z^+ is the upper bound of height considered in the model, in m, assuming $Z^+ = 1$ m;
T^+ is the near-surface air temperature at Z^+, in °C, assuming $T^+ = T_a$ (T_a is the ambient air temperature at ~2 m height, in °C); and
T_s is the surface temperature, i.e., T_Z at $Z = 0$, in °C.

Using the developed normalized dimensionless temperature and height, the normalized spatial profiles of near-surface air temperature can be developed. Some examples of normalized profiles of near-surface air temperature are presented in Figure 9.13, along with the corresponding original profiles.

9.4.2 Model Development and Obtaining Coefficient C through Regression on the Model

According to the trends shown by most of the normalized profiles presented in Figure 9.13, the following simple model is proposed for the normalized profiles of near-surface air temperature:

$$T_n = e^{C \cdot z_n} \tag{9.5}$$

where C is constant.

With the developed model for the normalized profiles of near-surface air temperature, the best-fitting constant C can be obtained through regression on the normalized profiles at various times. Some examples of measured and predicted normalized profiles of near-surface air temperature at various times are presented in Figure 9.14. The obtained constants C for each time are also shown in Figure 9.14. The constant C is different for the different times and is correlated to the wind speed at that time.

9.4.3 Correlation between Coefficient C and Wind Speed

As mentioned previously, the spatial profiles of near-surface air temperature are also influenced by the wind speed in addition to the ambient air temperature and surface temperature. To quantify the effect of wind speed on the profile, the obtained constants C at various times with various wind speeds are correlated to the wind speed at the corresponding time. The obtained data of constant C and wind speed WS (at ~2 m height) are plotted in Figure 9.15, along with the best-fitting line. The correlation between coefficient C and wind speed is then proposed as follows,

$$C = -5.13 + (-0.57^* WS) \tag{9.6}$$

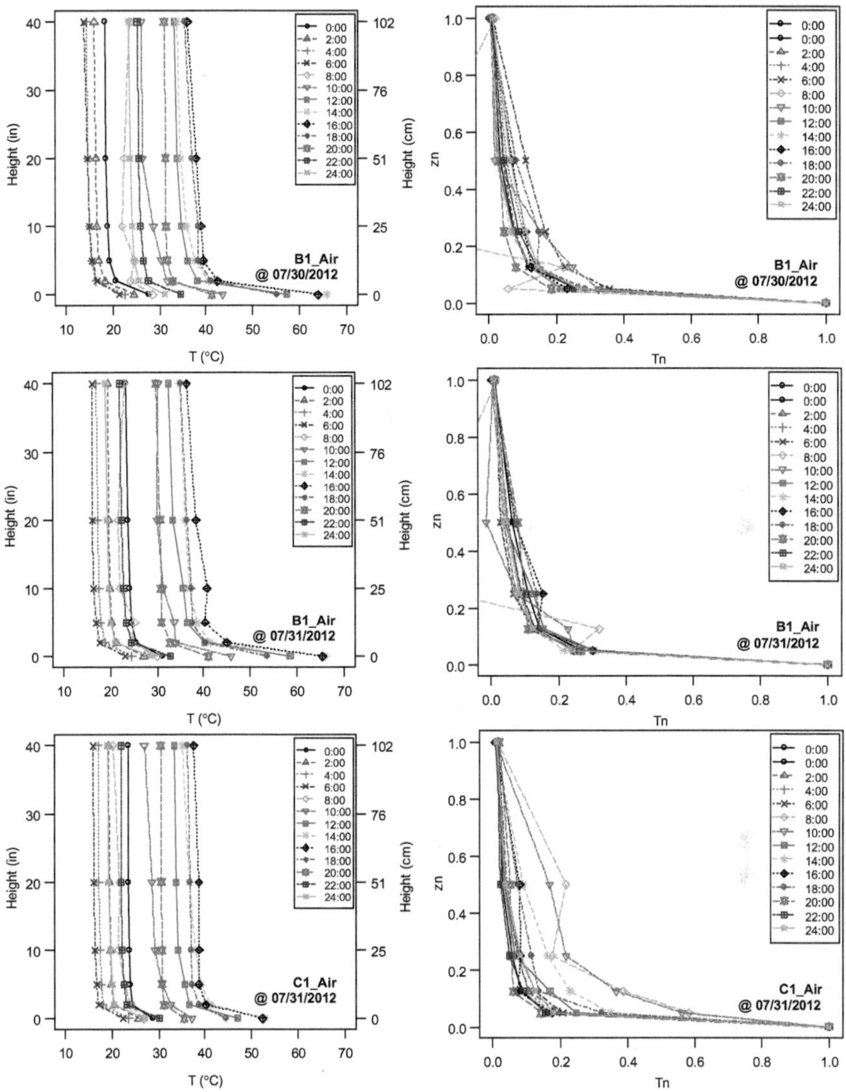

Figure 9.13 Examples of original and normalized profiles of near-surface air temperature (asphalt B1 and concrete C1, at various times). Dates are given as month/day/year.

9.4.4 Application of the Model

With the proposed model described by Eqs (9.3–9.6), the spatial profiles of near-surface air temperature could be predicted once the surface temperature, ambient air temperature, and wind speed are known. For instance, if

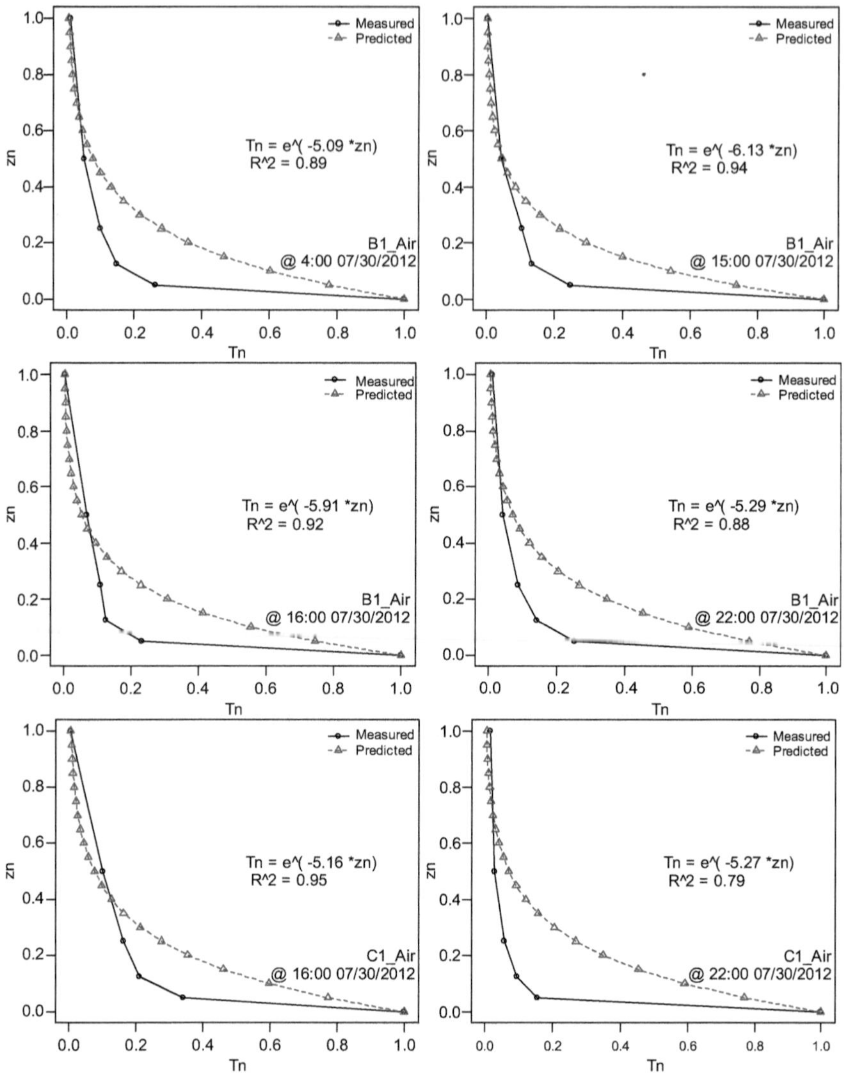

Figure 9.14 Examples of measured and predicted normalized profiles of near-surface air temperature (asphalt B1 and concrete C1, at various times). Dates are given as month/day/year.

the surface temperature, ambient air temperature, and wind speed (at ~2 m height) are known as $T_s = 65\ °\mathrm{C}$, $T^+ = T_a = 38\ °\mathrm{C}$, and $WS = 1.0$ m/s, then the normalized profile can be predicted as follows,

$$T_n = e^{C \cdot z_n} = e^{(-5.13 - 0.57 * WS) \cdot z_n} \tag{9.7}$$

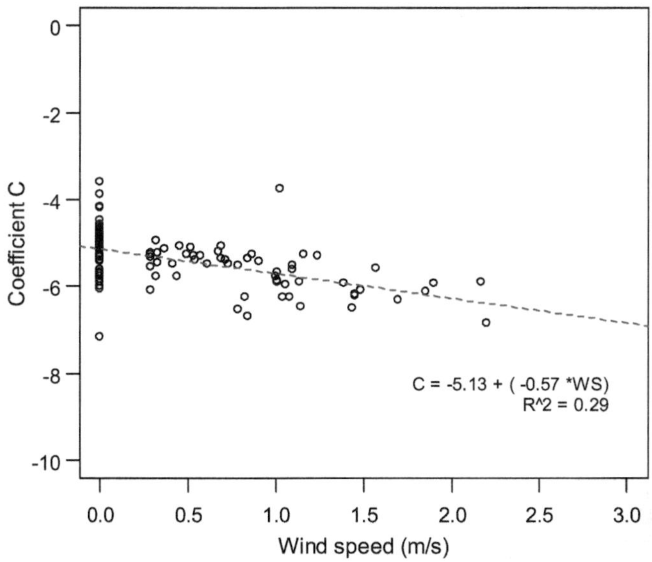

Figure 9.15 Correlation between coefficient *C* and wind speed at 2 m height for modeling normalized profiles of near-surface air.

$$T_n = \frac{T_z - T^+}{T_s - T^+} = e^{(-5.13-0.57*WS)\cdot\frac{Z}{Z^+}} \tag{9.8}$$

$$\begin{aligned} T_Z &= \left(T_s - T^+\right)\cdot e^{(-5.13-0.57*WS)\cdot\frac{Z}{Z^+}} + T^+ \\ &= (65-38)\cdot e^{(-5.13-0.57*1.0)\cdot\frac{Z}{1}} + 38 \\ &= 27e^{(-5.13-0.57)\cdot Z} + 38 \end{aligned} \tag{9.9}$$

The predicted original and normalized profiles of near-surface air temperature are presented in Figure 9.16. The contour plots of the profiles of near-surface air temperature are presented in Figure 9.17. The predicted profiles of near-surface air temperature can be used for the height-specific (spatial variation) evaluation of the thermal comfort and air quality (e.g., ground-level ozone) and other near-surface air-related issues.

9.5 SUMMARY AND CONCLUSIONS

The study in this chapter investigated the thermal interaction between pavement and near-surface air for various pavement types, measured the temperature profiles of near-surface air, and explored the factors affecting the profiles.

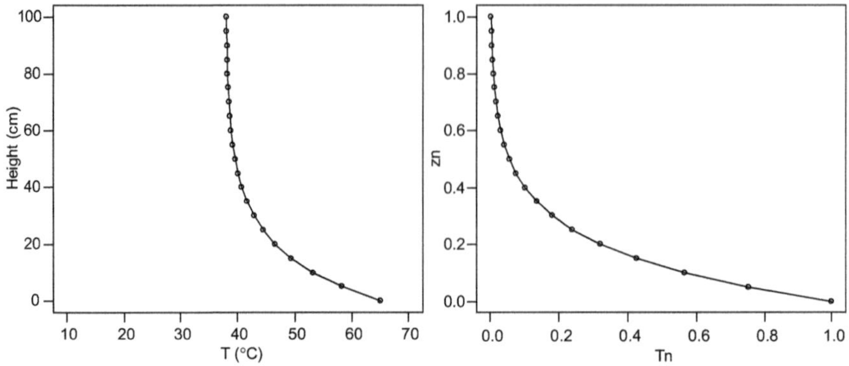

Figure 9.16 Examples of predicted original and normalized profiles of near-surface air temperature.

It is noted that the temporal profiles of near-surface air have a pattern similar to that of ambient air temperature for both asphalt and concrete pavements. For the spatial profiles, the near-surface air temperatures gradually decrease as the distance from the pavement surface increases, with greater slope (change rate) at 10 in (25 cm) from the pavement surface. In the range close to the pavement surface, the near-surface air temperatures are much higher than the ambient air temperature at ~2 m height, especially for the black asphalt pavement with higher surface temperature compared to higher reflectance concrete pavement. The high near-surface air temperatures would reduce the human thermal comfort, and they are more critical for babies and children, who are shorter and closer to the surface. In addition, the formation of ground-level ozone would be facilitated by these high near-surface air temperatures when smog (volatile organic compounds and nitrogen oxides) is emitted from vehicle exhaust pipes that are close to the pavement surface on roads or parking lots.

The wind speed will influence the shape of the temperature profile of near-surface air. Lower wind speed will make the spatial profile steeper owing to less heat dissipated by wind or airflow. The near-surface air temperatures at higher distances from the surface will be influenced by the pavement heat under lower wind speed. This further implies that the heat effects of pavements will be more critical in environments with lower wind speed, such as high-density urban areas, especially those with high-rise buildings that block prevailing winds.

It is revealed from the findings that during hot periods the near-surface air temperature diminishes gradually as the height above the surface

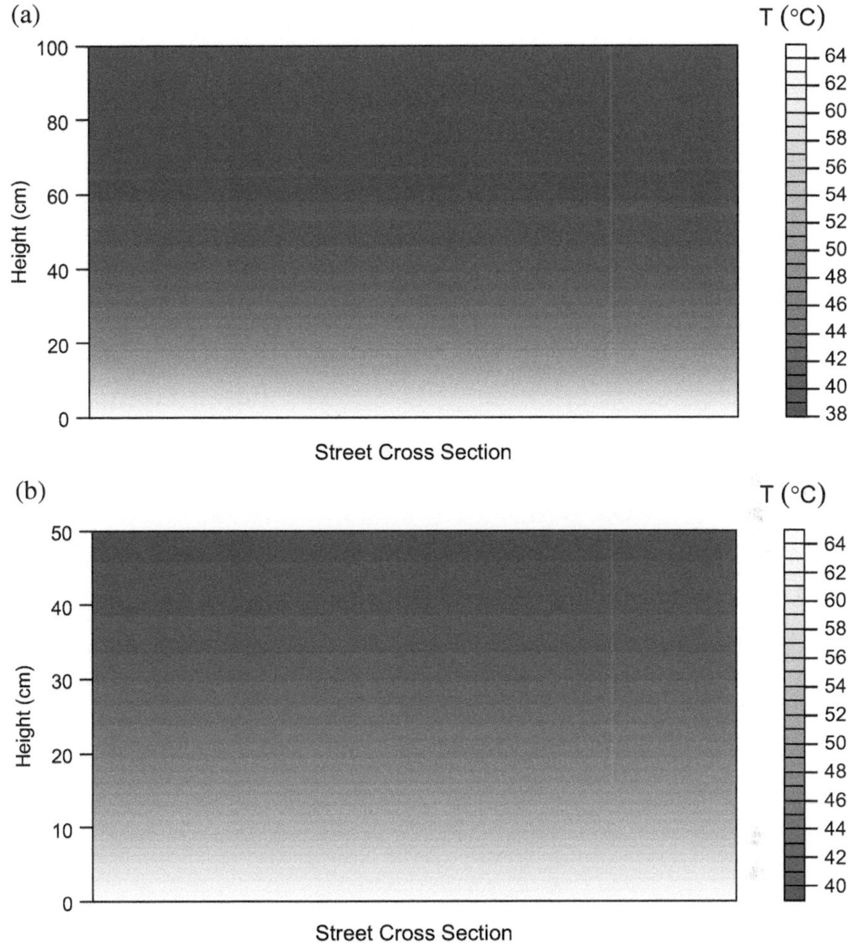

Figure 9.17 Examples of spatial contour plots of near-surface air temperatures. (a) Up to 1 m height. (b) Up to 0.5 m height.

increases, and the slope (change rate) also diminishes. The shape of the spatial profile is influenced by the surface temperature, ambient air temperature, and wind speed.

A simple model was developed to describe the spatial profile of near-surface air temperature, which characterizes the effects of surface temperature, ambient air temperature, and wind speed. The spatial profile of near-surface air temperature could be predicted once the surface temperature, ambient air temperature, and wind speed are known.

CHAPTER 10

Thermal Interaction between Pavement and Building Surfaces

Contents

10.1 INTRODUCTION

Pavement heat can influence near-surface air temperature profiles and consequently influence human thermal comfort and air quality in hot periods, as discussed in the previous chapter. In addition to the thermal impact on near-surface air, pavement heat could potentially also increase the temperature of the building surfaces as well as producing a reflection of solar radiation that could be absorbed by building surfaces, consequently increasing the energy use for cooling the building in hot periods.

The objective of the study presented in this chapter is to evaluate the thermal interactions between pavement and building walls through experiments and simulations, characterizing the impacts of various pavement types on building surface temperatures.

Pavement Materials for Heat Island Mitigation
ISBN 978-0-12-803476-7
http://dx.doi.org/10.1016/B978-0-12-803476-7.00010-6

10.2 EXPERIMENTAL MATERIALS AND METHODOLOGY

10.2.1 Experimental Setup

For simplification, two identical building walls (hollow wall box, 2 × 4 × 8 ft (0.6 × 1.2 × 2.4 m), made of commonly used half-inch-thick plywood, painted with white paint, Figure 10.1) were developed and put on some of the pavement sections described in Chapter 4 and some other existing pavements to evaluate the thermal interactions between pavement and building walls. Thermocouple wires were used to measure the temperatures on the walls at various heights above the pavement surface and

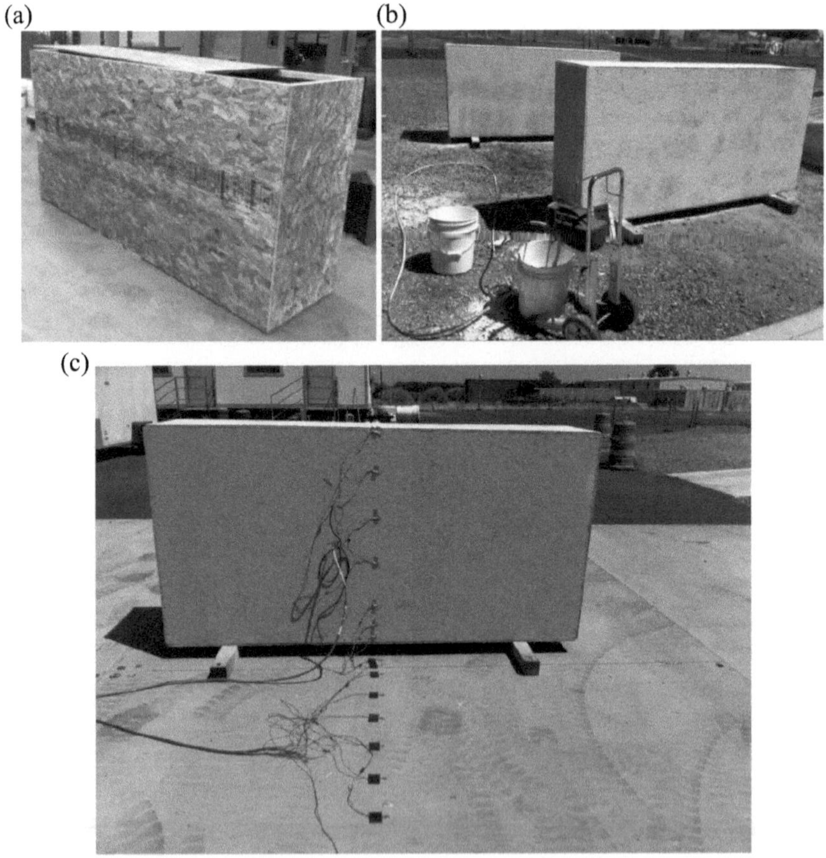

Figure 10.1 Preparation of building walls and example test setup with temperature sensors. (a) Raw wall. (b) Wall painting. (c) Example test setup with temperature sensors.

the temperatures on the pavement surface at various distances from the walls. Thermocouple sensor locations on the walls were 0, 2, 5, 10, 20, 40, and 48 in (0, 5, 12.5, 25, 50, 100, and 122 cm) above the pavement surface (Figure 10.1). Thermocouple sensor locations on the pavement were 5, 10, 20, 30, 40, 50, 60, and 80 in (12.5, 25, 50, 75, 100, 125, 150, and 200 cm) from the walls. Local weather data were also monitored using a portable weather station. Thermal images of walls and pavements were taken for some of the pavement sections at various times in a sunny hot day.

10.2.2 Pavement Sections and Measurement plan

Four pavement sections were chosen to conduct the experiments for measuring the temperature profiles of both wall and pavement surfaces. Two of them were asphalt pavements and the other two were concrete pavements. Two were on small sections and two were in large paved areas. For the first two large pavement sections (PA1 and PC1), the two walls were set up on the same sections in the same way, which provided the chance to directly compare the two walls to make sure they were thermally identical when installed under some conditions. For the last two sections (PA2 and PC2), the experiments were performed simultaneously for direct comparison with one wall on each section (Figure 10.2). The pavement sections and measurement dates are summarized in Table 10.1, along with the albedos (solar reflectivities) of both pavement sections and the two walls measured by an albedometer. The two walls had the same albedo value of 0.29.

10.3 EXPERIMENTAL RESULTS AND DISCUSSION

10.3.1 Examination of the Two Walls

Both walls were installed on the big asphalt section (see Figure 10.2(a)). As shown in Table 10.1, the two walls had the same albedo value of 0.29. Under the same conditions (weather and pavement), these two walls are expected to give the same or close surface temperatures. The temperatures at various heights on both walls in one day are presented in Figure 10.3. The temperatures on both walls at the same height are very close to each other. This implies that both walls are thermally identical when installed under some conditions.

(a)
(b)
(c)

Figure 10.2 Experimental setup for temperature profile measurement. (a) On asphalt pavement PA1. (b) On concrete pavement PC1. (c) On small asphalt (PA2 or B1) and concrete (PC2 or C1) pavements for the same period.

10.3.2 Example Results on Asphalt Pavement PA1 and Concrete Pavement PC1

The wall and pavement temperatures on asphalt pavement PA1 and concrete pavement PC1 were monitored at various periods from 2 through

Table 10.1 Summary of pavement sections and walls for experimental measurement

Section no.	Type	Size	Albedo	Experiment date (month/day/year)
PA1	Asphalt	~30 m diameter	0.06	7/2/2012–7/17/2012
PC1	Concrete	15 m wide × 45 m long	0.26	7/18/2012–7/27/2012
PA2 (B1)[a]	Asphalt	4 m wide × 4 m long	0.08	7/28/2012–8/20/2012
PC2 (C1)[a]	Concrete	4 m wide × 4 m long	0.28	7/28/2012–8/20/2012
Wall 1	Plywood	0.6 × 1.2 × 2.4 m	0.29	—
Wall 2	Plywood	0.6 × 1.2 × 2.4 m	0.29	—

[a]PA2 and PC2 are the experimental test sections B1 and C1 as described previously, respectively.

17 July 2012 and from 18 through 27 July 2012, respectively. The weather conditions did not change much during this summer period, especially on the sunny days. This makes it possible to compare the effects of asphalt and concrete pavement on near-surface air temperatures for clear days. Some results for PA1 and PC1 are presented below in Figures 10.4–10.7, as well as the example weather conditions during the experiment periods (see Figure 9.3).

It is noted that the temporal profiles (Figures 10.4 and 10.6) of both wall and pavement temperature have a pattern similar to that of ambient air temperature for both asphalt and concrete pavements. The difference is that the peak temperatures of wall and pavement came earlier than that of the ambient air (~14:00 vs ~16:00 h). The wall and pavement temperatures around noon were both much higher than the ambient air temperature on the sunny summer days. This might cause more energy use and high peak demand for cooling buildings with air conditioners during hot periods. While the nighttime pavement temperature stays higher than the ambient air temperature, the wall temperature drops and stays very close to the ambient air temperature during nighttime. This is because the pavement is quite thick (15 cm or thicker) and can absorb and store quite a large amount of heat during the daytime and maintain a high temperature with the stored heat during the nighttime. However, the walls used in the experiments were quite thin (the plywood layer of the wall was only ~1.3 cm (0.5 in), with air in the hollow wall box) and of low heat capacity and thermal conductivity, and consequently the thin plywood layer could not absorb

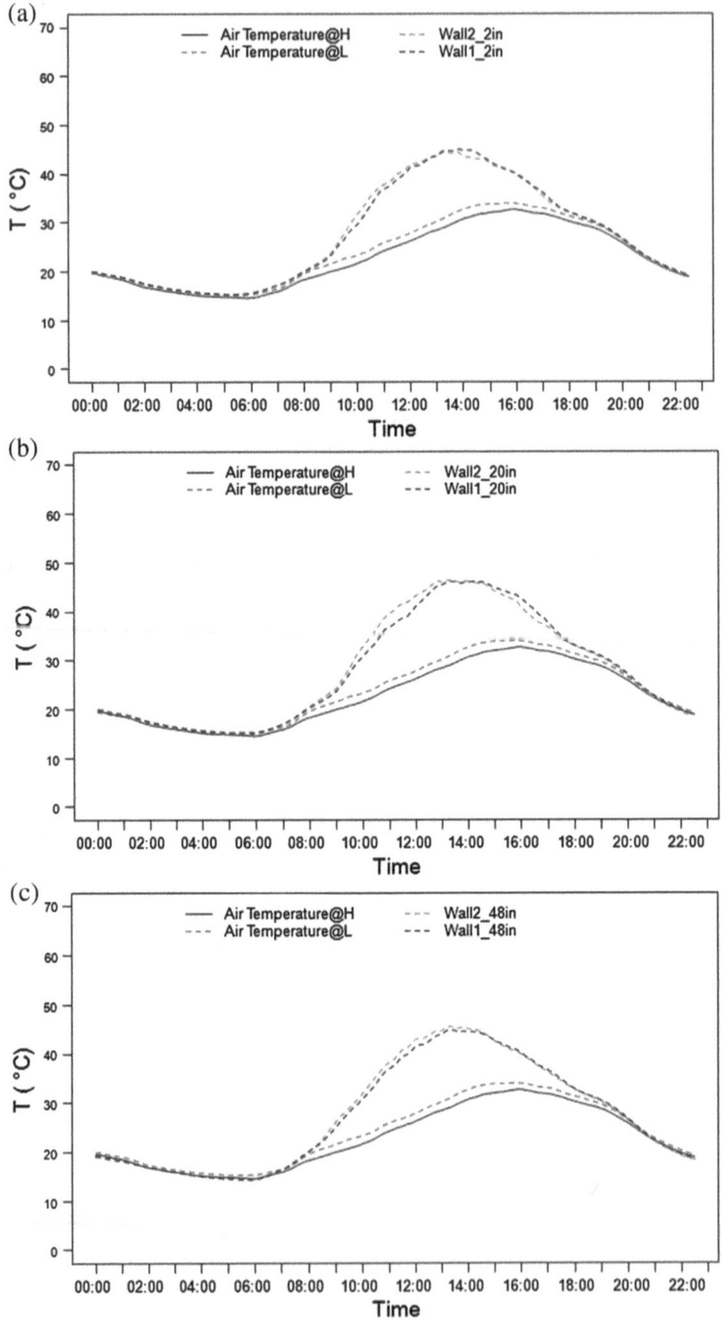

Figure 10.3 Temperatures on both walls under the same conditions. Air temperature@H and air temperature@L are ambient air temperatures measured at 1.7 m (67 in) and 0.3 m (11.8 in), respectively. (a) At 2 in (5 cm) height. (b) At 20 in (50 cm) height. (c) At 48 in (122 cm) height.

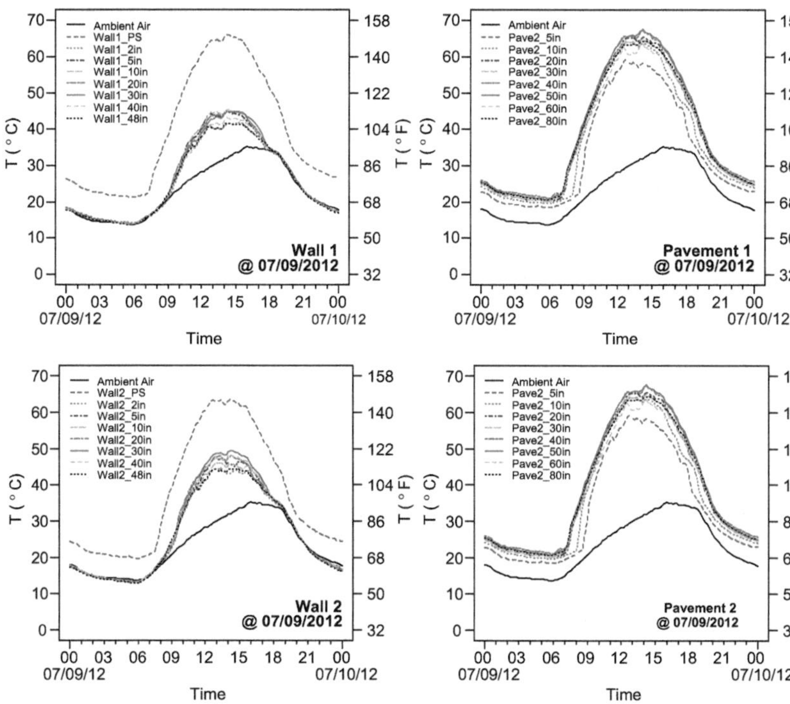

Figure 10.4 Example temporal profiles of wall and pavement temperatures on asphalt pavement PA1. *Ambient air is the ambient air temperature from a sensor on a portable weather station at 67 in (1.7 m). Walli_PS and Walli_xin are wall i surface temperatures at 0 and x in above the pavement surface, respectively. Pavei_xin are pavement i surface temperatures at x in from the wall.* Dates are given as month/day/year.

and store a large amount of heat during the daytime. During the nighttime, without solar radiation, the wall temperature dropped very quickly and stayed close to the ambient air temperature. This situation was for simulation and was different from a thick wall or a wall made of materials with a high heat capacity and high thermal conductivity, such as an aluminum wall surface.

For the spatial profiles (Figures 10.5 and 10.7), the wall temperatures at various heights on the wall were relatively constant (slightly higher at the bottom close to pavement surface) during the nighttime. However, during the daytime around noon the middle part of the wall presented a higher temperature under the same weather conditions, such as solar radiation, ambient air temperature, and wind speed. This indicated that there were some thermal interactions (through reflected and emitted radiation) between pavement and building wall. The thermal interactions increased

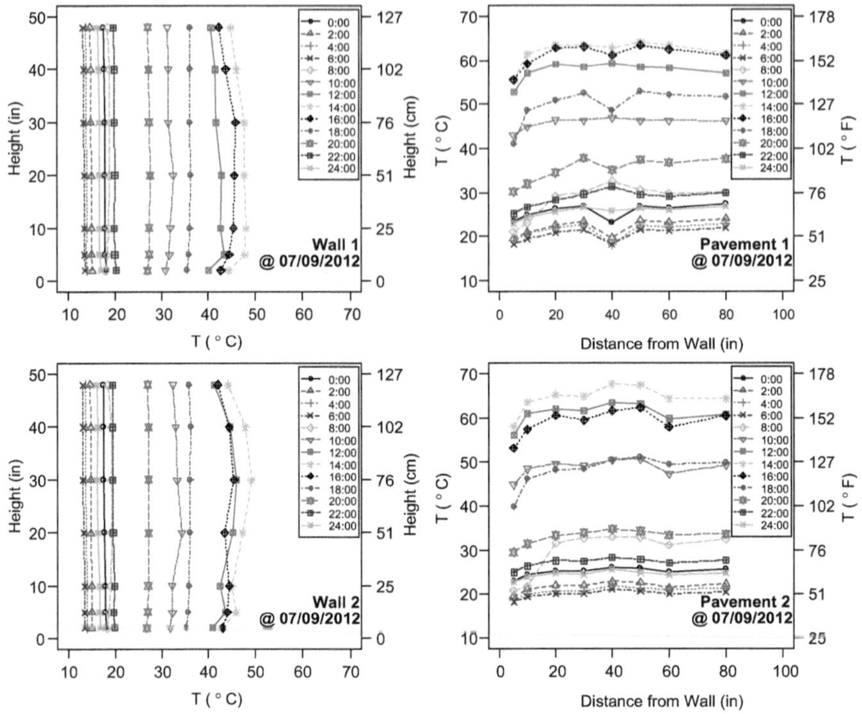

Figure 10.5 Example spatial profiles of wall and pavement temperatures on asphalt pavement PA1. Dates are given as month/day/year.

the wall temperature to some extent and potentially would contribute increased energy use for cooling buildings in hot periods. The nonuniform temperature of the pavement surface at various distances from the wall also confirmed the thermal interactions between pavement and building wall.

10.3.3 Temporal and Spatial Profiles of Wall Temperatures on Asphalt Pavement PA2 and Concrete Pavement PC2

The wall and pavement temperatures on asphalt pavement PA2 and concrete pavement PC2 were monitored simultaneously from 28 July through 20 August 2012. This made it possible to directly compare the effects of asphalt and concrete pavement on wall temperatures. Some results are presented in Figures 10.8 and 10.9, as well as the example weather conditions during the experiment period (see Figure 9.6).

In addition some observations similar to those for PA1 and PC1 presented above, there were some interesting observations obtained from the

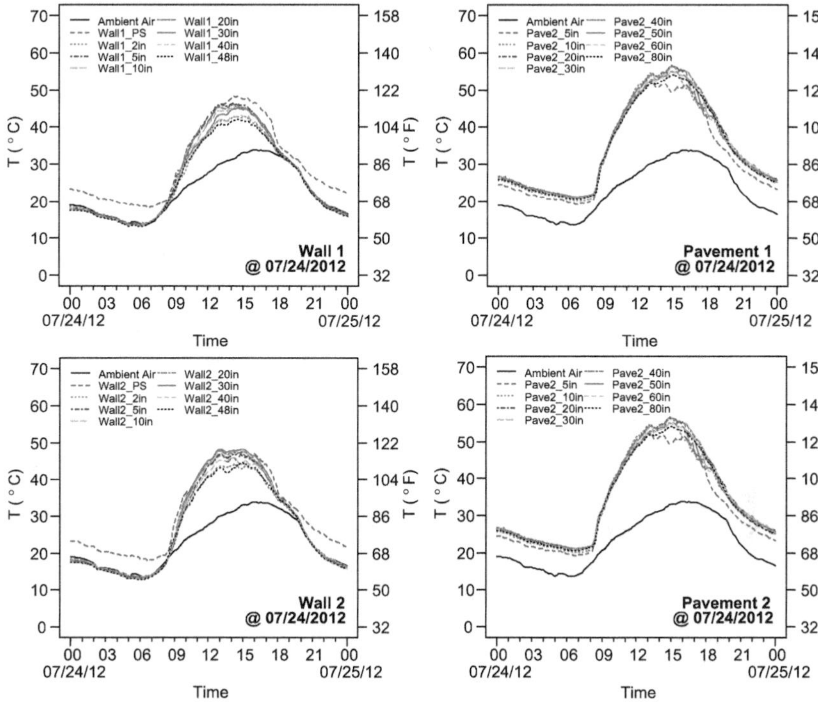

Figure 10.6 Example temporal profiles of wall and pavement temperatures on concrete pavement PC1. *Ambient Air is the ambient air temperature from a sensor on a portable weather station at 67 in (1.7 m). Walli_PS and Walli_xin are wall i surface temperatures at 0 and x in above the pavement surface, respectively. Pavei_xin are pavement i surface temperatures at x in from the wall.* Dates are given as month/day/year.

direct comparison between asphalt and concrete pavements under the same weather conditions.

The daytime surface temperatures on asphalt pavement (B1 in Figure 10.8) were much higher than those of the concrete pavement (C1 in Figure 10.8) (70 °C vs 55 °C), although they were under the same weather conditions. The nonuniform temperature along the wall implies that the thermal interaction between pavement and wall exists for both asphalt (B1 in Figure 10.9) and concrete (C1 in Figure 10.9) pavements, especially during the daytime and around noon with high solar radiation. Although the concrete pavement (C1) had a lower surface temperature than the asphalt pavement (B1), the temperatures in the middle part of the wall on the concrete pavement (C1 in Figure 10.9) around noon were higher and more nonuniform than those of the asphalt pavement (B1 in Figure 10.9). The reason for this is that, compared to the asphalt pavement (B1), the

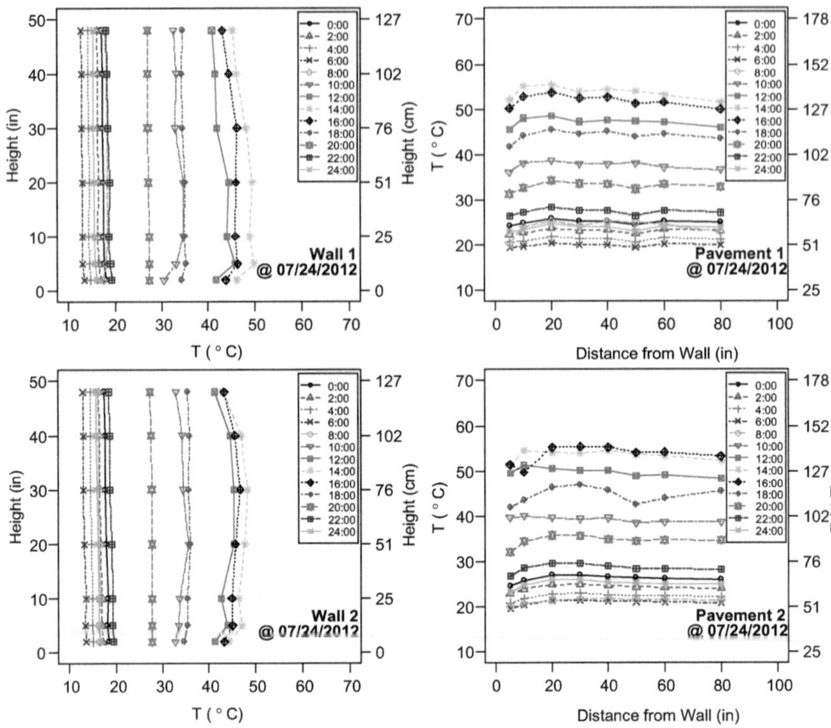

Figure 10.7 Example spatial profiles of wall and pavement temperatures on concrete pavement PC1. Dates are given as month/day/year.

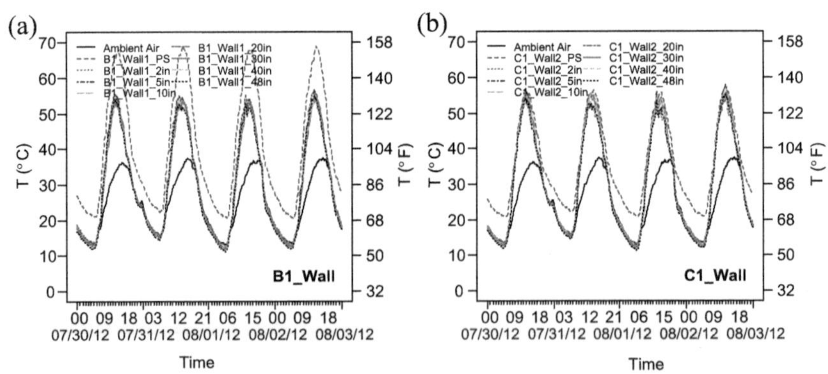

Figure 10.8 Example temporal profiles for wall temperature on asphalt PA2 (B1) and concrete pavement PC2 (C1). (a) Asphalt pavement PA2 (B1). (b) Concrete pavement PC2 (C1). *Ambient air is the ambient air temperature from a sensor on a portable weather station at 67 in (1.7 m).* Dates are given as month/day/year.

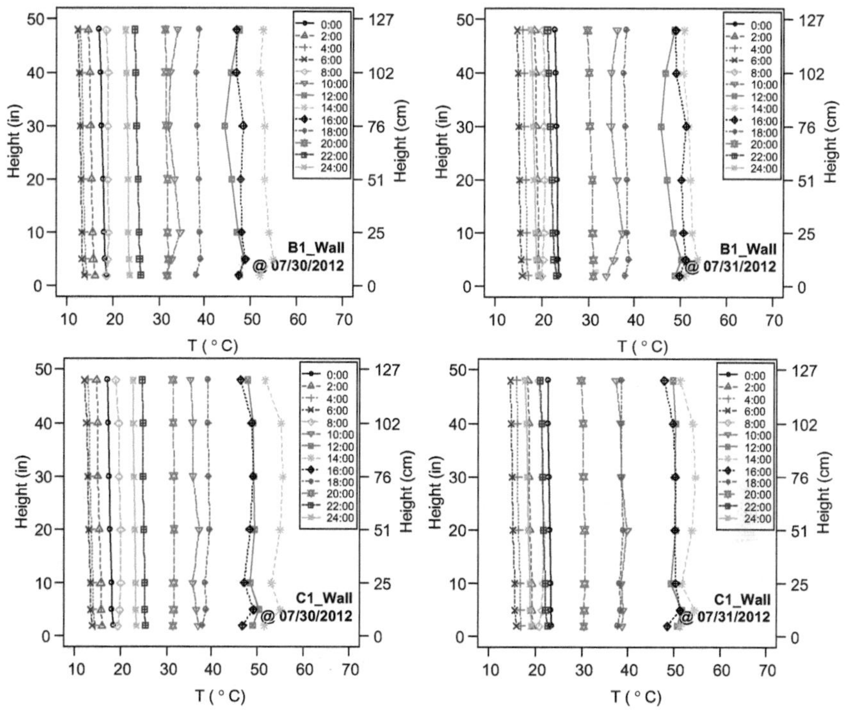

Figure 10.9 Example spatial profiles for wall temperatures on asphalt pavement PA2 (B1) and concrete pavement PC2 (C1) on 2 days. Dates are given as month/day/year.

concrete pavement (C1) has higher solar reflectivity (or albedo) and consequently reflects more solar radiation. Some of the reflected heat will hit and be absorbed by the building wall and heat it up. This implies that increased pavement albedo will reduce the pavement surface temperature; however, the temperature of the building wall will be increased by the energy reflected from the pavement surface and consequently would potentially increase the energy use for cooling the building in hot periods. This risk might not be a big issue in open areas but would tend to be more significant in high-density urban areas. Therefore, special attention should be given to the application of high-reflectance pavement for mitigating the heat island effect, especially in high-density urban areas.

10.3.4 Thermal Images of Walls and Pavements on B1 and C1

Thermal images of walls and pavements on B1 and C1 were taken on clear days in summer. Some example thermal images (13:00 h on 15 August

Figure 10.10 Optical and thermal images of walls and pavements on B1 and C1. (a) Optical images. (b) Thermal images. *Average temperatures on wall and pavement are shown in the thermal images.*

2012) are presented in Figure 10.10. This provides a more direct and obvious way to compare the thermal impacts of asphalt and concrete pavements on the walls.

Three thermal images were taken at each time (Figure 10.10(b)): one for the pavement and wall together (B1/C1-both), one for the pavement only (B1/C1-pavement), and one for the wall only (B1/C1-wall)). In addition, the optical images are also presented in Figure 10.10(a). The asphalt pavement (B1) presented higher pavement surface temperature than the concrete pavement (C1) (60 vs 45 °C). However, the average temperature of the wall on the concrete pavement (C1) was higher than that of the asphalt pavement (B1) (55 vs 52 °C). This verified the results obtained from the temperature sensors presented previously. Another more convincing and quantitative way of verifying the thermal impacts of different pavements on the wall temperature is to use a heat flux sensor (including both long- and short-wave radiation) to directly measure the heat flux on the wall surface at various heights and times and to directly compare the thermal impacts. However, such heat flux sensors are quite expensive (over $6000 per set of sensors). If more funding is available from any source, the direct measurement of heat flux (including both long- and short-wave radiation) on the wall surface at various heights and times can be performed and would definitely give more convincing and quantitative observations.

10.4 MODELING AND SIMULATION

A simple numerical model and simulation of the heat transfer and thermal interactions between pavement and wall were developed and performed to obtain some deeper insights regarding the thermal interactions for a typical summer climate in a dry and hot region of Sacramento, California.

10.4.1 Integrated Finite Element Method (FEM) Model

The model includes three parts: pavement (2 m long × 2 m wide × 2 m deep, 0.2 m surface layer, 0.3 m base, 1.5 m subgrade), bare soil (2 m long × 2 m wide × 2 m deep), and wall (2 m long × 0.6 m wide × 2 m high). The integrated FEM model is shown in Figure 10.11.

The parameters for temperature simulation are listed in Table 10.2. Detailed heat transfer model development and the parameters are presented in Chapter 11.

10.4.2 Boundary Conditions

The boundary conditions for this heat transfer model include the solar radiation heat flux on the pavement/soil/wall surfaces, the surface convection between all surfaces and air, the surface radiation of pavement/soil/wall, and the upper boundary of the near-surface air.

The solar radiation heat flux varies throughout a day, as defined by the user subroutine *DFLUX()* in *Abaqus*®. Surface convection, which is a function of the air temperature varying over time and the convection coefficient varying with wind velocity, is defined by the user subroutine *FILM()* in *Abaqus*®. The effective surface radiation is directly defined

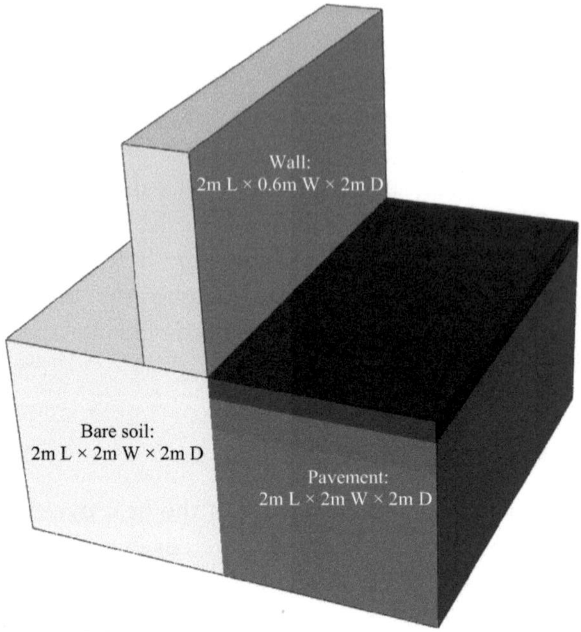

Figure 10.11 Integrated model for temperature simulation.

through surface radiation to the ambient atmosphere using the key word **Sradiate* in *Abaqus*®. All other boundary conditions are assumed to be regulated by heat insulation [168].

10.5 SIMULATION RESULTS AND DISCUSSION

Some simulation results are presented below, including temperature profiles and view factor. More simulations can be performed with this and/or a modified and validated model to explore the effects of various parameters (materials, climates, etc.)

10.5.1 Temperature Profiles

The temperature profiles were extracted for various times on a hot summer day in Sacramento, California. The results are presented in Figure 10.12. During the nighttime, the temperature on the wall surface was low and relatively uniform (Figure 10.12(a) and (c)). However, the daytime temperature on the wall surface was high and nonuniform (Figure 10.12(b)), and the lower middle part had higher temperature owing to the thermal

Table 10.2 Parameters for temperature simulation

Thermal and radiative properties and typical parameter values

Parameters	Wall (plywood)	Bare soil	Surface layer	Aggregate base	Subgrade soil
Thermal conductivity k^a (J·(h·m·°C)$^{-1}$)	432	3600	5000	4600	4200
Specific heat capacity c (J·(kg·°C)$^{-1}$)	2500	1100	950	920	1300
Density ρ (kg·m^{-3})	1600	1700	2200	1900	1800
Solar radiation absorptivity r_s	0.30	0.22	0.10	—	—
Thermal emissivity $\varepsilon = \alpha_s = 1 - r_s$	0.70	0.78	0.90	—	—
Heat convection coefficient h_c^a (J·(h·m^2·°C)$^{-1}$)	$h_c = 3600(3.7v_w + 6.1)$ (v_w, wind velocity, m/s)				
Absolute zero T^Z (°C)	−273				
Stefan–Boltzmann constant σ^a (J·(h·m^2·K^4)$^{-1}$)	2.041092×10^{-4}				

Typical summer climate data in dry and hot region of Sacramento, CA

Month	Daily peak air temperature T_a^{max} (°C)	Daily lowest air temperature T_a^{min} (°C)	Daily total solar radiation volume Q (MJ/m^2)	Daily effective sunlight hour c (h)	Daily average wind velocity v_w (m/s)
7 (July)	38	18	30.0	11	1.5

[a]Time is measured hourly and the related parameters are converted for the simulation analysis at intervals of 1 h.

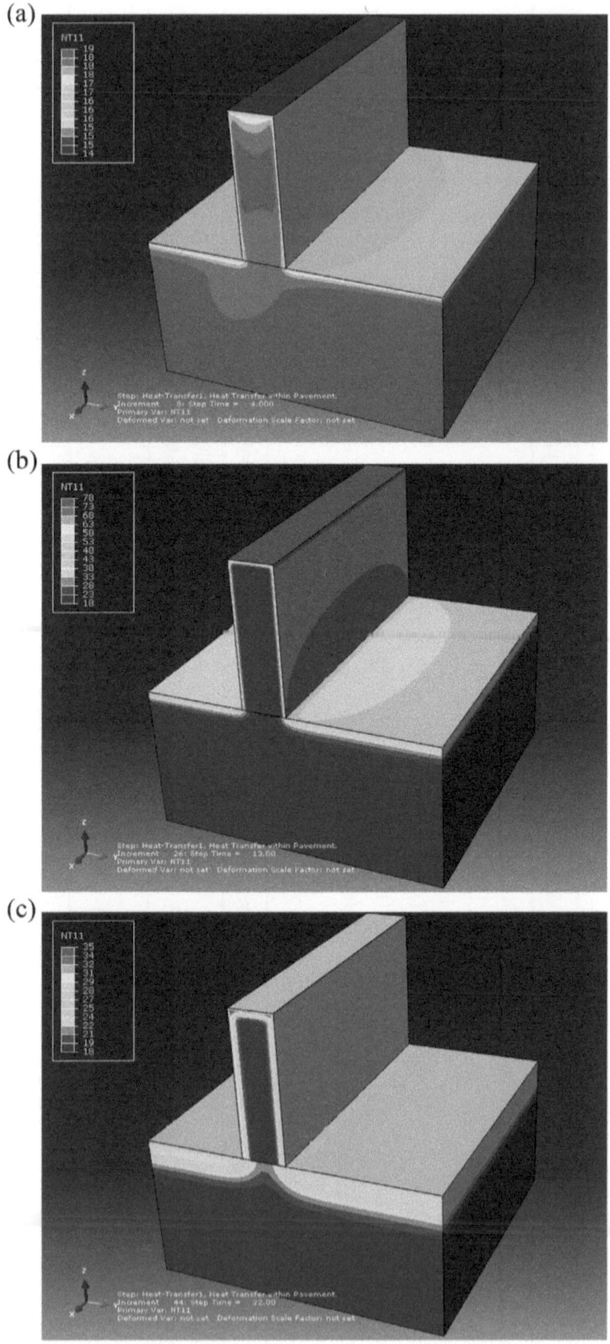

Figure 10.12 Temperature (in °C) contours at various times. (a) 4:00 h (b) 13:00 h (c) 22:00 h.

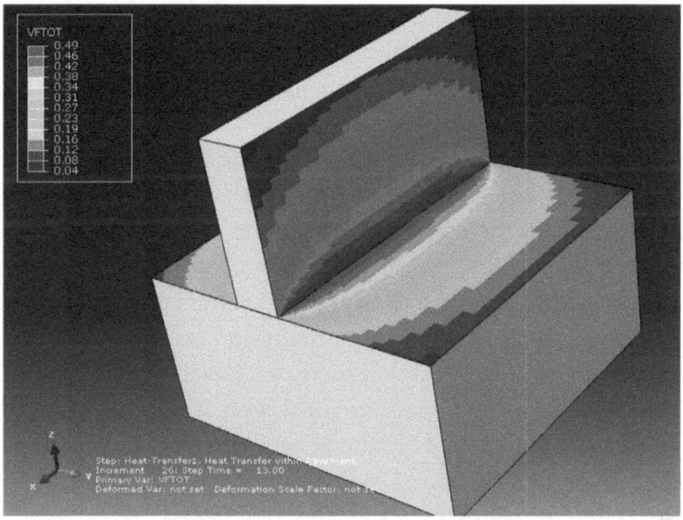

Figure 10.13 View factor contour.

interaction between pavement and wall from radiation. This in part verified the results from the experiments presented previously.

10.5.2 View Factor

The thermal interaction between pavement and wall includes reflected short-wave radiation and emitted long-wave radiation. The amount of the radiation impact between pavement and wall depends on the view factor (the proportion of the radiation that leaves one surface and strikes another surface) between the surfaces. The view factor between pavement and wall surfaces is simulated and presented in Figure 10.13. The view factor is higher in the lower and middle parts of the wall and is lower for other parts. This helps to explain the temperature difference on the wall obtained from both experimental and simulation results presented above.

10.6 SUMMARY AND CONCLUSIONS

This chapter preliminarily evaluated the thermal interaction between pavement and building walls through both experiments and simulations, to characterize the impacts of various pavement types on building surface temperatures.

Four pavement sections were selected to conduct the experiments for measuring the temperature profiles of both wall and pavement surfaces. Experimental results from temperature sensors and thermal images indicated that there are some thermal interactions (through reflected and emitted radiation) between pavement and building walls. The thermal interactions increase the wall temperature to some extent (~2–5 °C around noon) and potentially will contribute increased energy use for cooling the building in hot periods. The nonuniform temperature of the pavement surface at various distances from the wall also provided some confirmation that there are thermal interactions between pavements and building walls.

Experimental results implied that increased pavement albedo will reduce the pavement surface temperature; however, the temperature of the building wall will be increased by the reflected energy from the pavement surface and consequently would potentially increase the energy use for cooling the building in hot periods. This risk might not be a big issue in open areas but would tend to be more significant for high-density urban areas where walls are next to the pavement. Therefore, special attention should be given to the application of high-reflectance pavement for mitigating the heat island effect, especially in high-density urban areas. It also should be noted that the importance of thermal interactions between pavement and other surfaces depends on various factors, such as pavement size, distance of the walls from the pavement, the albedo of both pavement and wall, and the thermal resistance of the wall.

A simple numerical model and simulation for the heat transfer and thermal interaction between pavement and wall was developed and performed to obtain some deeper insights into the thermal interactions between pavements and walls for a typical summer climate in the dry and hot region of Sacramento, California.

Some simulation results were presented, including temperature profiles and view factor (the proportion of the radiation that leaves one surface and strikes another surface). More simulations can be performed with this and/or a modified and validated model to explore the effects of various parameters (materials, climates, etc.).

During the nighttime, the temperature on the wall surface is low and relatively uniform; however, the daytime temperature on the wall surface is high and nonuniform, and the lower middle part has higher temperature owing to the thermal interactions between pavement and wall. This in part verified the results from the experiments presented previously.

The view factor is higher in the lower and middle parts of the wall and is lower for other parts. This helps to explain the temperature difference on the wall obtained from both experimental and simulation results presented above.

Another more convincing and quantitative way of verifying the thermal impacts of various pavements on wall temperature is to use a heat flux sensor (including both long- and short-wave radiation) to directly measure the heat flux on the wall surface at various heights and times and to directly compare the thermal impacts. However, such heat flux sensors are quite expensive (over $6000 per set of sensors). If more funding is available from any source, the direct measurement of heat flux (including both long- and short-wave radiation) on the wall surface at various heights and times can be performed and would definitely produce quantitative and more convincing observations.

CHAPTER 11

Pavement Thermal Modeling: Development and Validation

Contents

Pavement Materials for Heat Island Mitigation
ISBN 978-0-12-803476-7
http://dx.doi.org/10.1016/B978-0-12-803476-7.00011-8

11.1 INTRODUCTION

The objective of this chapter is to develop an integrated local microclimate model that includes both the pavement structure and the near-surface air and considers coupled processes of radiation, convection, conduction, shading, and evaporation. The model will be numerically implemented and used to simulate the temporal and spatial distribution and variation of both pavement temperature and near-surface air temperature in the summer in a hot region (Sacramento, California) for validation against the measurements from the experimental test sections presented previously in this dissertation.

11.2 OVERVIEW OF THE INTEGRATED LOCAL MICROCLIMATE MODEL

As mentioned above, the major task of this chapter is to develop, numerically implement, and validate the local microclimate model to simulate the outdoor thermal environment, which will be used in the following tasks to analyze the effects of various cool pavements on the outdoor thermal environment.

To accomplish this specific task, several subtasks need to be undertaken, including (1) theoretical development of a local microclimate model, (2) numerical implementation of the model, and (3) model validation.

11.2.1 Theoretical Development of Local Microclimate Model

Urban climate models vary substantially in many aspects including physical basis, model scale, temporal and spatial resolution, input and output quantities, etc. The focus of this study will be on the effects of street design (mainly street layout and pavement type) on the street thermal environment and the corresponding thermal comfort of humans walking or cycling on the street. Hence, a street-level local microclimate model, which couples different processes in transient heat transfer, will be developed to describe the major processes in the near-ground atmosphere and to analyze the key meteorological factors and later their corresponding impacts on human thermal comfort within the pavement area, with a special emphasis on the zone near the ground surface where most human activities of interest to the study occur (up to 2-m above ground). Based on well-founded physical phenomena (i.e., the fundamental laws of heat transfer), the model will seek to reproduce the major processes in the near-ground atmosphere, including simulation of temperature, air flow, and radiation fluxes and shading and evaporation effects. This new integrated model will allow simulation of local microclimate dynamics on the basis of a 24-h daily cycle for various time frames (weeks, months, or years). It can be used for complex built environments including various street layouts, pavement types and structures, vegetation/trees, and building shapes and heights in various climate regions. In addition, the model can provide flexible spatial and temporal resolutions to balance the accuracy and computation cost for understanding the local microclimate at the street level for different seasons (hot summers and cold winters).

The local microclimate model to be developed should consider the following major items:

- Energy balance on the pavement surface;
- Thermal interactions between the pavement and other surfaces;
- Coupled processes of radiation, conduction, convection, shading, and evaporation.

To achieve the objectives of this study under controlled complexity, this simplified full three-dimensional street-level local microclimate model will include the following submodels.

Pavement/soil model, which will formulate and calculate the temperatures on the pavement surface and at various levels to a depth of about 2 m for the whole model area, including the major thermodynamic processes, such as conduction, convection, radiation, evaporation, and moisture diffusion, which vary according to the individual pavement/soil thickness and properties (e.g., thermal conductivity, specific heat, density, albedo, emissivity, convection coefficient, etc.) [169–173]. In addition, the evaporation model [174] for the evaporative cooling strategy and vegetation/trees (including shading) will be associated with this pavement/soil model.

Thermal interaction model, which will explicitly consider both reflected radiation and emitted radiation interactively from pavement surfaces and other surfaces (e.g., building surfaces), to more accurately evaluate the effectiveness of alternative design strategies for improving the thermal environment. The focus is on reflective pavements with high albedo installed in the context of high-density surroundings, since they will potentially influence the energy balance on the surfaces and thus influence the their own temperature and those of the surroundings.

Near-surface air model, which will consider changes in the air temperature, airflow (speed and direction), and radiation fluxes (including short-wave solar radiation, long-wave pavement and building surface radiation, etc.).

It will be based on the fundamental laws of fluid dynamics and thermodynamics (i.e., equations of conservation of mass, momentum, and heat) (e.g., Refs [175,176]), to simulate the main atmospheric variables that vary spatially and temporally.

11.2.2 Numerical Implementation of the Model

The numerical implementation will be achieved by numerically solving the corresponding partial differential equations governing each process through a coupled finite difference and finite element method [177,178]. All these processes will be numerically implemented using subroutines/plug-in packages integrated into the commercially available numerical simulation platform, *Abaqus*®, to limit the expensive but less meaningful efforts related to preprocessing and postprocessing of complex three-dimensional geometries (e.g., various street/building shapes and layouts, etc.) and some other fundamental numerical processes. The initial and boundary conditions of the developed model include the profiles (temperature, wind speed, or pressure) of inflow and outflow of the model, solar radiation, etc. These initial and boundary conditions of the local microclimate model will be given based on available local climate data, with partial calibration and validation using the field measurements conducted in this dissertation study.

11.2.3 Model Validation against Field Measurement

As described previously in this dissertation, nine test sections (4×4 m (13×13 ft) each section) were built at the University of California at Davis Advanced Transportation Infrastructure Research Center located in Davis, California. Some data from the field measurements on the test sections will be used to validate the developed model, as well as for empirically verifying the effectiveness of some cool pavement technologies and management strategies that were presented in previous chapters.

The thermal properties and radiative and evaporative characterization (e.g., thermal conductivity, heat capacity, solar reflectivity, evaporation rate, etc.) of the pavement materials used in these test sections and other commonly used pavement materials were obtained through laboratory testing and field measurements and are presented in previous chapters (Chapters 4–7) in this dissertation. These measured material properties, with some other secondary parameters from the literature, will be used for the implementation of the simulation model.

11.3 DEVELOPMENT OF A FRAMEWORK FOR THE GENERAL LOCAL MICROCLIMATE MODEL

11.3.1 Energy Balance on Pavement Surface

The energy balance method is usually used for microclimates in an urban canopy. The major processes involved in the energy balance are illustrated in Figure 11.1. The energy balance equation is based on the first law of thermodynamics, which states that the energy flowing into and out of any surface must be conserved. In the case of a pavement surface, this general equation is written as [2]:

$$\begin{aligned}&\text{Net Radiation} + \text{Anthropogenic Heat} - \\ &\quad(\text{Conduction/Storage} + \text{Convection} + \text{Evaporation}) = 0\end{aligned} \tag{11.1}$$

in which

$$\begin{aligned}\text{Net Radiation} &= \text{Incoming Radiation} - \text{Reflected Radiation} \\ &\quad - \text{Emitted Radiation}\end{aligned} \tag{11.2}$$

11.3.1.1 Conduction/Storage

Conduction/storage depends on three properties of materials: thermal conductivity, heat capacity, and density. Thermal conductivity determines

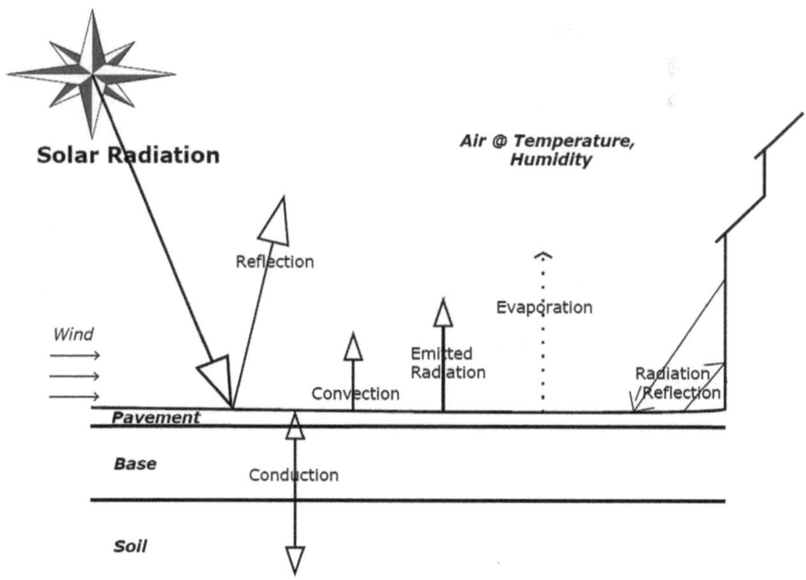

Figure 11.1 Energy balance on a pavement surface (same as Figure 1.7).

the speed and ease of conduction heat transfer, which occurs from the hot part to the cold part within a material owing to the interactions of microscale energy carriers (e.g., electrons or phonons in a solid; molecules in a gas or a liquid) [110]. Conduction heat transfer can be characterized by Fourier's law, which relates the heat flux in any direction to the temperature gradient in that direction [110]. For example,

$$q_{\text{cond}} = -k\frac{\partial T}{\partial z} \tag{11.3}$$

where q_{cond} is the conduction heat flux per unit area in the z direction and k is the thermal conductivity of the material.

Heat capacity c and density ρ determine the amount of energy needed to heat up the material to increase the temperature by a certain ΔT and then store it in the material. The internal specific heat energy U stored per unit volume of the material with temperature T could be described as

$$U = \rho c T \tag{11.4}$$

where ρ and c are the density and the specific heat capacity of the material, respectively.

Materials with high thermal conductivity are more able to direct heat into their depths. Materials with high heat capacity and density can store more heat in their bulk and also need more energy to heat them up.

The governing differential equation for a semi-infinite solid (e.g., pavement) can be derived from the principle of energy balance within a control volume of the semi-infinite solid that is differential in the z direction. The governing equation for conduction heat transfer in a semi-infinite solid is

$$\alpha\frac{\partial^2 T}{\partial z^2} = -\frac{\partial T}{\partial t} \tag{11.5}$$

where α is the thermal diffusivity, $\alpha = \frac{k}{\rho c}$.

11.3.1.2 Convection

Convection refers to heat transfer between a surface and a moving fluid (i.e., a liquid or gas) with macroscale bulk motion. In this study, it is the convection from/to pavement or building surfaces to/from the near-surface air above them. The equation that characterizes the rate of convective heat transfer (q_{conv}) per unit contact area is Newton's law of cooling [110]:

$$q_{\text{conv}} = h_c(T_s - T_\infty) \tag{11.6}$$

where h_c is the convective heat transfer coefficient and T_s is the surface temperature that is exposed to air with temperature T_∞. Note that the convection coefficient is not a material property, but rather a complex function of the geometry, fluid properties, and flow condition, as well as the roughness of the surface [110]. Generally, convection increases when wind speeds are higher, when air becomes more turbulent over a rougher surface, and when the temperature difference between the surface and the air is higher [2].

For the heat transfer problem within a stationary medium, such as in pavement, energy transport within the material of interest occurs entirely by conduction and is governed by Fourier's law (Eq. 11.5). Convection is considered only as the boundary condition for the relatively simple ordinary or partial differential equations that govern conduction problems. As mentioned above, convection is the transfer of energy to a moving medium, most often a liquid or gas flowing through a duct or over an object, in this case, air flowing over pavements driven by wind. The transfer of energy in a flowing fluid is due not only to conduction (i.e., the interactions between microscale energy carriers) but also to the enthalpy carried by macroscale flow. Enthalpy is the sum of the internal energy of the fluid and the product of its pressure and volume. The pressure–volume product is related to the work required to move the fluid across a boundary. The term of the internal energy can be addressed by heat transfer within the context of thermodynamics. The addition terms in the energy balance related to the fluid flow complicate the convection problems substantially and link the heat transfer problem to an underlying fluid dynamic problem. The complete solution to many convection problems related to the pavement–air intergraded system of interest therefore requires advanced computational fluid dynamic tools [110,179,180]. This will be detailed below.

11.3.1.3 Net Radiation

Net radiation, as listed above, encompasses three separate radiation processes taking place at the pavement surface: incoming radiation, reflected radiation, and emitted radiation.

Incoming radiation represents the amount of energy radiating from the sun to the (pavement) surface, including direct radiation and diffuse radiation. This obviously varies based on the latitude, the season, the time of the day (zero during night), the amount of cloud cover, and the atmospheric pollution levels. Incoming solar rate per unit area q_{sola} can be represented as

$$q_{\text{sola}} = I_{\text{dire}} + I_{\text{diff}} \tag{11.7}$$

where I_{dire} and I_{diff} are direct solar radiation and diffuse solar radiation, respectively.

Reflected radiation is the amount of solar energy reflected from a surface, based on the solar reflectance or albedo of the surface material. A surface with high solar reflectance, such as bright white materials (e.g., snow and ice), reflects most of the solar radiation that falls on it, whereas dark surfaces such as black asphalt pavement absorb most of the solar radiation. The reflected solar radiation rate per unit area q_{refl} can be represented as

$$q_{refl} = rq_{sola} \tag{11.8}$$

where r is the solar reflectivity or albedo of the material.

Emitted radiation is heat radiation emitted from a surface itself. This term is highly dependent on the temperature of the surface itself and the surroundings. A warmer surface would radiate more energy to its surroundings. For a pavement, the surface at temperature T_s interacts with the surroundings at temperature T_{surr} (e.g., air, other surfaces, etc.); thus the effective surface radiation rate per unit area q_{emit} can be represented as

$$q_{emit} = \sigma\varepsilon\left(T_s^4 - T_{surr}^4\right) \tag{11.9}$$

where σ and ε are the Stefan–Boltzmann constant (5.67×10^{-8} W/m^2 K^4) and the emissivity of the surface material, respectively. Here both T_s and T_{surr} must be expressed as absolute temperature (i.e., in unit K rather than °C) in the above equation.

11.3.1.4 Anthropogenic Heat

Anthropogenic heat represents "human-made" heat generated by buildings, vehicles, machinery, or people. In many areas, especially rural and suburban areas, the amount of anthropogenic energy is small compared to the other terms in the balance equation. In densely urban areas, the anthropogenic term might be larger and can be a significant influence on heat island formation. This item is usually small and thus can be neglected.

11.3.1.5 Shading from Solar Radiation

Shading, produced by trees or walls, will block the incoming solar radiation from hitting some surfaces, including paved surfaces and a human body surface. The sky view factor (*SVF*) could be used to characterize the shading. For a surface that is completely shaded, $SVF = 0$; for a surface that is barely shaded, $SVF = 1$. Some trees might partly allow solar radiation to pass through, and *SVF* on the ground surface or a human body under the trees could be somewhere between 0 and 1, e.g., 0.3. Then the effective incoming radiation incident on a shaded surface with the *SVF* can be described as

$$q_{\text{sola},e} = SVF \cdot q_{\text{sola}} \tag{11.10}$$

11.3.1.6 Evaporation

Evaporation is energy transmitted away from the pavement surface by the latent heat of water vapor owing to phase change (from liquid to gas). Water from moist soil or wet surfaces changes to vapor when heated by solar heat or other heat sources. Water vapor then rises into the atmosphere, taking the solar energy with it. The evaporation term also includes evapotranspiration, a more complicated process plants use to keep cool. During evapotranspiration, water is drawn from the soil by the roots of the plant and is evaporated through stoma on the plant's leaves. Both evaporation and evapotranspiration increase when there is more moisture available, when wind speeds are greater, and when the air is drier and warmer [2]. Some measurements of evaporation rate for various materials were conducted and presented in Chapter 7. The evaporation latent heat q_{evap} (W/m^2) can be described as follows:

$$q_{\text{evap}} = L \cdot ER \tag{11.11}$$

where *ER* is the evaporation rate, in $g/(s\ m^2)$ or $10^{-3} \times$ mm/s and *L* is the specific latent heat of water vaporization, in 2260 J/g.

11.3.2 Thermal Interactions between Pavement and Other Surfaces

From a thermodynamic perspective, thermal energy can be transferred across a boundary (i.e., heat transfer can occur) by three mechanisms: conduction, convection, and radiation. As mentioned above, conduction is the process in which energy exchange occurs owing to the interactions of molecular (or smaller)-scale energy carriers within a material. The conduction process is intuitive; it is easy to imagine energy carriers having a higher level of energy (represented by their temperature) colliding with neighboring particles and thereby transferring some of their energy to them. Convection is the process in which the surface of a solid (or liquid or gas) material exchanges thermal energy with a fluid. Although convection is commonly treated as a separate heat transfer mechanism, it is more properly viewed as conduction in a substance that is also undergoing motion.

The energy transfer by conduction and fluid motion is coupled, making convection problems more difficult to solve than conduction problems. However, convection is still an intuitive process since it can be explained by interactions between neighboring molecules with different energy levels. Radiation is a very different heat transfer process because energy is transferred without the benefit of any molecular interactions. Indeed, radiation energy exchange can occur over long distances through a complete vacuum. The solar energy that our Earth receives from the sun is a result of radiation exchange [110].

In addition, all substances emit energy in the form of electromagnetic radiation as a result of molecular and atomic activity; molecular electronic, vibrational, or rational transitions result in the emission of energy in the form of radiation. The characteristics and amount of radiation emitted by a substance are dependent on its temperature as well as its surface properties. Energy is exchanged between a system and its surroundings by radiation even when they are at the same temperature. In this case, however, the net energy exchange is zero. The rate at which the system is emitting radiation is equal to the rate at which it is absorbing the incident radiation that was emitted from its surroundings [110].

The net rate of radiation heat transfer to a surface is the difference between the rate of radiation that is emitted by the surface and the rate at which the radiation that is incident on the surface is absorbed. The amount of incident radiation is determined by the radiation emitted by other surfaces and their geometric orientation with respect to the surface of interest, which is mainly concerned with calculating view factors. View factors are dimensionless ratios that characterize the degree to which two surfaces "see" one another in terms of radiation amount and therefore how efficiently they exchange radiation [110]. For the pavement and a building surface with an included angle of θ, the average view factor will be $F_{1,2} = 1 - \sin(\theta/2)$. For the case of a horizontal pavement surface and a vertical building surface, $\theta = 90$, thus the average view factor is $F_{1,2} = 0.3$. This means 30% of the radiation emitted by a pavement surface will be the incident radiation hitting the building surface. The amount of this incident radiation that will be absorbed by the building surface is dependent on the reflectivity and transmissivity of the building surface [110].

To better understand the thermal interactions between pavements and building surfaces, several concepts need to be introduced, including specular and diffuse surfaces, reflected and emitted radiation and radiosity, and radiation exchange processes between surfaces.

11.3.2.1 Specular versus Diffuse Surface

Surfaces generally can be classified into two types according to the radiative characteristics: specular surface and diffuse surface [110]. For specular surfaces (Figure 11.2(a)), radiation is reflected only at the incident angle, e.g., a mirror or highly polished metal surface. For diffuse surfaces (Figure 11.2(b)), however, radiation reflected from the surface is angularly uniform and completely independent of the direction of the incident radiation, e.g., rough pavement/building surfaces or general surfaces with diffuse coating [110]. General pavement/building surfaces of interest in this study are diffuse surfaces and will reflect the radiation angularly uniformly in every direction.

11.3.2.2 Reflected and Emitted Radiation and Radiosity

Pavement, as a diffuse gray surface, has a constant emissivity ε at all wavelengths and emits radiation uniformly in all directions as does the reflected radiation [110]. The sum of the reflected and emitted radiation per unit area is called *radiosity* (Figure 11.3); it is the rate of radiation that is leaving surface i per unit area [110]. The radiosity will be an important variable to determine the energy balance on related surfaces and evaluate the effectiveness of high-albedo pavements for heat island mitigation in terms of human thermal comfort and building energy use.

11.3.2.3 Radiation Exchange between Surfaces

An example of the complete thermal interactions (radiation exchange) between relevant surfaces is illustrated in Figure 11.4. The amount of radiation exchange depends on the radiosity and view factor of each surface.

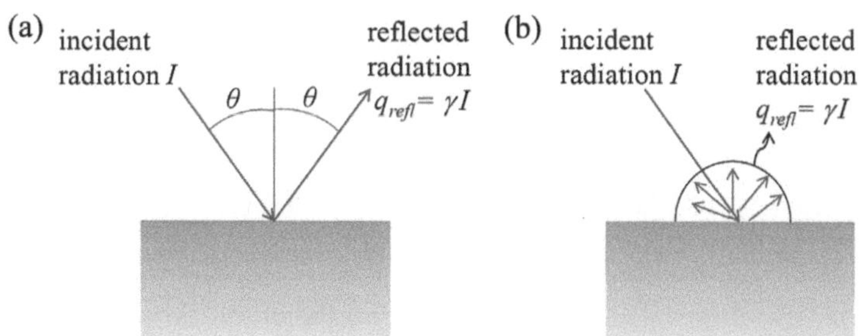

Figure 11.2 Specular versus diffuse surface. (a) Specular reflection. (b) Diffuse reflection.

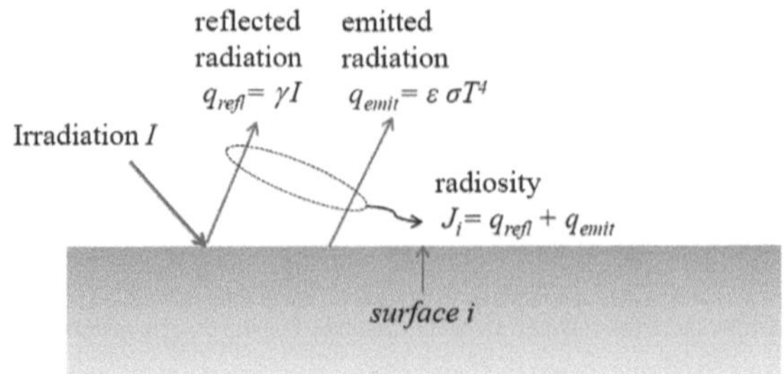

Figure 11.3 Reflected and emitted radiation and radiosity on a pavement surface.

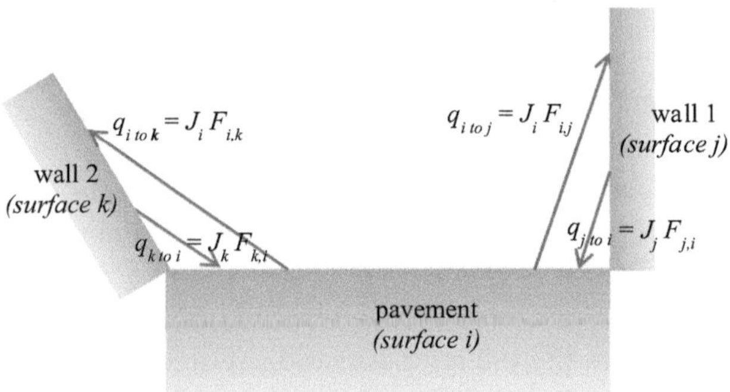

Figure 11.4 Thermal interactions between pavement and other surfaces. $F_{i,j}$ is the fraction of radiation that leaves surface i and subsequently hits surface j, i.e., the effective view factor.

To reduce the radiation of one surface to another, one method is to reduce the absorbed energy by increasing the reflected energy. However, this will increase the risk of more energy being absorbed by other surfaces unless most of the reflected energy is reflected directly back into far space and not reflected onto and absorbed by other building or vehicle surfaces or human bodies. Therefore, there should be an optimal balance of reflectivity among thermally interactive surfaces, especially in high-density urban areas.

11.4 SIMPLIFIED MODEL FOR THERMAL INTERACTIONS BETWEEN PAVEMENT AND NEAR-SURFACE AIR

To model the thermal interaction between pavement and near-surface air with a focus on the important pavement surface temperature, a simplified and specific model was extracted from the full model above to simulate the temperatures of pavement and near-surface air. The simulation results were compared with the results measured from the field experiments on test sections presented in previous chapters, in part to validate the model.

In addition to shading and evaporation, there are three primary thermal interactive processes between pavement and near-surface air: conduction, convection, and radiation [181,182]. The total heat energy that flows into the pavement is indicated as

$$q_T = q_s - q_r - q_c - q_e \tag{11.12}$$

where

q_T is the total energy flowing into the pavement;
q_s is the solar radiation absorbed by the pavement;
q_r is the effective surface radiation to the ambient atmosphere;
q_c is the convective heat loss to the ambient atmosphere; and
q_e is the evaporative heat loss from the pavement.

11.4.1 Conduction

Heat conduction is the transfer of heat by direct contact of particles of matter. Conduction defines the constitutive behavior of the system for heat transfer analysis (in terms of specific heat and conductivity). It determines the total internal energy of the heat transfer system and drives the system to reach a heat energy balance. The pavement conduction heat transfer equation is defined as follows [181,183]:

$$\frac{\partial^2 T_i}{\partial z^2} = \frac{\rho_i c_i}{k_i}\frac{\partial T_i}{\partial t} = \alpha_i \frac{\partial T_i}{\partial t} \tag{11.13}$$

where

T_i is the temperature of the ith layer of material, °C;
z is the depth from the surface, m;
t is the time of day, 0–24 h;
ρ_i is the density of the ith layer of material, kg/m^3;
c_i is the heat capacity (or specific heat) of the ith layer of material, J/(kg °C);
k_i is the thermal conductivity of the ith layer of material, W/(m °C); and
α_i is the thermal diffusivity of the ith layer material, m^2/s, $\alpha_i = k_i/\rho_i c_i$.

Heat transfer by conduction occurs between the layers of pavement as well as in the air above the pavement:

$$q_{\text{cond}} = -k \frac{\partial T_1}{\partial z}\Big|_{z=0} \tag{11.14}$$

11.4.2 Convection

Convection is the transfer of thermal energy by the movement of molecules from one part of a material to another. It is one of the boundary conditions for the heat transfer system, which can be specified as surface heat flux per area or as volumetric heat flux per volume. Convection determines the heat exchange between pavement and near-surface air.

The rate of convective heat transfer per area q_c is given by [181,183]:

$$q_c = h_c\left[T_1\big|_{z=0} - T_a\right] \tag{11.15}$$

where h_c is the convection (film) coefficient, $T_1|_{z=0}$ is the surface temperature, and T_a is the ambient air temperature.

If the hourly data of ambient air temperature are not available or a reduced data set is needed to save storage space even if they are available, the ambient air temperature changes over a day can be calculated from the daily peak air temperature and daily lowest air temperature, usually available from general weather record data, using the following formula [184]:

$$T_a = \overline{T_a} + T_m\left[0.96 \sin \omega(t - t_0) + 0.14 \sin 2\omega(t - t_0)\right] \tag{11.16}$$

where

$\overline{T_a}$ is the daily average air temperature, °C, $\overline{T_a} = \frac{1}{2}(T_a^{\max} + T_a^{\min})$;

T_m is the daily air temperature vibration amplitude, °C, $T_m = \frac{1}{2}(T_a^{\max} - T_a^{\min})$;

$T_a^{\max}$ and $T_a^{\min}$ are the daily peak air temperature and daily lowest air temperature, respectively;

t is the time of day, 0–24 h;

t_0 is the initial phase position, $t_0 = 11$ (peak at 16:00 h, see Section 8.3.3 in Chapter 8); and

w is the angular frequency, $w = 2\pi/24$ rad.

11.4.3 Radiation

Radiation is the transfer of heat energy through empty space via electromagnetic waves. All objects with a temperature above absolute zero radiate energy at a rate equal to their emissivity multiplied by the rate at which energy would radiate from them if they were a black body.

There are two radiation processes in the pavement heat transfer system: solar radiation to pavement and effective surface radiation to the ambient atmosphere (including pavement radiation both to atmosphere and in the opposite direction).

Similar to the ambient air temperature above, the hourly data of solar radiation flux might not be available sometimes. Even if they are available, sometimes a reduced data set might be needed for some situations. In these cases, solar radiation flux to the pavement at various times over a day $q(t)$ can be calculated from the daily total solar radiation volume (peak at 13:00 h, see Section 4.3.3 in Chapter 3) using the formula shown below [184]:

$$q(t) = \begin{cases} 0 & 0 \le t < t_{s0} - \dfrac{c}{2} \\ q_0 \cos m\omega(t - t_{s0}) & t_{s0} - \dfrac{c}{2} \le t \le t_{s0} + \dfrac{c}{2} \\ 0 & t_{s0} + \dfrac{c}{2} < t \le 24 \end{cases} \tag{11.17}$$

where

q_0 is the peak solar radiation at noon, $q_0 = 0.131m \cdot Q$, in J/h;
Q is the daily total solar radiation volume, in J;
m is the distribution coefficient for solar radiation, $m = 12/c$;
c is the daily effective sunlight hour, h;
t is the time of day, 0–24 h;
t_{s0} is the peak position, $t_{s0} = 13$ (peak at 13:00 h, see Section 8.3.3 in Chapter 8); and
w is the angular frequency, $w = 2\pi/24$ rad.

The effective incoming radiation $q_e(t)$ incident on a shaded surface with the SVF can be described as

$$q_e(t) = SVF \cdot q_e(t) \tag{11.18}$$

The solar radiation absorbed by pavement q_s can be calculated as

$$q_s = \alpha_s \cdot q_e(t) \tag{11.19}$$

where α_s is the solar radiation absorptivity of the pavement surface ($=1 - r$; r is the solar reflectivity).

Equation (11.17) is not smooth and continuous, but rather results in discontinuities across time in temperature calculations. Therefore, the Fourier series ($k = 30$ yields an accurate estimate) is used to address this problem as below [185]:

$$q(t) = \frac{a_0}{2} + \sum_{k=1}^{\infty} a_k \cos \frac{k\pi(t - 12)}{12} \tag{11.20}$$

where $a_0 = \frac{2q_0}{m\pi}$;

$$a_k = \begin{cases} \dfrac{q_0}{2m} & k = m \\ \dfrac{q_0}{\pi}\left[\dfrac{1}{m+k}\sin(m+k)\dfrac{\pi}{2m} + \dfrac{1}{m-k}\sin(m-k)\dfrac{\pi}{2m}\right] & k \neq m \end{cases} \tag{11.21}$$

The effective surface radiation to the ambient atmosphere, a boundary condition for the pavement heat transfer system, can be defined as [168,181]:

$$q_r = \varepsilon\sigma\left[\left(T_1|_{z=0} - T^Z\right)^4 - \left(T_a - T^Z\right)^4\right] \tag{11.22}$$

where

q_r is the effective surface radiation, W/(m^2 °C);
ε is the thermal emissivity;
σ is the Stefan–Boltzmann constant, 5.6697 × 10^{-8} W/(m^2 K^4);
$T_1|_{z=0}$ is the temperature at the pavement surface, °C;
T_a is the temperature of the air atmosphere, °C; and
T^Z is absolute zero, −273 °C.

11.4.4 Evaporation

Evaporation is energy transmitted away from the pavement surface by the latent heat of water vapor owing to phase change (from liquid to gas). Water from moist soil or wet surface changes to vapor when heated by solar heat or other heat sources. Some measurements of evaporation rate for various materials were conducted and presented in Chapter 7. The evaporation latent heat q_e (W/m^2) can be described as follows:

$$q_e = L \cdot ER \tag{11.23}$$

where

ER is the evaporation rate, in g/(s m^2) or 10^{-3} × mm/s and
L is the specific latent heat of water vaporization, in 2260 J/g.

11.5 MODEL VALIDATION

Some data from the field measurements on the test sections are used to partly validate the developed model. More detailed and complete validation could be performed with more data from field measurements of the test sections and from other different and complex contexts (climates, surroundings). The validation includes both asphalt and concrete pavements for 10 typical hot summer days in a row. Some validation results, with a focus on the key surface temperatures, are presented below.

11.5.1 Pavement Structures and Model Parameters

Two asphalt pavement sections and two concrete pavement sections (B1 and B3, C1 and C3, with different structures and materials; see Chapter 4) were used for validation of the simplified model for 10 sunny days in summer (20–30 July 2012). The key material properties for the surface asphalt layer (albedo, thermal conductivity, heat capacity, density) are from the experimental measurements presented previously (Tables 4.4, 6.7, and 8.1). Some other secondary material properties (e.g., thermal properties of the base materials and subgrade soil that will not heavily influence the results) are from the literature (e.g., [181,182,184,186,187]). Some key parameters for temperature simulation for validation are summarized here in Table 11.1. Other information (e.g., numerical implementation, boundary conditions, etc.) can be found in Section 12.1 of Chapter 12.

11.5.2 Weather Data

The local hourly weather data (ambient air temperature, solar radiation and wind speed) for the 10 sunny days in summer (20–30 July 2012) were available and used for the simulation, as shown in Figure 11.5.

11.5.3 Validation Results for Asphalt Pavements

The simulated and measured results for the two asphalt pavements (B1 and B3) are presented in Figure 11.6. The simulated and measured results generally agree very well with one another. The comparison of the simulated and the measured results for the asphalt pavements is presented in Figure 11.7. This indicates that the developed model can be used to simulate the temperatures for asphalt pavements under various weather conditions.

11.5.4 Validation Results for Concrete Pavements

The simulated and measured results for the two concrete pavements (C1 and C3) are presented in Figure 11.8. The simulated and measured results generally agree well with one another, with relatively larger errors at the low temperatures for C3. The comparison of the simulated and the

Table 11.1 Parameters for temperature simulation for validation

Layer	Material	Thickness (cm)	Albedo	Thermal conductivity (W/(m °C))	Heat capacity (J/(kg °C))	Density (kg/m^3)	Thermal emissivity
B-1 surface	Asphalt-D	10	0.09	1.73	852	2399	0.80
B-3 surface	Asphalt-O	20	0.08	1.23	798	2200	0.80
C-1 surface	Concrete-D	10	0.29	1.83	1001	2257	0.80
C-3 surface	Concrete-O	20	0.26	1.29	761	2050	0.80
Base	Aggregate	15	—	1.40	920	1800	—
Subgrade	Soil	—	—	1.30	860	1700	—
Air	Air	200	—	0.00257	1006	1.205	—

D, dense graded; O, open graded; U, uncompacted; C, compacted.

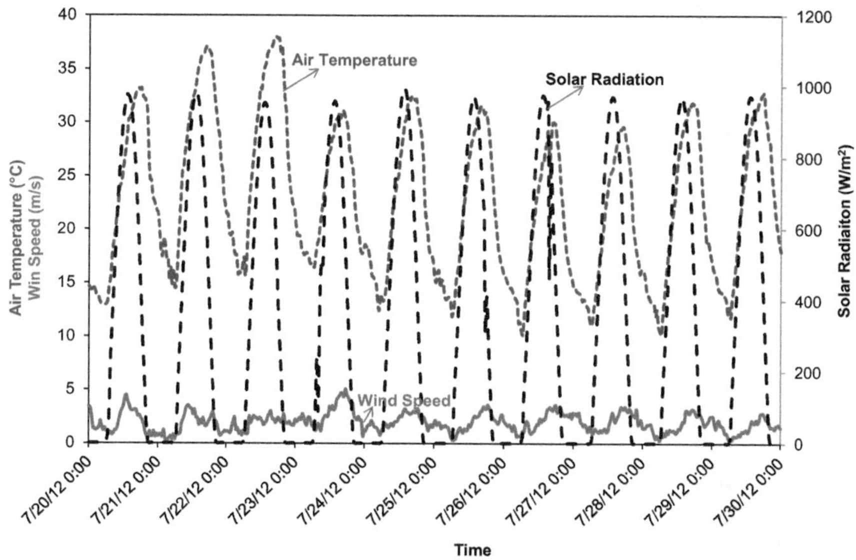

Figure 11.5 Weather data for the 10 sunny days in summer (20–30 July 2012). Dates are given as month/day/year. 0:00 indicates midnight.

measured results for the concrete pavements is presented in Figure 11.9. This indicates that the developed model can be used to simulate the temperature for concrete pavements for various weather conditions.

11.6 SUMMARY AND CONCLUSIONS

This chapter developed a framework for a general local microclimate model. This integrated local microclimate model includes both pavement structure and near-surface air and considers coupled processes of radiation, convection, conduction, shading, and evaporation. A simplified model was extracted from the general model to simulate the temporal and spatial distribution of temperatures of pavement and near-surface air. It can be used for various pavement structures under various climate conditions for various time frames.

The model was numerically implemented using the finite element method and the finite difference method and used to simulate the temporal and spatial distribution and variation of both pavement temperature and near-surface air temperature in the summer in a hot region (Sacramento, California) for validation against the field measurements from experimental test sections presented previously in this dissertation. The model was

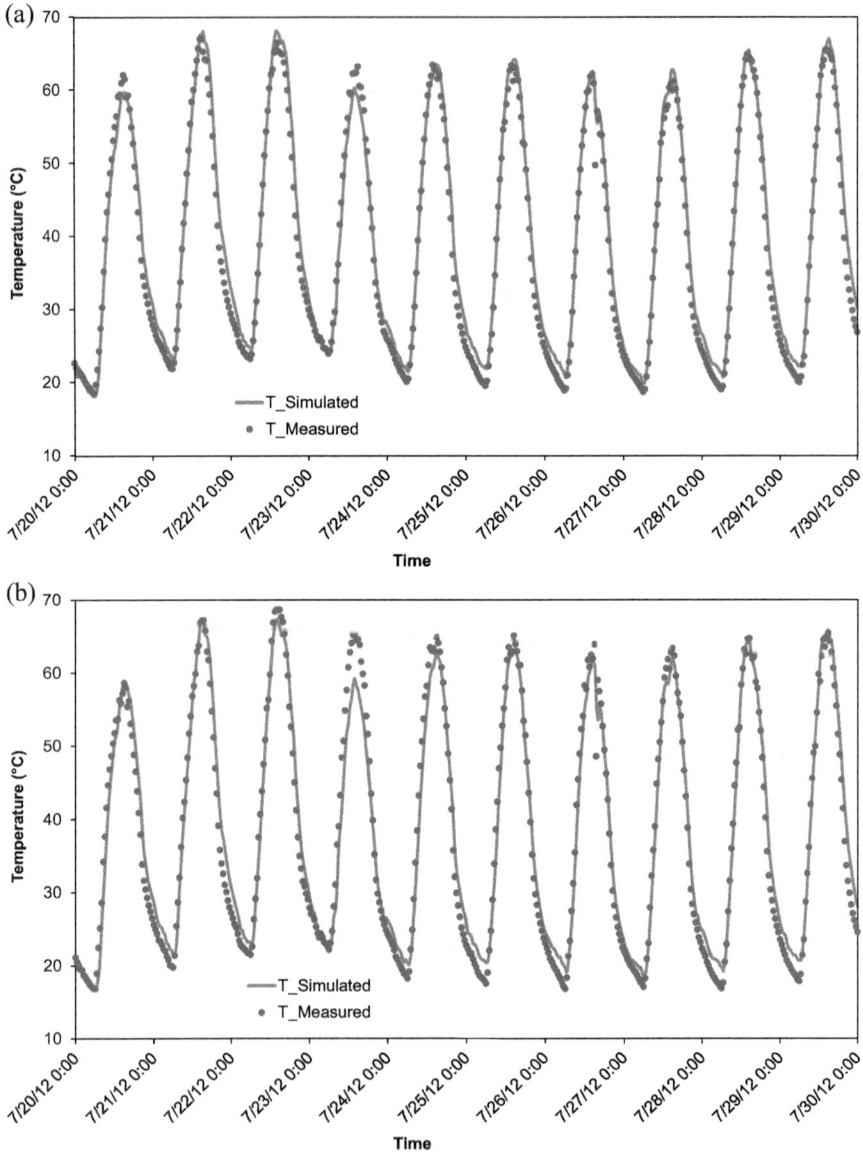

Figure 11.6 Simulated and measured surface temperatures for asphalt pavements. Dates are given as month/day/year. 0:00 indicates midnight. (a) B1. (b) B3.

(a)

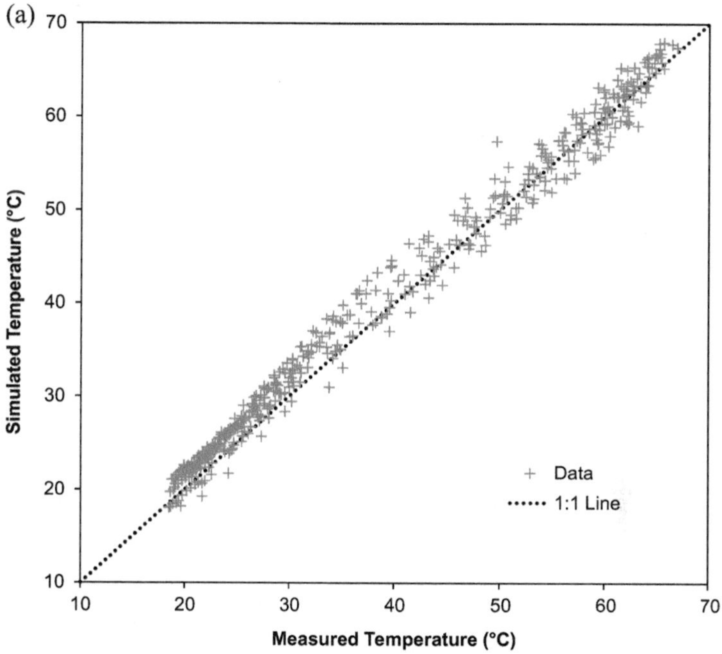

(b)

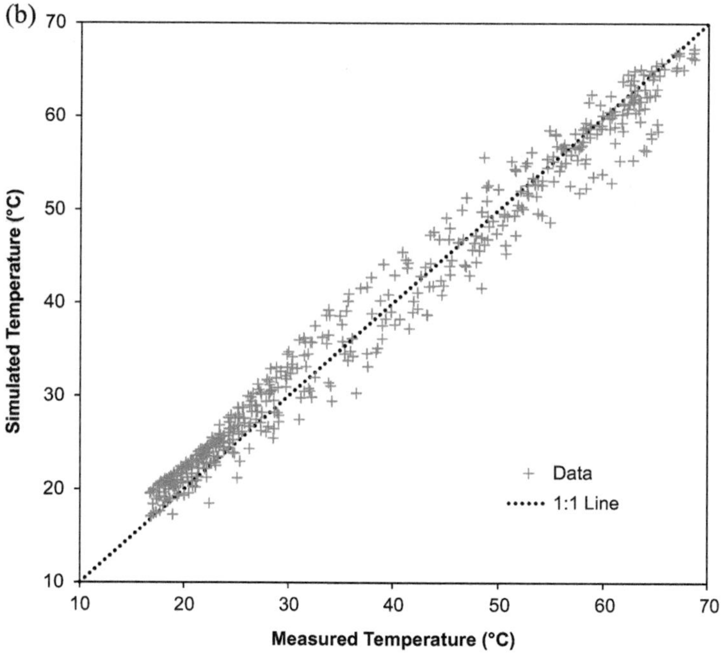

Figure 11.7 Comparison of simulated and measured results for asphalt pavements. (a) B1. (b) B3.

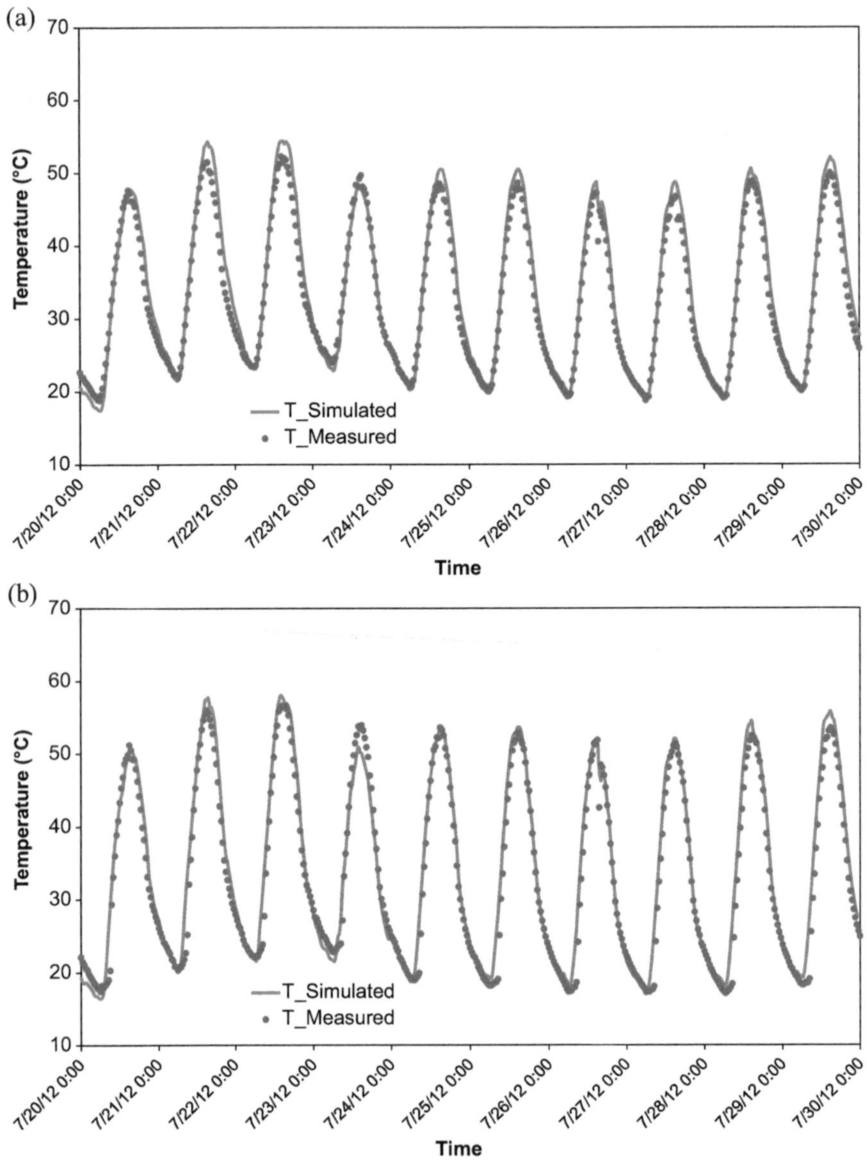

Figure 11.8 Simulated and measured surface temperatures for concrete pavements. Dates are given as month/day/year. 0:00 indicates midnight. (a) C1. (b) C3.

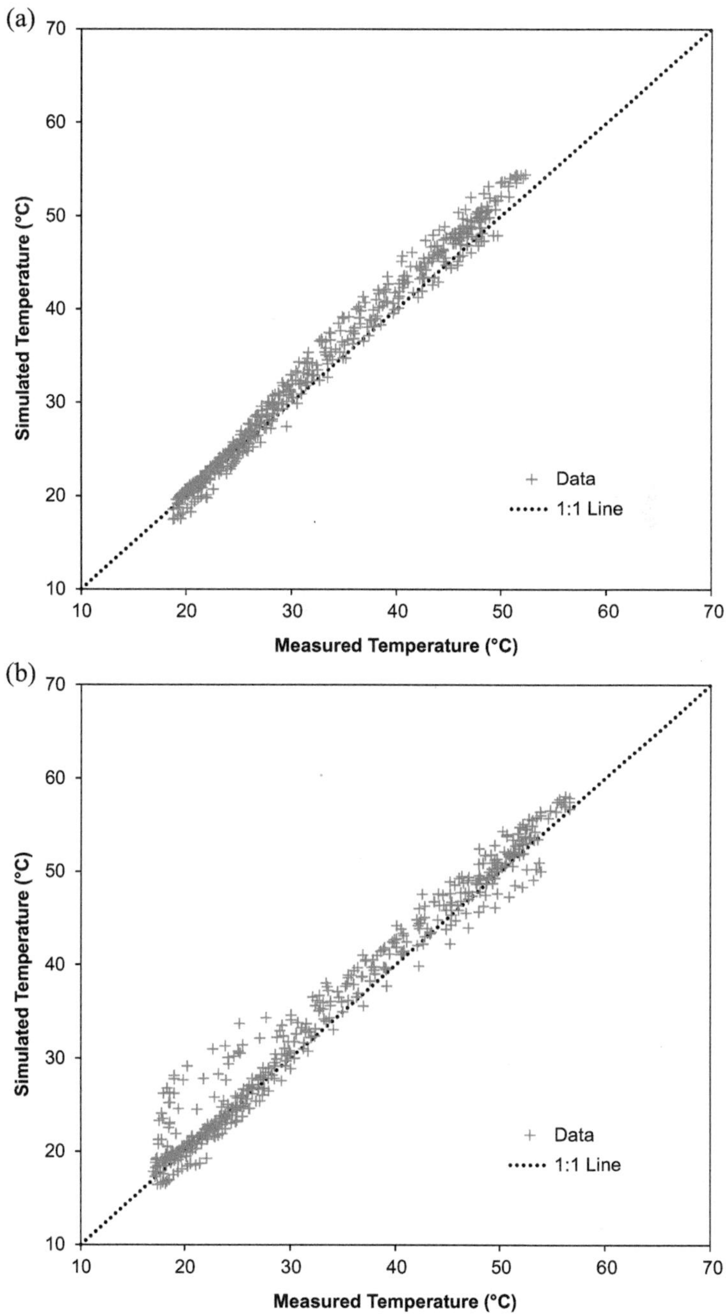

Figure 11.9 Comparison of simulated and measured results for concrete pavements. (a) C1. (b) C3.

validated against field measurements for both asphalt and concrete pavements (with different materials and structures) and under various weather conditions. The simulated and measured results generally agree well with one another for both asphalt and concrete pavements with different materials and structures and under various weather conditions. This indicates that the developed model can be used to simulate the temperature for asphalt and concrete pavements for various weather conditions.

CHAPTER 12

Simulation of Thermal Behavior of Design and Management Strategies for Cool Pavement

Contents

This chapter uses the validated model to run the temperature simulation for various factors and then identifies the significant factors affecting thermal behavior. Depending on the identified significant factors affecting the thermal behavior, effective potential alternative design strategies are then proposed.

12.1 SIMULATION USING THE SIMPLIFIED MODEL

The validated model was used first to run some simulations for a typical pavement in a hot climate; then based on the simulation, a sensitivity analysis on the model was performed for key model factors, which is presented in the following section.

Pavement Materials for Heat Island Mitigation
ISBN 978-0-12-803476-7
http://dx.doi.org/10.1016/B978-0-12-803476-7.00012-X

12.1.1 Pavement Structure and Integrated Local Modeling

A typical asphalt pavement structure (as shown in Figure 12.1) was chosen for the numerical simulation of the temperature field. The simplified integrated local model (as shown in Figure 12.2) was used for the heat transfer analysis between pavement and near-surface air. This integrated modeling includes the pavement structure and the near-surface air (~2 m above the pavement surface).

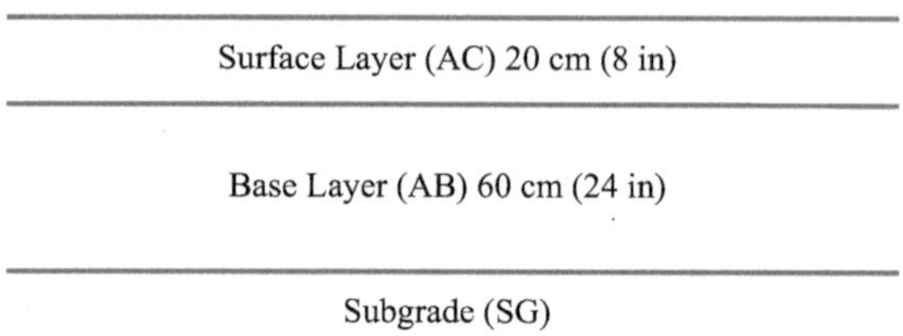

Figure 12.1 Typical asphalt pavement structure for temperature simulation.

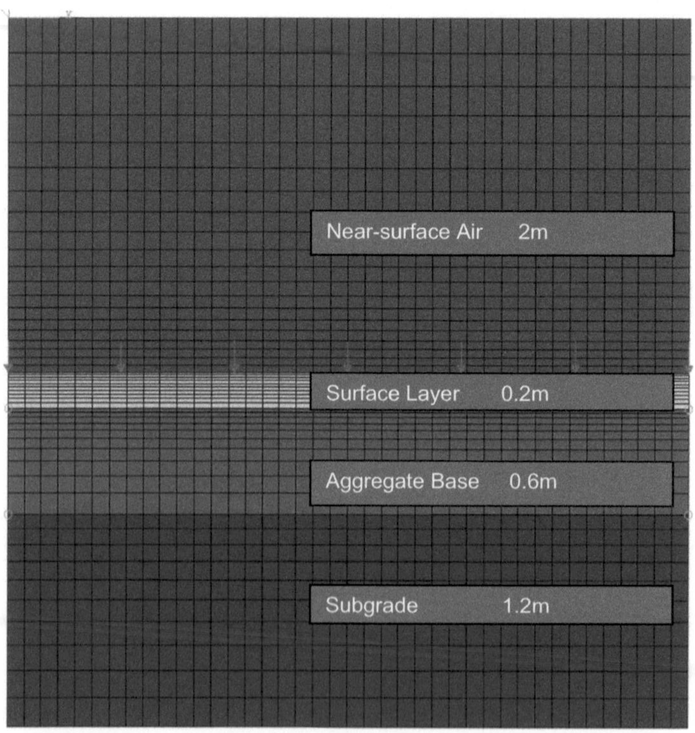

Figure 12.2 Integrated modeling for temperature simulation.

12.1.2 Thermal Property Parameters and Climate Conditions

The thermal properties and the typical parameter values for temperature simulation are shown in Table 12.1 and were determined from experimental measurements presented previously (Tables 4.4, 6.7, and 8.1). Some other material properties (e.g., thermal properties of the base materials and subgrade soil that will not heavily influence the results) are from the literature (e.g., [181,182,184,186,187]). The hot climate conditions in summer (July) in Sacramento, California [188], as shown in Table 12.1, were used to set up the boundary conditions (convection and radiation) for the integrated local model implemented in the finite element analysis software *Abaqus*® [168].

12.1.3 Initial and Boundary Conditions

The initial condition includes the predefined temperature field for the whole model, which was set at 25 °C.

The boundary conditions for this simplified heat transfer model include the solar radiation heat flux on the pavement surface, the surface convection between pavement surface and air, the surface radiation of pavement and upper boundary of near-surface air, and the temperature boundary for the upper boundary of near-surface air.

The solar radiation heat flux (calculated from Eq. (11.20)) varies throughout a day, as defined by the user subroutine *DFLUX()* in *Abaqus*®. Heat loss by evaporation is integrated into the surface heat flux with *DFLUX()* in *Abaqus*®. Surface convection, which is a function of the ambient air temperature (calculated from Eq. (11.16)) varying over time and the convection coefficient varying with wind velocity, is defined by the user subroutine *FILM()* in *Abaqus*®. The effective surface emitted radiation is directly defined through surface radiation to the ambient atmosphere using the key word **Sradiate* in *Abaqus*®. The temperature boundary for the upper boundary of near-surface air at 2 m height was set as the ambient air temperature. All other boundary conditions are assumed to be regulated by heat insulation [168].

12.2 EXAMPLE SIMULATION RESULTS

Using the integrated model developed and validated above (Figure 12.2), the temperatures of pavement and near-surface air under the typical climate

Table 12.1 Parameters for temperature simulation in Sacramento, California

Thermal properties and typical parameter values

Parameters	Air	Asphalt concrete	Aggregate base	Subgrade
Thermal conductivity k^a $(J\cdot(h\cdot m\cdot°C)^{-1})$	92.5	5400	5040	4680
Density $\rho/(kg\cdot m^{-3})$	1.205	2200	1800	1700
Specific heat capacity c $(J\cdot(kg\cdot°C)^{-1})$	1006	950	920	860
Surface solar radiation absorptivity α_s	0.90			
Thermal emissivity ε	0.80			
Heat convection coefficient $h_c{}^a$ $(J\cdot(h\cdot m^2\cdot°C)^{-1})$	$h_c = 3600(3.7v_w + 6.1)$ (v_w, wind velocity, m/s)			
Absolute zero T^Z (°C)	−273			
Stefan–Boltzmann constant σ^a $(J\cdot(h\cdot m^2\cdot K^4)^{-1})$	2.041092×10^{-4}			

Hot summer climate data in hot region of Sacramento, California

Month	Daily peak air temperature T_a^{max} (°C)	Daily lowest air temperature T_a^{min} (°C)	Daily total solar radiation volume Q (MJ/m²)	Daily effective sunlight hour c (h)	Daily average wind velocity v_w (m/s)
7 (July)	38	18	30.0	11	4

[a]Time is measured in hours and the related parameters are converted for the simulation analysis at intervals of 1 h.

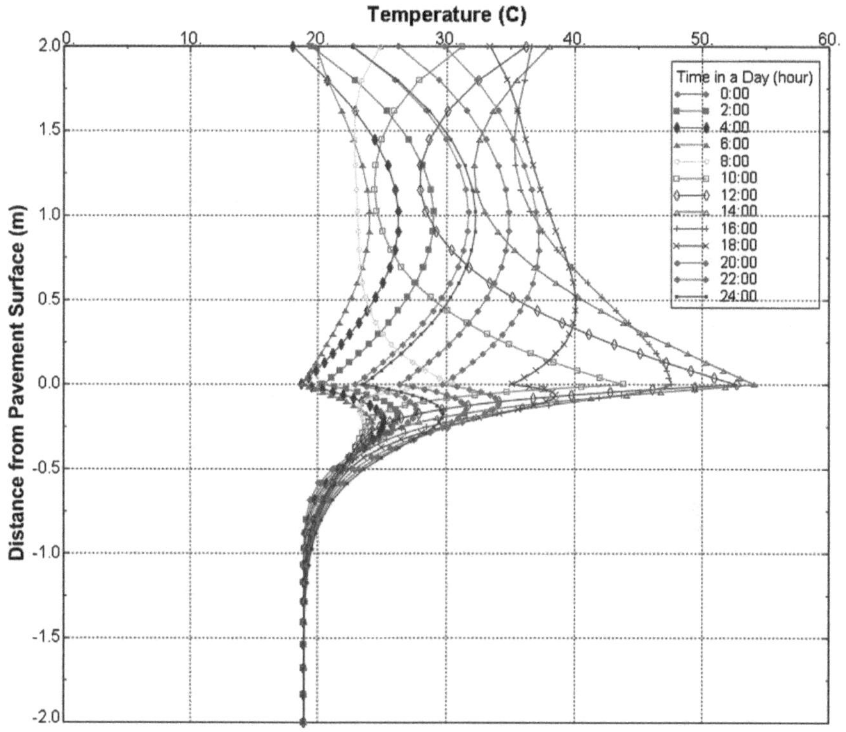

Figure 12.3 Temperature of whole model over depth at various times.

conditions in July in Sacramento, California (Table 12.1), were simulated and the results are shown in Figures 12.3 and 12.4.

12.2.1 Temperature Over Depth

The lowest surface temperature occurs at ~4:00 h (4:00 in Figure 12.3). The maximum temperature for the whole model reaches to approximately 55 °C, which occurs on the pavement surface at ~14:00 h (14:00 in Figure 12.3). At this time, away from the pavement surface, the temperatures both down in the pavement and up in the near-surface air decrease with increasing distance from the pavement surface.

During the noon and afternoon hours (e.g., 12:00, 14:00, and 16:00 in Figure 12.3), the highest temperatures occur on the surface, while the highest temperatures of both pavement and near-surface air occur at some distance from the surface during nighttime (e.g., 18:00 and 20:00 in Figure 12.3). This dynamic is driven by daytime solar radiation. Also, the temperature gradient is higher in the area close to the surface than some

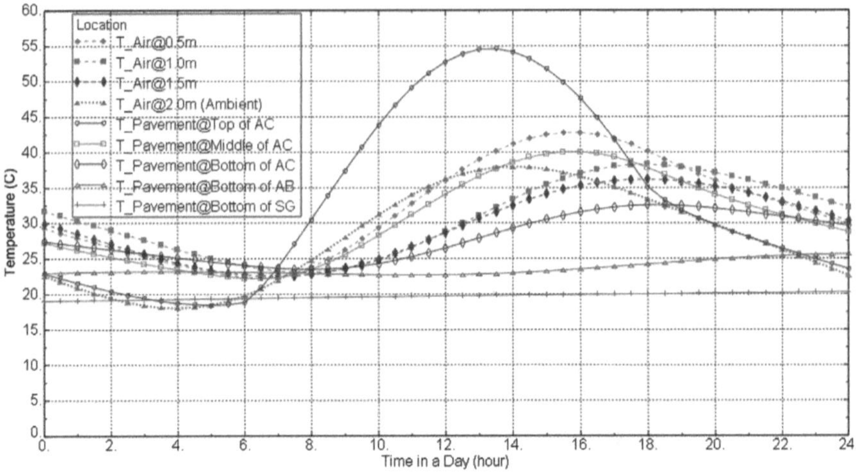

Figure 12.4 Temperature over time at various locations.

distance away from the pavement surface. This high temperature gradient is driven by heat transfer between the pavement and the air around the surface. The resultant higher temperature level and temperature gradient are adverse to both pavement life and human comfort and health in areas near pavement surfaces.

12.2.2 Temperature Over Time

The temperature fluctuations of pavement and near-surface air at various depths/heights over one day are shown in Figure 12.4, along with ambient air temperatures from weather data (T_Air@2.0m (Ambient)) for comparison.

Figure 12.4 shows that the area close to the pavement surface, i.e., the top of the asphalt concrete (AC) surface, and the near-surface air at a height of 0.5 m above the surface have higher peak temperatures during the daytime and higher fluctuations over the day. The temperatures in the area some distance away from the pavement surface (i.e., the bottom of the aggregate base (AB) and the subgrade (SG) and the near-surface air at 1.5 m above the surface) change little over time.

Also, the temperatures of both pavement and near-surface air at most depths and at most times of the day are higher than the ambient air temperature (T_Air@2.0m (Ambient)). Especially around noon, the highest temperatures of the pavement surface and near-surface air (0.5 m high) are approximately 20 and 10 °C higher than the ambient air temperature, respectively. Even during the night, the pavement temperatures at most pavement depths (Figure 12.4) are still much higher than the ambient air temperature. The heat energy absorbed by and stored in the pavement is released into the near-surface air and then into the ambient atmosphere during the nighttime, increasing the temperature of the near-surface air as well as of the ambient atmosphere. This process of heat exchange occurring during the nighttime can have negative impacts on building energy use and human comfort and health in hot climates but can be a positive in cold climates.

12.3 SIMULATION-BASED SENSITIVITY ANALYSIS USING THE SIMPLIFIED MODEL

The effects of various parameters on pavement and near-surface temperature were investigated using the simplified model, to explore the specific effects of each factor and identify the significant factors. The factors used for sensitivity analysis include surface material thermal properties, solar radiation absorptivity, thermal emissivity of pavement, and the heat convection coefficient, as well as solar radiation, wind speed, and evaporation rate. In addition to the base case (C0 in Table 12.2), four additional values (C1–C4 in Table 12.2) for each factor were chosen based on the experimental measurements presented previously and some literature for the simulation of temperature fields.

For sensitivity analysis, in each case only one factor was changed in the simulation run, as shown in Table 12.2. All the other parameters were kept constant with the values shown in Table 12.1, which was set as the reference base case. The temperature distributions along the depth in the upper pavement (0.2 m AC + 0.6 m AB, −0 on the y axis) and in the near-surface air (~1.0 m above the surface, +0 on the y axis) were extracted at 14:00 h (at 14:00 in Figures 12.5–12.9) and 4:00 h (at 4:00 in Figures 12.5–12.9) to compare the effects of each factor on both daytime high temperatures and nighttime low temperatures.

12.3.1 Thermal Conductivity

As the thermal conductivity of the pavement surface layer material increases, the heat exchange between the pavement and the near-surface air is facilitated by conduction. The increased thermal conductivity reduces the high near-surface air temperature at 14:00 h during daytime (at 14:00 in Figure 12.5) with more heat flux from the air into the pavement, while

Table 12.2 Factors for sensitivity analysis

Factor	C0 (Base case)	C1 (Low case)	C2 (Slightly low case)	C3 (Slightly high case)	C4 (High case)
Thermal conductivity k ($J \cdot (h \cdot m \cdot °C)^{-1}$)	5400	3240	4320	6480	7560
Specific heat capacity c ($J \cdot (kg \cdot °C)^{-1}$)	950	570	760	1140	1330
Density ρ ($kg \cdot m^{-3}$)	2200	1320	1760	2640	3080
Solar radiation absorptivity α_s	0.9	0.5	0.6	0.7	0.8
Thermal emissivity ε	0.8	0.5	0.6	0.7	0.9
Heat convection coefficient, slope A^a	3.7	2.2	3.0	4.4	5.2
Daily total solar radiation Q (MJ/m^2)	30	18	24	36	42
Daily average wind velocity (m/s)	4	2	3	5	6
Evaporation rate (mm/h)	0	0.2	0.5	1	1.5

All material properties are of the surface.
[a] $h_c = 3600(A \cdot v_w + 6.1)$ (v_w, wind velocity, m/s).

increasing the low near-surface air temperature at 4:00 h during the nighttime (at 4:00 in Figure 12.5) with more heat flux from the pavement into the air. This provides an opportunity to moderate the near-surface air temperature through easier heat exchange between the pavement and the near-surface air.

The increased thermal conductivity of the pavement surface layer facilitates heat flow into deep pavement layers from upper pavement layers. This results in higher temperatures of deep pavement layers at both the high temperature peak time (14:00 h) and the low temperature time (4:00 h), but a slightly reduced temperature in the upper pavement layer immediately close to and on the pavement surface. The ways to increase the thermal conductivity include using a high-conductivity aggregate, carbon filler, steel fiber, etc.

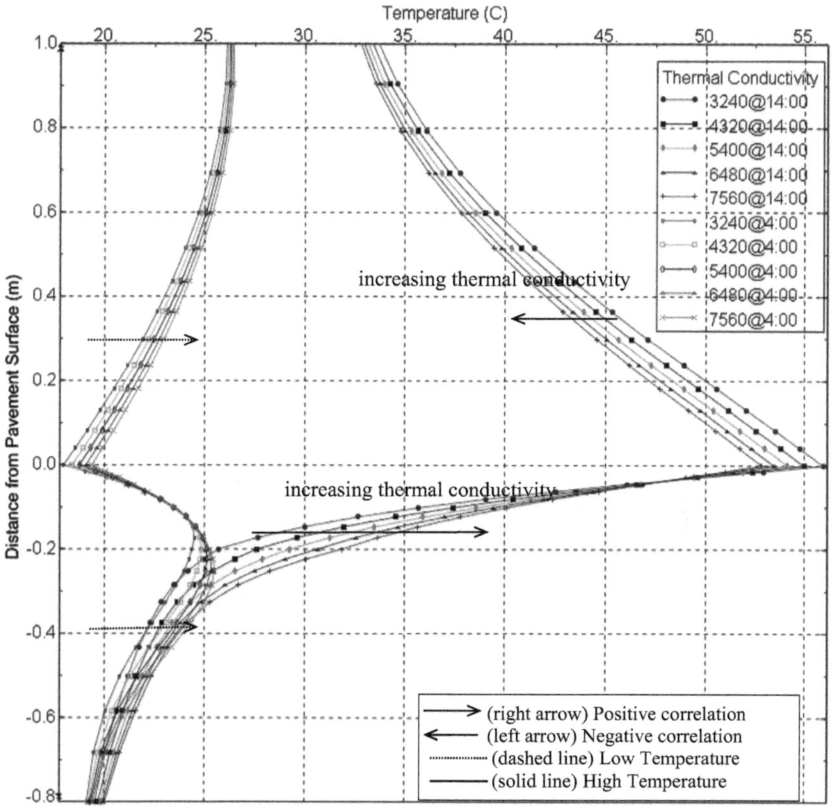

Figure 12.5 Temperature versus thermal conductivity.

12.3.2 Specific Heat

Similar to the heat conductivity, the increased specific heat of the surface layer can reduce the daytime high near-surface air temperature (negative correlation) and increase the nighttime low near-surface air temperature (positive correlation). Unlike thermal conductivity, which has a positive correlation with both high and low pavement temperature, the increased specific heat not only increases the nighttime low pavement temperature but also reduces the daytime high pavement temperature, moderating the temperatures. The ways to increase the specific heat include using aggregate and filler with high specific heat or adding some phase-change admixture.

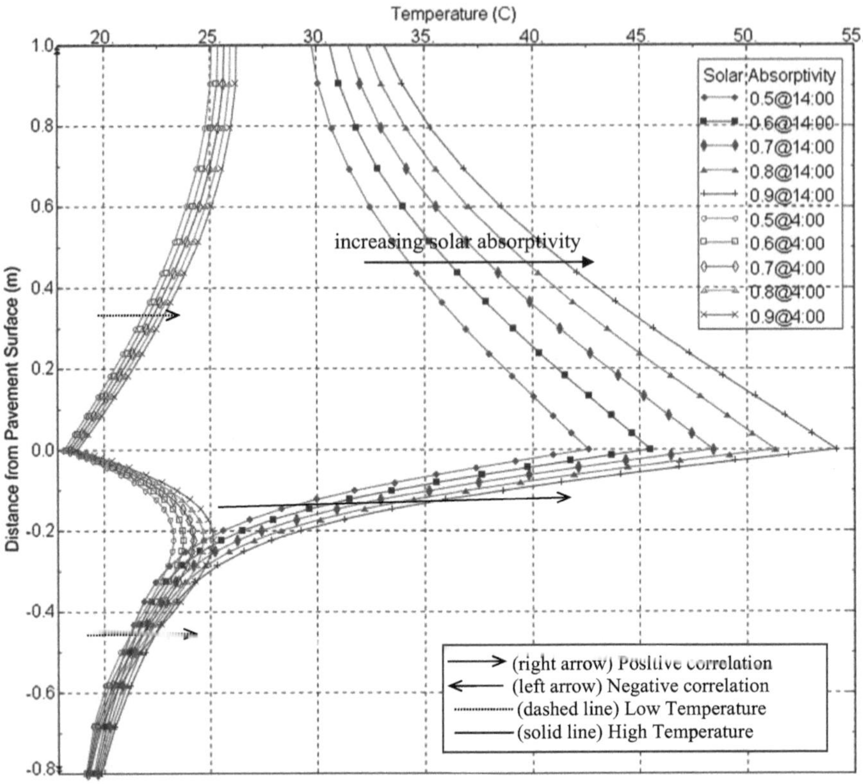

Figure 12.6 Temperature versus solar radiation absorptivity (= 1 − albedo).

12.3.3 Density

The density also shows similar effects on the temperature for both the pavement and the near-surface air as with specific heat above. The mechanism is that more energy is needed because of an increased total mass of the surface layer to reach the energy balance. Using materials with higher density for pavement can moderate the temperature for both the pavement and the near-surface air, but the effect is limited, as in the case of specific heat.

12.3.4 Solar Absorptivity

Figure 12.6 shows that the daytime high temperature of both the pavement surface and the near-surface air substantially decreases as the solar absorptivity (which equates to 1 − reflectivity) of the pavement surface goes down, with less solar energy absorbed by the pavement. Also, the drop in the daytime high temperatures of both pavement and near-surface air is more

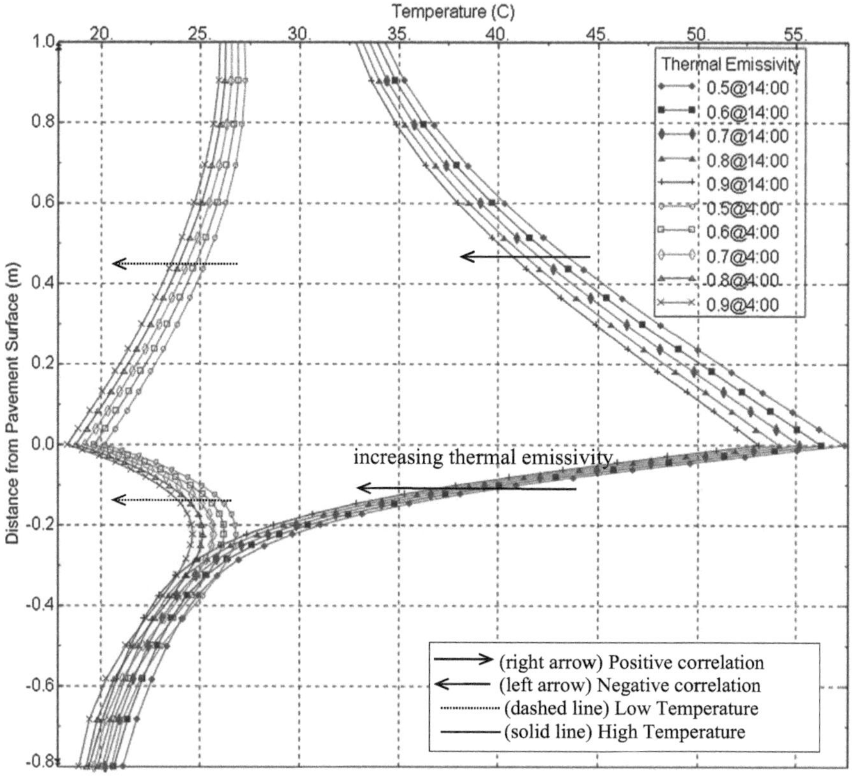

Figure 12.7 Temperature versus thermal emissivity.

pronounced than the drop in the nighttime low temperature, especially the pavement surface temperature. In Figure 12.6, the highest temperature of the pavement surface (14:00 h) drops by about 12 °C (from 54 to 42 °C) when the solar absorptivity decreases from 0.9 to 0.5 (reflectivity increases from 0.1 to 0.5). However, the lowest temperature of the pavement surface (4:00 h) remains roughly constant. This implies that, using materials with low absorptivity (high reflectivity) such as light-colored aggregate or coating on pavement could reduce the high pavement temperatures but would probably not have a strong effect on the lowest pavement temperatures. Adding this characteristic would probably improve pavement properties under high-temperature conditions while not increasing the possibility of pavement low-temperature distresses (such as cracking).

From the near-surface air at 0.5 m height (Figure 12.6) it can be seen that the daytime high temperature (14:00 h) drops by about 7 °C (from 40 to 33 °C) as the solar absorptivity goes down from 0.9 to 0.5. However,

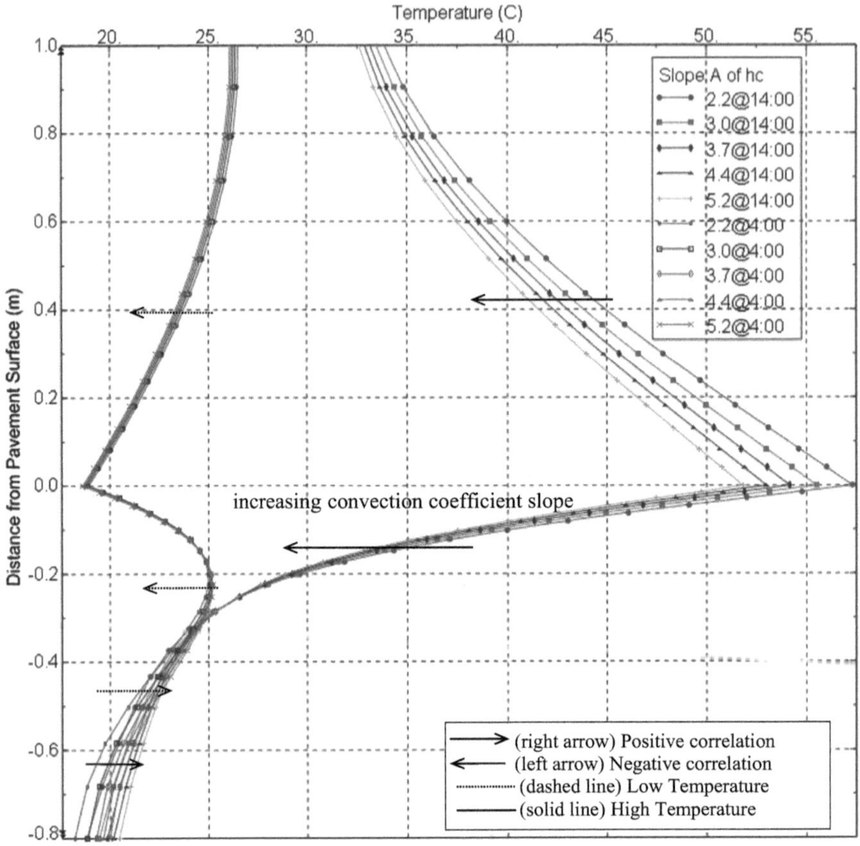

Figure 12.8 Temperature versus convection coefficient slope.

the low temperature of the near-surface air (4:00 h) also does not change much, by only about 2 °C. Reducing the near-surface air temperature, by reducing solar absorptivity with high reflectivity pavement materials, would probably contribute to mitigating the urban heat island effect, thus potentially reducing energy consumption for cooling by people living and driving in urban areas during summer and increasing the thermal comfort of people walking/driving on and/or living near pavements in hot seasons and a hot climate. However, on the side of heat flux of radiation, the reflected radiation might hit the building surfaces and human bodies and be absorbed by these surfaces. This will increase the chance of causing negative impacts of the high-reflectivity pavement materials in terms of human comfort and building energy use, as mentioned previously in the experiments.

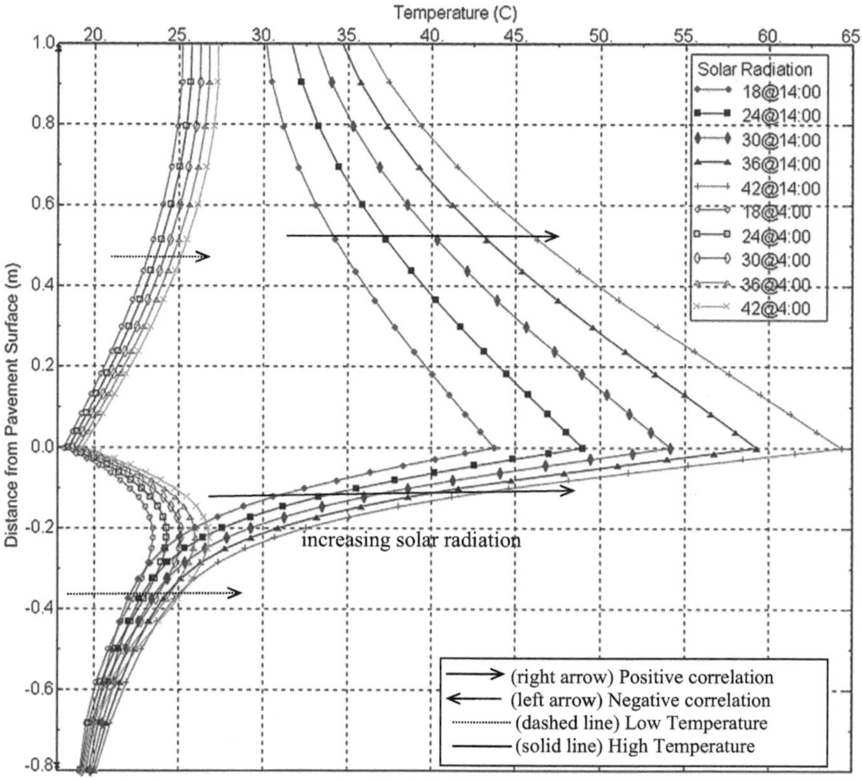

Figure 12.9 Temperature versus solar radiation.

12.3.5 Thermal Emissivity

The temperature distributions of pavements with various thermal emissivities and the corresponding near-surface air temperatures are shown in Figure 12.7.

Higher thermal emissivity results in lower temperatures in both the pavement (−0 in Figure 12.7) and the near-surface air (+0 in Figure 12.7), with more heat energy emitted/radiated to the near-surface air and then to the whole ambient atmosphere. Similar to solar absorptivity, thermal emissivity has a greater effect on the pavement surface temperature than on the near-surface air temperature. An increase in thermal emissivity from 0.5 to 0.9 causes the highest temperature of pavement surface (at 14:00 h) to drop by about 4 °C (from 57 to 53 °C) and the high near-surface air temperature at 0.5 m height to drop by about 2 °C (from 42 to 40 °C).

Although the effect of thermal emissivity is not as marked as that of solar absorptivity, using materials with high thermal emissivity can still reduce the temperatures of both the pavement surface and the near-surface air, resulting in cooler pavement and bringing all the benefits of it. However, increased thermal emissivity also reduces the nighttime low temperatures (at 4:00 h) of both pavement and near-surface air. This might have a negative effect on building energy use, thermal comfort, and pavement life during nighttime or winter time and in cold regions. In addition, effectively changing the thermal emissivity might be difficult for pavement materials. Nano-materials might provide the opportunity to modify the thermal emissivity of surface materials.

12.3.6 Convection Coefficient–Slope *A*

The heat convection coefficient, the most important parameter for the convection process, is dependent on not only the wind velocity (the speed of air convection near the pavement surface) but also the pavement surface characteristics (roughness) [189,190]. Faster wind velocities correspond to larger convection coefficients and lead to a reduction in the temperatures of both the pavement surface and the near-surface air during summer. Rough surfaces have a higher convection coefficient than smooth ones under constant wind velocity, owing to more effective surface area exposed to air and the enhanced air turbulence over the pavement [3].

Therefore, the convection coefficients (given by slope *A* in Table 12.2) were used to simulate the effect of convection on the temperatures of pavement and near-surface air. The results are shown in Figure 12.8.

Higher convection coefficients imply lower pavement temperatures and near-surface air temperatures. This is because more heat energy is transferred to the ambient atmosphere through convection as air moves over the hot pavement. The daytime high temperature of the pavement surface drops approximately by 5 °C (from 57 to 52 °C) as the slope of the convection coefficient increases from 2.2 to 5.2, compared with a drop of about 3 °C (from 42 to 39 °C) for the near-surface air temperature at 0.5 m height.

Therefore, pavements can be built with rougher surfaces, such as permeable pavements, chip seals, etc., to increase their effective surface area and create air turbulence over the pavement, thus reducing the temperatures of both pavement and near-surface air. However, "while this roughness can increase convection and cooling, it may also reduce a surface's net solar reflectance" [3]. More empirical and modeling studies are needed to investigate and confirm the detailed effects of pavement surface roughness on heat convection and pavement temperature.

12.3.7 Wind Speed

Wind speed shows an effect on the temperature near the pavement surface that is similar to that of slope A of the convection coefficient. Increased wind speed significantly reduces the daytime high temperature of near-surface air almost without impact on the nighttime low temperature. Also, increased wind speed enhances convection between the pavement and the air, which can moderate the pavement temperature. A reduced daytime high temperature and increased nighttime low temperature in pavement can be explained by this mechanism. Therefore, enhancing natural ventilation, especially in high-density urban areas, could improve the thermal condition of pavements and near-surface air.

12.3.8 Solar Radiation

As shown in Figure 12.9, solar radiation has a significant impact on the temperature of both the pavement and the near-surface air, especially during the daytime high temperature period. When total solar radiation decreases by about 50% (from 42 to 18), the high temperature on the pavement surface decreases by about 20 °C (from 64 to 44 °C). However, the nighttime lowest temperatures do not decrease much. This suggests that shading of some solar radiation using trees, canopies (e.g., photovoltaic panels) [65], or buildings could help to significantly reduce the daytime high temperature of both pavements and the near-surface air, potentially without impact on nighttime low temperatures.

12.3.9 Evaporation

Evaporation of water can consume some solar radiation/heat absorbed by the pavement through latent heat loss, which is insensible (unable to be perceived by the senses), unlike the sensible reflected and emitted radiation loss. The insensible latent heat loss will not influence the energy balance of the surroundings (e.g., building surface or human bodies, etc.) while reducing the effective solar radiation absorbed by the pavement and consequently reducing the pavement temperature. Therefore, enhancing evaporation from the pavement will influence the pavement thermal behavior and consequently effectively improve the thermal environment. However, as indicated by the findings from evaporation rate and field measurement of temperature presented in Chapters 7 and 8, to effectively reduce the pavement temperature through evaporative cooling, moisture needs to be available on or just below the pavement surface. Otherwise, the evaporation rate will be very low and the consequent effect of evaporative cooling will be minor.

12.3.10 Summary of Sensitivity Analysis

A sensitivity analysis based on simulation was conducted on some variables. Findings on the specific effects of some factors are listed in Table 12.3. The table shows complex correlations between these factors and temperatures of pavement and near-surface air. Findings from the sensitivity analyses based on the simulation indicate that temperatures are highly sensitive to solar reflectivity (1 − absorptivity) and solar radiation and shading. Temperatures show relatively high sensitivity to wind speed, convection coefficient, evaporation, thermal conductivity, and thermal emissivity. Temperatures present low sensitivity to specific heat and density. This implies that increasing the solar reflectivity and reducing the incident solar radiation (e.g., shading, etc.) could be effective ways to reduce the pavement temperature. Other strategies, such as enhancing ventilation and evaporation and using thermal-resistance materials, also can help reduce the temperature to some extent. In addition, the potential benefits provided by these potential cool pavement strategies might vary with climate. Different climate regions might need different strategies for heat island mitigation.

12.4 SUMMARY AND CONCLUSIONS

The validated model was used to simulate temperature for a typical pavement structure under typical climate conditions in summer (July) in Sacramento, California, a location with high summertime temperatures. Sensitivity analysis based on the simulation was conducted on some variables. Findings on the specific effects of some factors show complex correlations between these factors and pavement and near-surface air temperatures.

Findings obtained from the sensitivity analysis indicate that temperatures are highly sensitive to solar reflectivity (1 − absorptivity) and solar radiation and shading. Temperatures show relatively high sensitivity to wind speed, convection coefficient, evaporation, thermal conductivity, and thermal emissivity. Temperatures present low sensitivity to specific heat and density. This implies that increasing solar reflectivity and reducing incident solar radiation (e.g., shading, etc.) could be effective ways to reduce pavement temperatures. However, application of high-reflectivity pavement materials might increase the chance of causing negative impacts in terms of human

Table 12.3 Effect of factors on temperature of pavement and near-surface air (summer in hot region)

	Pavement surface and deeper layers		Near-surface air		
Factors	**Daytime high temperature**	**Nighttime low temperature**	**Daytime high temperature**	**Nighttime low temperature**	**Order of sensitivity**
Solar absorptivity[a]	+	+	+	+	Very high
Shading	−	−	−	−	Very high
Solar radiation	+	+	+	+	Very high
Evaporation rate	−	−	−	−	High
Thermal conductivity	−/+*	+/−*	−	+	High
Wind velocity	−/+*	−/+*	−	−	High
Convection coefficient	−/+*	−/+*	−	−	High
Thermal emissivity	−	−	−	−	High
Density	−	+/−*	−	+	Low
Specific heat	−	+/−*	−	+	Low

Note: + denotes positive correlation; − denotes negative correlation; *−/+ denotes a correlation for upper layer and lower layer, respectively.
[a]Solar absorptivity = 1 − solar reflectivity.

thermal comfort and building energy use owing to the reflected solar radiation, which will be explored in the following chapters.

According to the identified significant factors affecting the pavement thermal performance, potentially effective strategies include technologies such as high-reflectance and thermal-resistance pavement materials with low thermal conductivity, i.e., they conduct heat very poorly and leave the heat only near the surface, e.g., porous materials for permeable pavements, and management strategies such as shading, ventilation, and evaporation, and the combined use of these technologies and management strategies. The effectiveness these strategies at improving the thermal environment at the street level when applied to a specific context can be evaluated using the model developed and validated previously in this study.

It should also be noted that while the technologies (e.g., modifying thermal variables, etc.) that cool down the pavement and near-surface air potentially provide a positive effect on pavement life, building energy use, and human thermal comfort during hot times (especially in daytime during hot summers) and in hot climates, they might also lead to negative effects in terms of the same factors during cold seasons (e.g., nighttime in the winter) or in cold climates. Therefore, comprehensive consideration of both positive and negative effects of cool pavements should be given to ensure that the overall net benefits are positive. These will be addressed in the following chapters with respect to human thermal comfort and building energy use.

Any effects on the larger scale of the microclimate of cities are outside the scope of this work, but should be considered before making any decision based on the results presented in this dissertation, which is focused on near-surface effects of individual paved areas.

CHAPTER 13

Impacts of Pavement Strategies on Human Thermal Comfort

Contents

The objectives of this chapter are to investigate the impacts of various pavement technologies and management strategies on human outdoor thermal comfort.

Some important concepts will be first introduced for a better understanding of human thermal comfort, such as mean radiant temperature (MRT), thermal comfort index, and energy balance of human body.

Pavement Materials for Heat Island Mitigation
ISBN 978-0-12-803476-7
http://dx.doi.org/10.1016/B978-0-12-803476-7.00013-1

13.1 MEAN RADIANT TEMPERATURE

13.1.1 Definition

The MRT is defined as the uniform temperature of an imaginary enclosure (or environment) in which the radiant heat transfer from the human body is equal to the radiant heat transfer in the actual nonuniform enclosure (or environment) [191,192].

MRT is the most important parameter governing human energy balance, especially on hot sunny days. MRT also has a strong influence on thermophysiological comfort indexes such as physiological equivalent temperature (PET) or predicted mean vote (PMV) (discussed in Section 13.3) [191,192].

13.1.2 Difference between Mean Radiant Temperature and Air Temperature

Radiant heat can play a significant role in achieving thermal comfort, promoting a healthier environment and lowering building heating costs in the winter. To understand how radiant heat makes these benefits possible we must first clarify the difference between ambient air temperature and the mean radiant temperature. The ambient air temperature is a measure of the average air temperature in the environment, while the mean radiant temperature is a measure of the net radiant heat gain and loss in the environment.

Most people are familiar with how fluctuations in air temperature affect their perception of comfort, but relatively few people are conscious of how their comfort is affected by mean radiant temperature differences. Living human skin has extraordinarily high emissivity and absorptivity, making it very sensitive to radiant heat loss and gain. Radiant heat loss or gain is not dependent on the air temperature. For example, when you stand in the sun on a cold winter day you feel radiant heat gain from the sun even though the air temperature is cold. On the other hand, when you open the freezer door on a hot summer day you feel radiant heat loss to the interior of the freezer. In both cases your thermal comfort is being largely affected by difference in radiant heat gain or loss, not air temperature.

Further, it is important to understand that you can experience heat gain from a warm surface and heat loss to a cold surface at the same time. In other words, your skin can simultaneously absorb heat from a warm surface and emit heat to a cold surface. Depending on the strength of the radiation of heat from the warm surface to your skin and the absorption or heat from your skin to the cold surface you will feel either a net gain or a net loss of heat energy. The mean radiant temperature measures this combined net radiant heat loss and gain [191,192].

13.1.3 Calculation of Mean Radiant Temperature

To calculate the MRT (T_{mrt}, in °C), the relevant properties and dimensions of the radiating surfaces and the sky view factors as well as the posture of the human body (e.g., seated or standing, etc.) need to be known. The entire surroundings of the human body are divided into n thermal surfaces with the temperatures T_i (in °C) and emissivity ε_i, to which the view factors VF_i are to be assigned as weighting factors ($i = 1, 2, 3, \ldots, n$) [191,192]:

$$T_{mrt} = \left[\frac{1}{\sigma}\sum_{i=1}^{n}\left(E_i + \alpha_{hb}\frac{D_i}{\varepsilon_{hb}}\right)VF_i + F_{hb}\alpha_{hb}\frac{SVF_{hb}I}{\sigma\varepsilon_{hb}}\right]^{0.25} - 273 \quad (13.1)$$

where

E_i is the emitted long-wave radiation from each surface, which can be calculated as (σ is the Stefan–Boltzmann constant, 5.67×10^{-8} W/m^2 K^4)

$$E_i = \varepsilon_i\sigma(T_i + 273)^4 \quad (13.2)$$

D_i is the diffuse short-wave radiation from each surface, mainly the diffusely reflected global radiation. It can be calculated as (I is the total global radiation; SVF_i is the sky view factor of surface i; α_i is the absorptivity of the surface i for the short-wave radiation)

$$D_i = (1 - \alpha_i)SVF_iI \quad (13.3)$$

α_{hb} is the absorptivity of the human body surface for the short-wave radiation (standard value 0.7); ε_{hb} is the emissivity of the human body surface (standard value 0.97); and SVF_{hb} is the sky view factor of human body. F_{hb} is the human body surface projection factor, which is a function of the incident radiation direction and the body posture. For practical application in human biometeorology, it is generally sufficient to determine F_{hb} for a rotationally symmetric person standing up or walking. F_{hb} ranges from 0.308 for 0° of the solar angle and 0.082 for 90° [191,192].

More details associated with determining the view factors VF_i are discussed in ASHRAE Fundamentals 2001 and radiation literature. In the case of large flat surfaces without any restriction of the horizon, for instance, a large paved parking lot, the problem of determining VF_i is reduced to an upper and a lower hemisphere with a view factor of 0.5 for each [191,192].

13.2 SHADING

Shading, produced by trees or walls, will block solar radiation from hitting some surfaces, including paved surfaces and human body surfaces. The sky view factor (SVF) could be used to characterize shading. For a surface that is completely shaded, the $SVF = 0$; for a surface that is barely shaded, the $SVF = 1$. If some trees might partly allow solar radiation passing through, the SVF on the ground surface or human body under the trees could be somewhere between 0 and 1, e.g., 0.3.

13.3 THERMAL COMFORT INDEX

Generally, thermal comfort is influenced by a large number of factors, such as the surrounding thermal environment (temperature, humidity, radiation flux, air flow, etc.) and people's activity, clothing, and perception about how hot an area is. Assessing comfort outdoors is not simple, because of the complexity and methodological differences observed in the related literature, which make any comparison with available results difficult. Generally, comfort can be assessed by means of comfort indices. There are a large number of indices referred to in the literature, such as PMV, index of thermal stress (ITS), perceived temperature (PT), operative temperature (OP), MRT, standard effective temperature (SET), and PET (e.g., [89–103]).

Compared to other thermal comfort indices such as ITS, PT, and OP, which are more suitable for indoor thermal environments, the SET and PET are more rational for evaluating outdoor thermal environments. Both SET and PET are based on the human body energy balance, for which a two-node model (skin and core nodes) is used to model the thermal conditions of the human body in a physiologically relevant way. Both SET and PET are defined as the air temperature at which, in a typical indoor setting, the heat budget of the human body is balanced with the same core and skin temperatures as under the complex outdoor conditions to be evaluated. To better understand these two indices, human body energy balance modeling should be first introduced and understood.

13.4 HUMAN BODY ENERGY BALANCE MODELING

The heat exchange between the human body and the thermal environment is illustrated in Figure 13.1. The total energy gain or loss of the human body can be described by the following heat balance equation [191,192]:

$$M + W + C + R + E_{sw} + C_{res} + E_{res} = S \tag{13.4}$$

M is the metabolic rate (W/m^2), W is the rate of mechanical work (W/m^2), C is the sensible heat gain or loss by convection (W/m^2), R is the sensible heat gain or loss by emitted radiation (W/m^2), E_{sw} is the total evaporative latent heat loss from the skin by sweat (W/m^2), C_{res} is the convective heat gain or loss by respiration (W/m^2), E_{res} is the evaporative heat loss by respiration (W/m^2), and S (W/m^2) is the total storage of heat flow in the body. All the terms are positive (body heating) when heat is produced in the body and gained from the environment and negative (body cooling) when heat is lost to the environment.

In the heat balance equation, the storage of heat flow in the body, S, means body heating when positive and cooling when negative. When S is

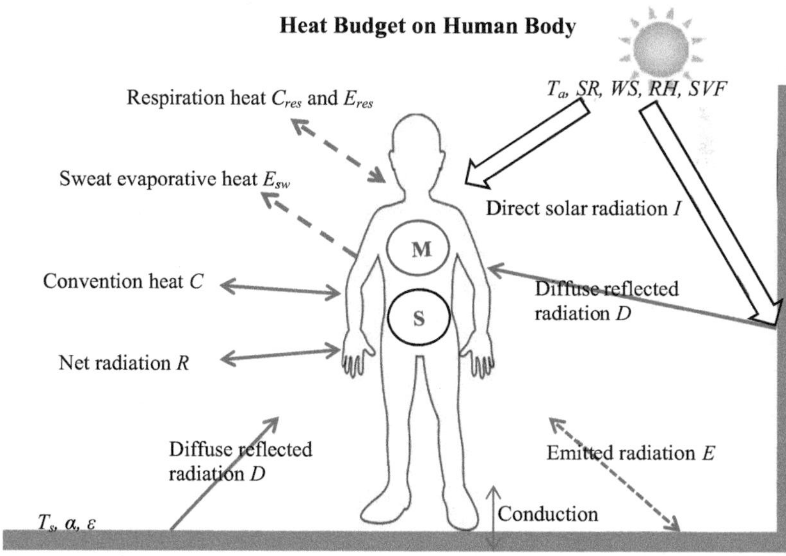

Figure 13.1 Illustration of heat budget on the human body. T_s, α, and ε, are temperature, albedo, and emissivity of pavement or other vertical surfaces, respectively. T_a, *SR*, *WS*, *RH*, and *SVF* are air temperature, total solar radiation, wind speed, relative humidity, and sky view factor, respectively.

equal to 0 the amount of heat produced in the body and gained from the environment is the same as that lost to the environment, and the body temperature is in a steady state.

For the two-node model, the energy balance at the core and skin nodes may be written as follows:

$$S = S_{cr} + S_{sk} \tag{13.5}$$

$$S_{cr} = M + W + (C_{res} + E_{res}) - H_{c-s} \tag{13.6}$$

$$S_{sk} = H_{c-s} + (C + R + E_{sk}) \tag{13.7}$$

$$S_{cr} = \frac{m_{cr}c}{A_{hb}}\frac{dT_{cr}}{dt} \tag{13.8}$$

$$S_{sk} = \frac{m_{sk}c}{A_{hb}}\frac{dT_{sk}}{dt} \tag{13.9}$$

where S_{cr} is the net heat flow to (positive, heating) or from (negative, cooling) the core (in W/m^2); S_{sk} is the net heat flow to (positive, heating) or from (negative, cooling) the skin (in W/m^2); m_{cr} and m_{sk} are the masses of the body core and the skin, respectively (in kg); c is the body heat capacity (in J/kg °C); A_{hb} is the body surface area (in m^2); T_{cr} and T_{sk} are the transient temperatures of the body core and the skin, respectively (in °C); and t is the exposure time for assessment of the human body in the thermal environment (in s).

13.4.1 The Mass and Body Surface Area

The mass at the core and skin nodes is calculated using the fraction of the total mass, also called the effective shell thickness (α), following Gagge et al. [191,192]:

$$m_{co} = (1 - \alpha)Wt \tag{13.10}$$

$$m_{sk} = \alpha Wt \tag{13.11}$$

where Wt is the total body weight (kg).

The body surface area is calculated from DuBois [191,192]:

$$A_{hb} = 0.203Ht^{0.725}Wt^{0.425} \tag{13.12}$$

where Ht is the body height (m).

13.4.2 Thermal Signals of the Human Body

The thermoregulation processes are controlled through feedback thermal signals by deviations in the skin, core, and body temperatures. The thermal signals for the skin (TS_{sk}, in °C), core (TS_{cr} in °C), and body (TS_{hb} in °C) are calculated following Gagge et al.:

$$TS_{sk} = T_{sk} - 34.1 \tag{13.13}$$

$$TS_{cr} = T_{cr} - 36.6 \tag{13.14}$$

$$TS_{hb} = \alpha TS_{sk} + (1 - \alpha) TS_{cr} \tag{13.15}$$

The numeric values (34.1 and 36.6) in the above equations are the set points for the skin and core nodes in the two-node model. The thermal signals from the skin and core are either warm (positive) or cold (negative) signals to control vasodilation, vasoconstriction, blood flow, and shivering, while the body thermal signal is used only for warmth to regulate sweating along with the skin thermal signal for warmth.

13.4.3 Skin Blood Flow

The skin blood flow (L/m^2 h) is calculated following Stolwijk and Hardy:

$$v_{bl} = (6.3 + 75 TS_{cr})/(1 - 0.5 TS_{sk}) \tag{13.16}$$

The skin fraction of the total body mass (α) is a factor that increases with vasoconstriction and decreases with vasodilation. It is calculated in the two-node model following Gagge et al. [192]:

$$\alpha = 0.0418 + \frac{0.7425}{v_{bl} + 0.5854} \tag{13.17}$$

13.4.4 Sensible Heat Exchange from the Skin Surface

The heat exchange by convection (C, W/m^2) and radiation (R, W/m^2) from the clothed body are calculated from the following equations:

$$C = F_{cl} h_c (T_a - T_{cl}) \tag{13.18}$$

$$R = F_{cl} h_r (T_{\text{mrt}} - T_{cl}) \tag{13.19}$$

where h_c is the convective heat transfer coefficient (W/m^2 °C), T_{cl} is the clothed body surface temperature (°C), T_a is the air temperature (°C), h_r is the radiant heat transfer coefficient (W/m^2 °C), T_{mrt} is the mean radiant temperature (°C), and F_{cl} is the clothing surface area increase factor from body surface area.

The clothing surface area factor F_{cl} is calculated in the model from (ISO 7730:2005):

$$F_{cl} = \begin{cases} 1.00 + 1.290 I_{cl}, & \text{if } I_{cl} \leq 0.078 \text{ m}^2\text{K/W} \\ 1.05 + 0.645 I_{cl}, & \text{if } I_{cl} > 0.078 \text{ m}^2\text{K/W} \end{cases} \tag{13.20}$$

where I_{cl} is the clothing insulation, in m^2 K/W (1 clo = 0.155 m^2 K/W).

The convective heat transfer coefficient is calculated as follows:

$$h_c = \begin{cases} 3.3, & \text{if } v_w \leq 0.1 \text{ m/s} \\ 8.6 v_w^{0.53}, & \text{if } v_w > 0.1 \text{ m/s} \end{cases} \tag{13.21}$$

where v_w is relative wind speed (or air velocity), in m/s.

The radiant heat transfer coefficient is calculated as follows:

$$h_r = r_{er} \varepsilon_{hb} \sigma \frac{(T_{mrt} + 273)^4 - (T_{cl} + 273)^4}{T_{mrt} - T_{cl}} \tag{13.22}$$

where r_{er} is the ratio of the body's effective radiation area to the whole body area, which is 0.73 for standing, 0.7 for sitting, and 0.9 for lying down. Other variables are defined previously.

13.4.5 Latent Heat Exchange from the Skin Surface by Sweat

Heat exchange by evaporation at the skin surface is mainly through the sweat secretion process and controlled by the feedback thermal signal for warmth from the skin and body nodes. The sweat evaporation heat exchange (W/m^2) is calculated as follows:

$$E_{sw} = w h_e (P_a - P_{sk,s}) \tag{13.23}$$

where w is the total skin wetness including wetness due to regulatory sweating (w_{sw}) and diffusion through the skin (w_{dif}):

$$w = w_{sw} + w_{dif} = w_{sw} + 0.06(1 - w_{sw}) = 0.06 + 0.94 w_{sw} \tag{13.24}$$

h_e is the evaporative heat transfer coefficient (W/m^2 °C). It is calculated from:

$$h_e = F_{pcl} \kappa h_c \tag{13.25}$$

where κ is the Lewis ratio and equals 2.2 °C/mm Hg at sea level. h_c is the convective heat transfer coefficient as defined previously. F_{pcl} is the permeation efficiency factor for water vapor evaporated from the skin surface through clothing to the ambient air and is calculated as:

$$F_{pcl} = \frac{1}{(1 + 0.92 I_{cl} h_c)} \tag{13.26}$$

P_a (mm Hg) is the ambient vapor pressure, which is calculated as:

$$P_a = P_{a,s} \cdot RH \tag{13.27}$$

where RH is the relative humidity of ambient air. $P_{a,s}$ (mm Hg) is the saturated vapor pressure at ambient air temperature T_a and is calculated as:

$$P_{a,s} = e^{\left(18.67 - \frac{4030.18}{T_a + 235}\right)} \tag{13.28}$$

$P_{sk,s}$ (mm Hg) is the saturated vapor pressure at mean skin temperature T_{sk} and is calculated as:

$$P_{a,s} = e^{\left(18.67 - \frac{4030.18}{T_{sk} + 235}\right)} \tag{13.29}$$

The regulatory sweat signal or sweat rate (R_{sw}, g/m^2 s) is calculated through body and skin signals for warmth by the following equation [191]:

$$R_{sw} = 0.047\, TS_{hb} e^{(TS_{sk}/10.7)} \tag{13.30}$$

where the body and skin thermal signals are set to zero if negative.

The regulatory signal is then used to estimate the wetness due to regulatory sweat (w_{sw}) as follows [192]:

$$w_{sw} = \min(L \cdot R_{sw} / E_{sk,\max}, 1) \tag{13.31}$$

where L is the latent heat of water evaporation, 2260 J/g. $E_{sk,\max}$ (W/m^2) is the maximum evaporation heat loss when $w = 1$ in Eq. (13.23).

13.4.6 Heat Exchange through Respiration

The dry and latent heat exchange (W/m^2) through respiration are estimated as follows [191]:

$$C_{res} = 0.0014M(T_a - 34) \tag{13.32}$$

$$E_{res} = 0.0023M(P_a - 44) \tag{13.33}$$

13.4.7 Heat Transfer from Core to Skin

The heat transfer from core to skin node, including conduction through tissue and blood flow, is estimated as follows [192]:

$$H_{c-s} = (K + c_b v_{bl})(T_{cr} - T_{sk}) \tag{13.34}$$

where K is the conductance of body tissues (5.28 W/m^2 °C) and c_b is the blood thermal capacity (1.163 J/L °C).

13.4.8 Metabolic Rate

By the oxidation of the constituents of food (carbohydrates, fat, or proteins), energy is transformed into heat in the body. The metabolic heat production rate (M) is primarily dependent on physical activity. Some examples of metabolic rates are listed in Table 13.1 (ISO 8996).

With the system of the equations above, all relevant heat fluxes and thermophysiological body parameters can be calculated for any given climatic condition considering all relevant meteorological parameters. Also, comfort indices, such as SET and PET, can be calculated from these models for steady state or dynamic state.

The ASHRAE SET index is defined as the equivalent air temperature of an isothermal environment at 50% relative humidity in which a subject, while wearing standardized clothing for the activity concerned, would have the same heat stress (skin temperature T_{sk}) and thermoregulatory strain (skin wetness, w) as in the actual environment being evaluated. The isothermal environment refers to the environment at sea level in which the air temperature is equal to the mean radiant temperature and the air velocity is zero.

The PET was developed by a German research group headed by Peter Hoppe and is recommended as a thermal index by the German Association of Engineer's VDI guidelines (German guidelines for urban and regional planners). PET is defined as the equivalent air temperature at which, in a typical indoor setting ($T_{mrt} = T_a$; $VP = 12$ hPa; $v = 0.1$ m/s), the heat balance of the human body is maintained with core and skin temperatures equal to those under the actual complex conditions being assessed.

Both SET and PET have a thermophysiological background and therefore they give the real effect of the sensation of climate on a human

Table 13.1 Metabolic rates for various activities

Activity	Metabolic rate (met)	Metabolic rate (W/m^2)
Reclining	0.8	46
Seated, relaxed	1.0	58
Standing, light activity	1.6	93
Standing, medium activity	2.0	116
Walking on level ground, 2 km/h	1.9	110
Walking on level ground, 3 km/h	2.4	140
Walking on level ground, 5 km/h	3.4	200

1 metabolic unit = 1 met = 58 W/m^2.

being. Moreover, they both have the unit degree Celsius (°C) and can therefore be more easily related to common experience and interpreted by planners for design purposes. Therefore, both SET and PET can be used to evaluate the outdoor thermal environment. For this study, the PET is selected as the main thermal index to evaluate the effects of various pavement technologies and management strategies on the outdoor thermal environment. The thermal comfort model and thermal index are implemented and calculated using *R* programming.

13.5 EXAMPLE CALCULATION OF PHYSIOLOGICAL EQUIVALENT TEMPERATURE

To demonstrate the heat balance model of a human body, the heat flux, body temperature, and PET were calculated for a typical hot outdoor condition and illustrated in Figure 13.2. In this case, an adult of 1.80 m height, of 75 kg weight, and wearing light clothes of 0.5 clo (1 clo = 0.155 m^2 K/W), is walking at the speed of 2 km/h on a street paved with black asphalt and without shading. The weather conditions are $T_a = 38$ °C, $RH = 50\%$, and $v_w = 0.5$ m/s. The mean radiant temperature (T_{mrt}) of the surroundings is 55 °C, which is very hot. The clothing temperature T_{cl} is up to 41.85 °C, the mean skin temperature T_{sk} is

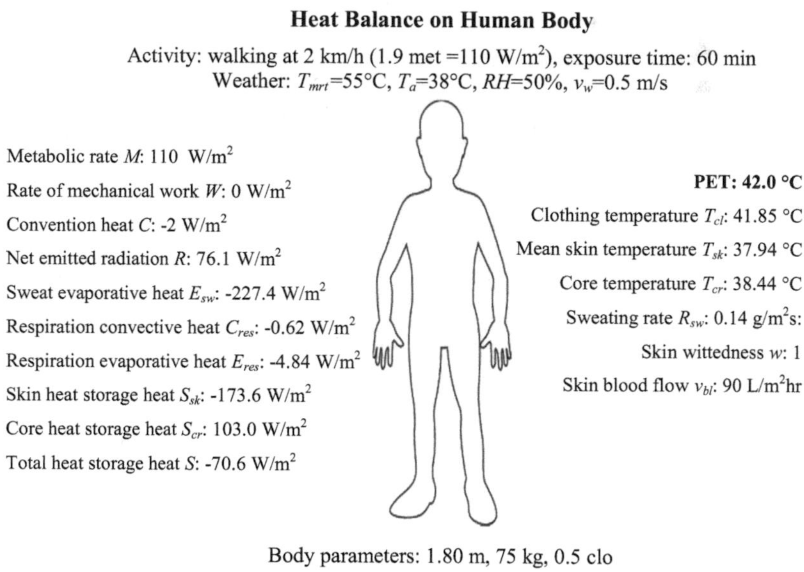

Figure 13.2 Illustration of energy balance model on a human body.

37.94 °C, and the core temperature T_{cr} is 38.44 °C. The sweating rate R_{sw} is estimated at 0.14 g/m^2 s. The PET the adult feels is 42.0 °C for this assumed context, which is very hot.

The thermal sensation or subjective thermal perceptions are related to PETs. Previous studies (e.g., [193–196]) indicated that occupant thermal sensations and preferences vary for various regions owing to the differences in behavioral adjustment, physiological acclimatization, and psychological habituation or expectations. This may lead to different thermal comfort ranges. Table 13.2 lists the thermal sensation classification for various regions. The PET ranges for Taiwan and western/middle European countries are from some survey studies conducted in these regions [194–196]. The comparison of the thermal sensation classifications for these two regions demonstrates that the thermal comfort range in western/middle European countries (18–23 °C PET) is lower than that of Taiwan (26–30 °C PET). People living in western/middle European countries cannot tolerate temperatures as high as those tolerated by people living in Taiwan, with a hot climate. The climate in Davis, California, is not as hot as that in Taiwan but is hotter than that in western/middle European countries. Therefore, the thermal sensation classification is estimated for Davis, California, based on those for Taiwan and western/middle European countries, as listed in Table 13.2. A more accurate local thermal sensation classification for Davis,

Table 13.2 Thermal sensation classifications for various regions

Thermal sensation	PMV	PET range for Taiwan[a] (°C PET)	PET range for western/middle Europe[b] (°C PET)	PET range for Davis, CA (°C PET)
Very hot	>3.5	>42	>41	>40
Hot	2.5–3.5	38–42	35–41	35–40
Warm	1.5–2.5	34–38	29–35	30–35
Slightly warm	0.5–1.5	30–34	23–29	26–30
Neutral	−0.5 to 0.5	26–30	18–23	19–26
Slightly cool	−1.5 to −0.5	22–26	13–18	16–19
Cool	−2.5 to −1.5	18–22	8–13	12–16
Cold	−3.5 to −2.5	14–18	4–8	6–12
Very cold	<−3.5	<14	<4	<6

[a]Lin and Matzarakis [196].
[b]Matzarakis and Mayer [194,195].

California, could be obtained from an outdoor field study and survey (e.g., [196,197]) to establish the correlation between thermal sensation and PET for local occupants.

13.6 EVALUATION OF OUTDOOR THERMAL ENVIRONMENT USING PHYSIOLOGICAL EQUIVALENT TEMPERATURE

A large flat paved area (e.g., parking lot, playground, etc.) is used to evaluate and compare the effects of various pavement technologies and management strategies on the outdoor thermal environment for three climates (Sacramento and Los Angeles in California and Phoenix in Arizona). The pavement technologies and management strategies to be evaluated include high reflectance, evaporation, and both high reflectance and evaporation, as well as shading. The summer (July) and winter (January) climate data for those three locations are used for assessment.

13.6.1 Climate Data in the Three Chosen Regions

The air temperature data across a year for the three regions, namely Sacramento and Los Angeles in California and Phoenix in Arizona, are presented in Figure 13.3. The Sacramento and Phoenix areas have inland climates with large air temperature fluctuations across the seasons, whereas the Los Angeles region has a coastal climate with much smaller air temperature fluctuations across the seasons. Phoenix has a very hot climate during the summer. Los Angeles, with the coastal climate, is cool during summer. Sacramento has a cold winter compared to the other two regions. The typical summer (July) and winter (January) climate data for these three locations are listed in Table 13.3 and were used for assessment.

13.6.2 Pavement Strategies and Surface Temperatures

The scenarios of pavement technologies and management strategies to be evaluated include high reflectance, evaporation, and both high reflectance and evaporation, as well as shading. The parameters for each pavement scenario are listed in Table 13.4.

For the assumed large flat paved area in an open area, the surroundings of a human body include the pavement ground and the sky with the view factor VF of 0.5 for each (see Eq. (13.1)). The first critical step is to estimate the mean radiant temperature (T_{mrt}) for this surrounding. The pavement surface temperature plays an important role in estimating the MRT.

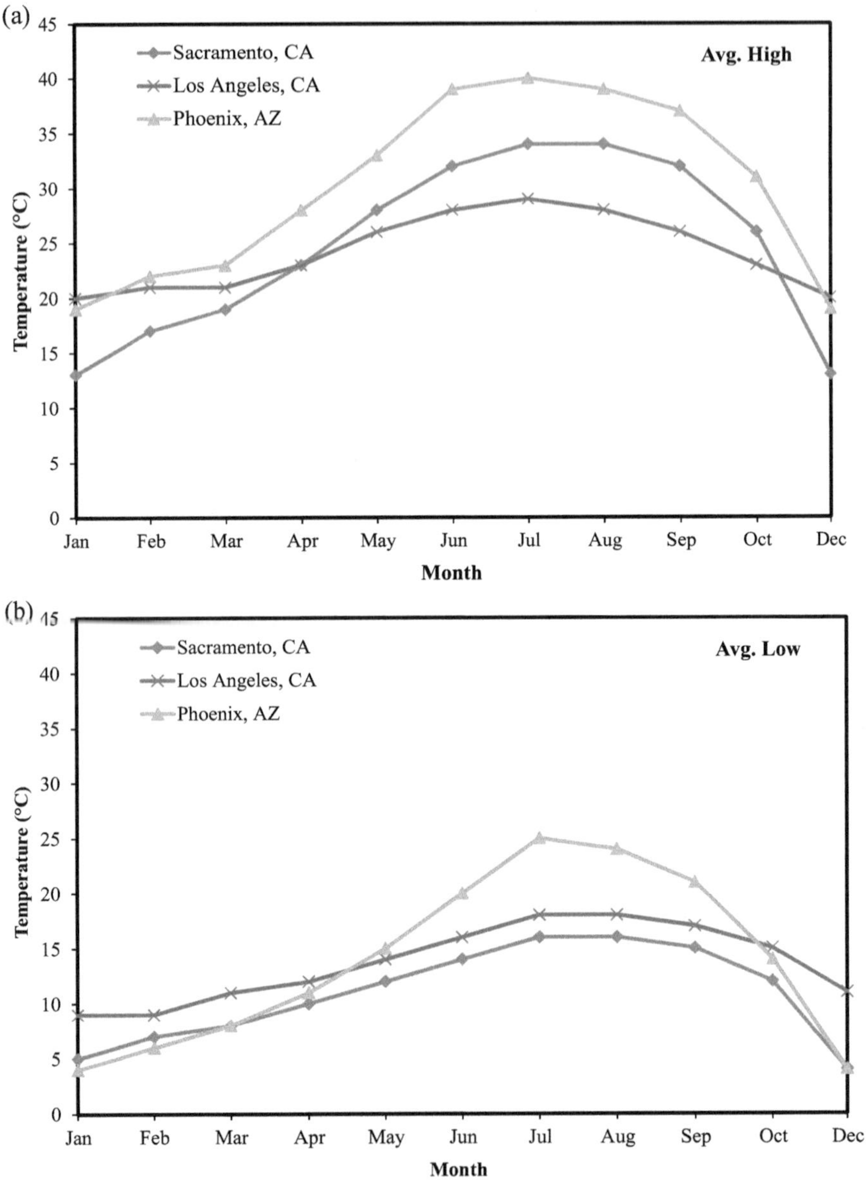

Figure 13.3 Comparison of the climate data in three regions: Sacramento and Los Angeles in California and Phoenix in Arizona. (a) Average high air temperature. (b) Average low air temperature.

Table 13.3 Typical summer and winter climate data for three regions

Season	Daily peak air temperature T_a^{max} (°C)[a]	Daily low air temperature T_a^{min} (°C)[a]	Daily total solar radiation volume Q (MJ/m^2)[b]	Daily effective sunlight hour c (h)[b]	Daily average wind velocity v_w (m/s)[c]
Sacramento, California					
Summer (July)	34	16	28.3	11	4.0
Winter (January)	13	5	6.3	8	3.2
Los Angeles, California					
Summer (July)	29	18	22.6	10	2.8
Winter (January)	20	9	9.7	8	2.2
Phoenix, Arizona					
Summer (July)	40	25	27.4	11	3.2
Winter (January)	19	4	11.4	9	2.4

Note: data obtained from:
[a]http://www.weather.com/weather/wxclimatology.
[b]http://rredc.nrel.gov/solar/old_data/nsrdb.
[c]http://rredc.nrel.gov/solar/old_data/nsrdb/1991-2005/hourly/list_by_state.html#C.

Table 13.4 Pavement scenarios used for analysis

Parameter	Pavement scenario: Baseline	High reflectance	Evaporation	High reflectance + evaporation	Shading
r	0.1	<u>0.5</u>	0.1	<u>0.5</u>	0.1
ER (mm/h)	0.00	0.00	<u>1.0</u>	<u>1.0</u>	0.00
SVF	1	1	1	1	<u>0</u>

Changed parameter is underlined for each scenario. r, solar reflectivity; ER, evaporation rate; SVF, sky view factor.

To calculate the pavement surface temperatures, the simplified model developed and validated in Chapter 11 was used for the three different climates (Table 13.3) and five different pavement scenarios (Table 13.4). A typical pavement structure (10 cm surface +15 cm base) was used for analysis and simulation. The other parameters for estimating the pavement temperatures are listed in Table 13.5 and are approximately typical mid-range values (Table 6.1).

Example results of the pavement surface temperatures for the three climates are presented in Figure 13.4.

Table 13.5 Parameters for temperature simulation in Sacramento, California

Parameters	Thermal properties and typical parameter values: Air	Asphalt concrete	Aggregate base	Subgrade
Thermal conductivity k^a $(J\cdot(h\cdot m\cdot°C)^{-1})$	92.5	6228	5040	4680
Density ρ $(kg\cdot m^{-3})$	1.205	2257	1800	1700
Specific heat capacity c $(J\cdot(kg\cdot°C)^{-1})$	1006	852	920	860
Solar radiation absorptivity α_s	0.90			
Thermal emissivity ε	0.80			
Heat convection coefficient h_c^a $(J\cdot(h\cdot m^2\cdot°C)^{-1})$	$h_c = 3600(3.7v_w + 6.1)$ (v_w, wind velocity, m/s)			
Absolute zero T^Z (°C)	−273			
Stefan–Boltzmann constant σ^a $(J\cdot(h\cdot m^2\cdot K^4)^{-1})$	2.041092×10^{-4}			

[a]Time is measured in hours and the related parameters are converted for the simulation analysis at intervals of 1 h.

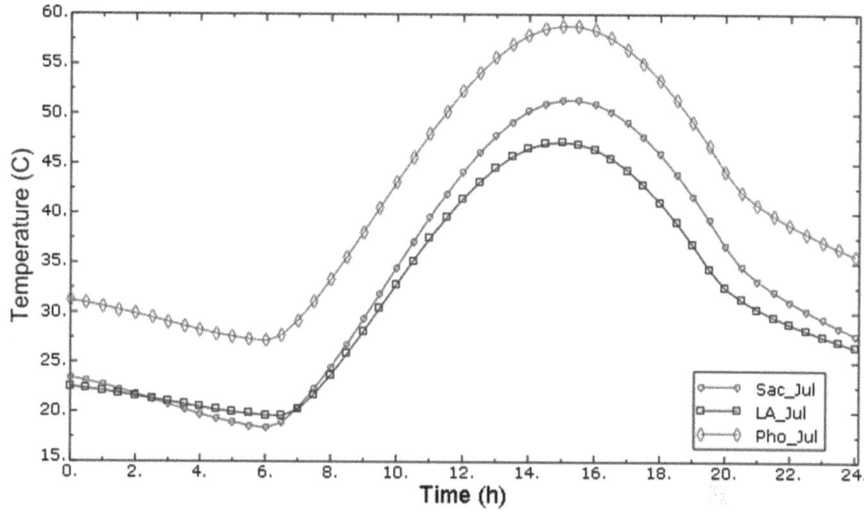

Figure 13.4 Example results of the pavement surface temperatures for the three climates (baseline, summer). Sac, Sacramento; LA, Los Angeles, Pho, Phoenix.

13.6.3 Mean Radiant Temperature and Physiological Equivalent Temperature

The maximum pavement surface temperatures were extracted from modeling and simulation of pavement temperatures to analyze the thermal environment along with the corresponding weather data. Then the information on the thermal environment was used to calculate the corresponding mean radiant temperature T_{mrt} and thermal comfort index PET using the thermal comfort model for the three locations for both summer and winter.

13.6.3.1 Results for Summer

The results for summer are summarized in Table 13.6 and illustrated in Figures 13.5 and 13.6. Using the high-reflectance pavement will definitely reduce the pavement surface temperatures (T_s) for all three locations. However, using the high-reflectance pavement would increase the MRT owing to the increased reflected radiation hitting the human body. The increased MRT would produce a higher PET compared to the baseline by 7.8, 8.4, and 5.3% for Sacramento and Los Angeles in California and Phoenix in Arizona, respectively. This indicates that using high-reflectance pavement will reduce the pavement surface temperature and consequently might help decelerate the formation of ground-level ozone

Table 13.6 Temperature results for various pavement scenarios in three regions (summer)

Pavement scenario	Parameter altered	New parameter value	Max. surface temperature T_s (°C)	Mean radiant temperature T_{mrt} (°C)	PET (°C)	PET Change from baseline (%)
Sacramento, California						
Baseline	—	—	51	47	37.0	0.0
High reflectance	*r*	0.5	42	58	39.9	7.8
Evaporation	*ER* (mm/h)	1.5	37	41	35.5	−4.1
High reflectance + evaporation	*R* + *ER* (mm/h)	0.5 + 1.5	34	55	39.2	5.9
Shading	*SVF*	0	31	16	28.5	−23.0
Los Angeles, California						
Baseline	—	—	47	40	32.4	0.0
High reflectance	*r*	0.5	38	50	35.1	8.3
Evaporation	*ER* (mm/h)	1.5	33	35	31.0	−4.3
High reflectance + evaporation	*R* + *ER* (mm/h)	0.5 + 1.5	30	47	34.3	5.9
Shading	*SVF*	0	27	12	23.5	−27.5
Phoenix, Arizona						
Baseline	—	—	59	52	42.4	0.0
High reflectance	*r*	0.5	49	62	44.7	5.4
Evaporation	*ER* (mm/h)	1.5	44	46	41.0	−3.3
High reflectance + evaporation	*R* + *ER* (mm/h)	0.5 + 1.5	40	58	43.8	3.3
Shading	*SVF*	0	37	21	35.3	−16.7

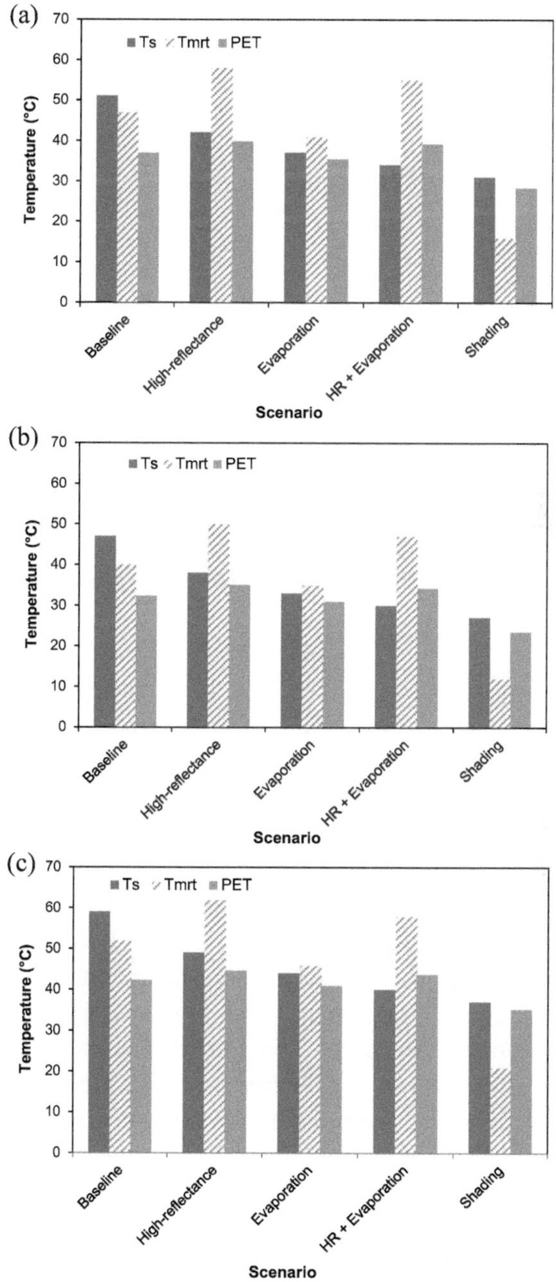

Figure 13.5 Calculated results for various pavement scenarios at three regions: pavement surface temperature T_s, mean radiant temperature T_{mrt}, and PET (summer). (a) Sacramento, CA. (b) Los Angeles, CA. (c) Phoenix, AZ.

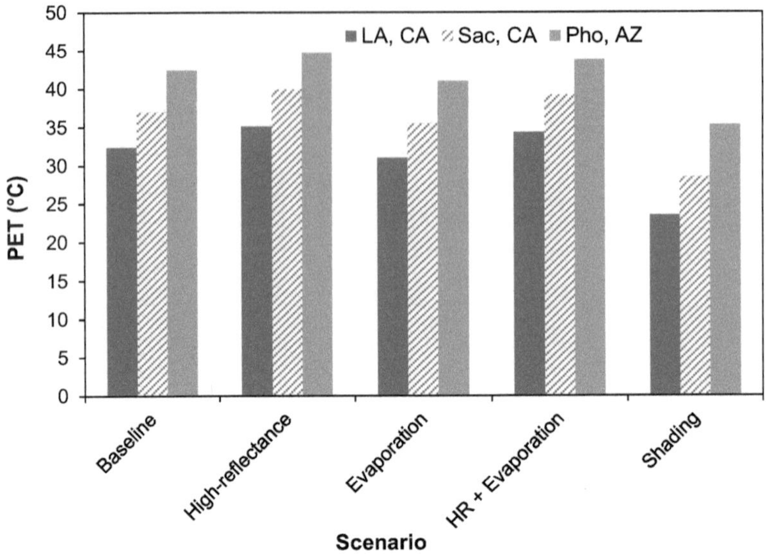

Figure 13.6 Comparison of PET for various pavement scenarios at three regions (summer).

and help improve the air quality; however, increasing the pavement reflectance will increase the risk of reducing human thermal comfort during hot periods by approximately 5–8% in the three locations analyzed in this study. Therefore, high-reflectance pavements should be applied only after very careful examination of the potential negative impacts of reflected radiation.

Enhancing the evaporation from pavement will help reduce both the pavement surface temperature (T_s) and the MRT (T_{mrt}) and also help reduce the PET and improve the human thermal comfort in hot periods by 4.1, 4.3, and 3.3% for Sacramento and Los Angeles in California and Phoenix in Arizona, respectively. This implies that enhancing the evaporation from pavement will be an effective way to reduce both the pavement surface and the near-surface air temperatures and the PET, helping improve human thermal comfort in hot periods by approximately 3–4% in the three locations analyzed in this study.

Owing to the increased solar reflectance, the high-reflectance plus evaporation strategy will also produce an increased MRT and a higher PET compared to the baseline by approximately 3–6% in the three locations analyzed in this study, although it will reduce the pavement surface temperature.

Shading alone will considerably reduce both the pavement surface temperature and the MRT and thus significantly reduce the PET and improve human thermal comfort in hot periods by 23, 27.5, and 16.7% for Sacramento and Los Angeles in California and Phoenix in Arizona, respectively. This indicates that enhancing the shading with trees or other ways over pavement will be a very effective strategy to reduce both the pavement surface temperature and the PET, helping to improve human thermal comfort in hot periods by approximately 17–28% in the three locations analyzed in this study.

13.6.3.2 Results for Winter

The results for winter are summarized in Table 13.7 and illustrated in Figures 13.7 and 13.8. As mentioned above, using the high-reflectance pavement will definitely reduce the pavement surface temperatures (T_s) for all three locations. However, using the high-reflectance pavement would increase the MRT (T_{mrt}) owing to the increased reflected radiation hitting the human body. This is not desired in hot summer but is desired in cold winter. The increased MRT would produce a higher PET compared to the baseline by 15, 10.4, and 12.6% for Sacramento and Los Angeles in California and Phoenix in Arizona, respectively. This indicates that using high-reflectance pavement will reduce the pavement surface temperature; however, increasing the pavement reflectance will increase the human thermal comfort during cold periods by approximately 10–15% in the three locations analyzed in this study. Therefore, high-reflectance pavements can be applied to cold regions to improve the human thermal comfort in winter.

Enhancing the evaporation from pavement will help reduce the PET and improve the human thermal comfort in hot summer periods as shown previously. However, pavement evaporation will also reduce the PET and lower the human thermal comfort in the already cold winter period by 8.7, 6.6, and 6.8% for Sacramento and Los Angeles in California and Phoenix in Arizona, respectively. This implies that enhancing the pavement evaporation might reduce human thermal comfort in the winter in cold regions by approximately 7–9% in the three locations analyzed in this study.

Owing to the increased solar reflectance, the high reflectance plus evaporation strategy will also produce an increased mean radiant temperature and a higher PET compared to the baseline by approximately 7–12% in the three locations. This is desired to improve human thermal comfort in winter in cold regions.

Table 13.7 Results for various pavement scenarios in three regions (winter)

Pavement scenario	Parameter altered	New parameter value	Max. surface temperature T_s (°C)	Mean radiant temperature T_{mrt} (°C)	PET (°C)	PET change from baseline (%)
Sacramento, California						
Baseline	—	—	18	9	12.7	0.0
High reflectance	*r*	0.5	15	14	14.6	15.0
Evaporation	*ER* (mm/h)	1.5	13	6	11.6	−8.7
High reflectance + evaporation	*R* + *ER* (mm/h)	0.5 + 1.5	12	13	14.2	11.8
Shading	*SVF*	0	11	−4	8.1	−36.2
Los Angeles, California						
Baseline	—	—	30	21	21.2	0.0
High reflectance	*r*	0.5	24	27	23.4	10.4
Evaporation	*ER* (mm/h)	1.5	21	17	19.8	−6.6
High reflectance + evaporation	*R* + *ER* (mm/h)	0.5 + 1.5	19	25	22.7	7.1
Shading	*SVF*	0	17	3	14.6	−31.1
Phoenix, Arizona						
Baseline	—	—	29	21	20.6	0.0
High reflectance	*r*	0.5	22	28	23.2	12.6
Evaporation	*ER* (mm/h)	1.5	19	17	19.2	−6.8
High reflectance + evaporation	*R* + *ER* (mm/h)	0.5 + 1.5	17	26	22.4	8.7
Shading	*SVF*	0	16	2	13.6	−34.0

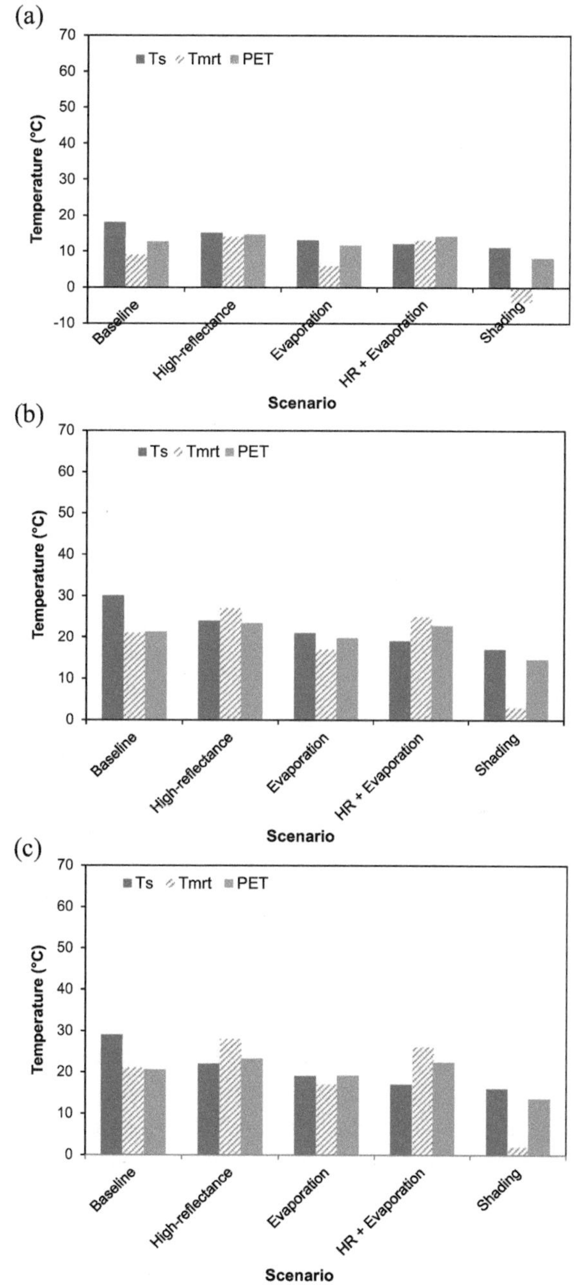

Figure 13.7 Calculated results for various pavement scenarios at three regions: pavement surface temperature T_s, mean radiant temperature T_{mrt}, and PET (winter). (a) Sacramento, CA. (b) Los Angeles, CA. (c) Phoenix, AZ.

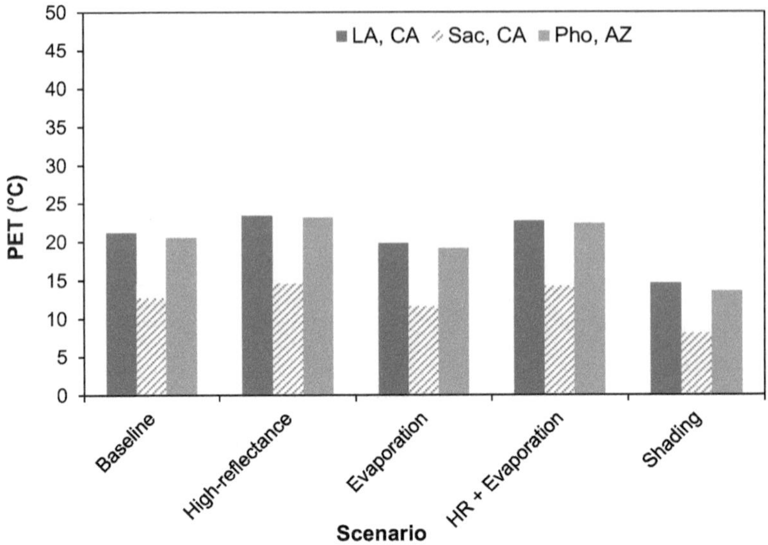

Figure 13.8 Comparison of PET for various pavement scenarios at three regions (winter).

As shown previously, shading alone can considerably reduce both the pavement surface temperature and the mean radiant temperature and thus significantly reduce the PET and improve human thermal comfort in hot summer. It also significantly reduces the PET and thus lowers human thermal comfort in cold winter by 36.2, 31.1, and 34% for Sacramento and Los Angeles in California and Phoenix in Arizona, respectively. This indicates that enhancing the shading with trees or other ways over pavement could reduce both the pavement surface temperature and the PET, lowering human thermal comfort in cold periods by approximately 31–36% in the three locations analyzed in this study.

Using cool pavement strategies might improve human thermal comfort during the hot summer periods. However, it also might cause some negative impacts during cold winter times. The results obtained for winter did demonstrate this concern. For some areas such as Sacramento, California, it is hot in summer and cold in winter. Some cool pavement strategies used to improve the summer thermal environments might make the PET slightly colder in winter, as shown in Figures 13.7 and 13.8. On the other hand, some cool pavement strategies such as reflective pavement could improve human thermal comfort in cold winter while reducing human thermal comfort in hot summer. Therefore, a strategy that can help reduce the summer hot temperatures but will not reduce the winter cold temperature is desirable, such as evaporation only in summer and deciduous tree shading only in summer, etc.

13.7 SUMMARY AND CONCLUSIONS

The study in this chapter investigated the impacts of various pavement technologies and management strategies on human outdoor thermal comfort. Some concepts and models related to human thermal comfort index, mean radiant temperature, and the human body energy balance model were introduced in this study. The physiological equivalent temperature was selected as a human thermal comfort index to evaluate and compare the effects of various pavement strategies on the outdoor thermal environment for three climates (Sacramento and Los Angeles in California and Phoenix in Arizona). The pavement strategies evaluated include high-reflectance pavement, evaporation cooling (permeable pavement), both high reflectance and evaporation, and shading along with a baseline for control (conventional pavement). The summer (July) and winter (January) climate data for those three locations were used for assessment.

Using high-reflectance pavement or a high-reflectance plus evaporation strategy will definitely reduce pavement surface temperatures for all three locations. However, using high-reflectance pavement would increase MRT because of the increased reflected radiation hitting the human body. The increased MRT would produce a higher PET compared to the baseline (conventional pavement). This indicates that using high-reflectance pavement will reduce the pavement surface temperature and consequently might help decelerate the formation of ground-level ozone and help improve the air quality; however, increasing the pavement reflectance (especially by a large amount) will increase the risk of reducing human thermal comfort during hot periods by approximately 3–8% in the three locations. Enhancing the pavement evaporation will help reduce both pavement surface temperature and MRT and also help reduce PET and improve human thermal comfort in hot periods by approximately 3–4% in the three locations. Shading alone will considerably reduce both pavement surface temperature and MRT and thus significantly reduce PET and improve human thermal comfort in hot periods by approximately 17–28% in the three locations. This indicates that enhancing the shading with trees or other ways over pavement will be a very effective strategy to reduce both the pavement surface temperature and the PET, helping improve human thermal comfort during hot periods.

Using cool pavement strategies might produce some benefits during hot summer periods, such as improving human thermal comfort. However, it also might cause some negative impacts during cold winter times. For some areas such as Sacramento, California, where it is hot in summer and cold in winter, some cool pavement strategies (e.g., pavement evaporation or shading) used to improve the summer thermal environments might make the cold winter slightly colder. On the other hand, some cool pavement strategies such as reflective pavement could improve human thermal comfort in cold winter while possibly reducing human thermal comfort in hot summer. Therefore, a strategy that can help reduce the summer hot temperatures but will not reduce the winter cold temperature is desirable, such as enhanced evaporation only in summer and tree shading only in summer. This chapter provides useful insights for the use of cool pavement strategies in urban areas to improve the thermal environment and mitigate heat island effects. More pavement strategies can be evaluated using the model developed in this study to examine the effect on human thermal comfort for existing or new built environments in urban areas with different climates.

CHAPTER 14

A Model Framework for Evaluating Impacts of Pavement Strategies on Building Energy Use

Contents

14.1 OBJECTIVE AND SCOPE

The objective of this chapter is to develop a framework and preliminary model for evaluating the impacts of pavement strategies on building energy use, which can be used to provide a first-order preliminary estimate of the effectiveness of cool pavements at reducing building energy use if the model is validated with measured data. The primary pavement data needed for the model are cooling and heating degree hours near the intake to the heating, ventilation, and air conditioning (HVAC). Data measured from the nine test sections in Davis, California (see Figure 4.3), were used to provide a preliminary indication of pavement type effect. Other input variables and a rigorous check of the building energy use model are outside the scope of this dissertation and are recommended for future improvement studies. The results from this type of study can provide insights for designers and policymakers on the effectiveness of applying cool pavement strategies for mitigating near-surface heat island effects, from the viewpoint of building energy demand.

The preliminary model developed in this study is a simplified one based on near-surface air temperatures, assuming that they are controlled by the pavement surrounding the building, and did not consider the radiation interactions between building and pavements. The intention of this model

Pavement Materials for Heat Island Mitigation
ISBN 978-0-12-803476-7
http://dx.doi.org/10.1016/B978-0-12-803476-7.00014-3

is to provide a first-level estimate of the potential benefit from energy savings for cooling and heating in buildings on an annual basis.

14.2 PRELIMINARY MODEL

Energy use for cooling and heating in buildings is determined by a large number of factors, such as the climate, building size, efficiency of the cooling and heating system, ventilation, insulation, roof type, orientation, thermal preference of the occupants, time duration of using cooling and heating, etc. [5,16,17,106,107]. With simplification but without loss of generality and first-order accuracy, this model is can be used to estimate the building energy use for cooling and heating based on the first law of thermodynamics, i.e., energy balance.

Assumptions for the simplified model include:

1. The building is a one-floor residential building simplified as a rectangular box with an average height of 2.5 m.
2. Air is continually being pumped, at a certain rate, from the outside at an average height of 2.5 m.
3. Corrected factors are used for considering loss of cooling and intrusion of heat related to insulation.
4. The air conditioning (A/C) works for cooling or heating when the near-surface air temperatures surrounding the building are higher or lower, respectively, than the thermostat set-point temperatures (comfort range) for cooling and heating.
5. The thermostat set-point temperatures (comfort range) for cooling and heating are 26 and 18 °C, respectively.
6. Near-surface air temperatures surrounding the building are controlled by the pavement surrounding the building.

The total energy cost and total energy use for cooling and heating a building are determined using a simplified model developed for this study, which is defined by the following equations:

$$C = C_c + C_h = p_c E_c + p_h E_h \tag{14.1}$$

$$E = E_c + E_h \tag{14.2}$$

$$E_c = A_c \frac{1}{\mu_c} m_c c \cdot \mathrm{CDH} \tag{14.3}$$

$$E_h = A_h \frac{1}{\mu_h} m_h c \cdot \mathrm{HDH} \tag{14.4}$$

$$m_c = (1 + r_c)\rho AH \tag{14.5}$$

$$m_h = (1 + r_h)\rho AH \tag{14.6}$$

$$\text{CDH} = 30\sum_{m=1}^{12}\sum_{t=1}^{24}\left(T^s_{mt} - T_c\right) \quad \text{for} \quad T^s_{mt} > T_c, \\ t = 1,2,\ldots,24; m = 1,2,\ldots,12 \tag{14.7}$$

$$\text{HDH} = 30\sum_{m=1}^{12}\sum_{t=1}^{24}\left(T_h - T^s_{mt}\right) \quad \text{for} \quad T^s_{mt} < T_h, \\ t = 1,2,\ldots,24; \; m = 1,2,\ldots,12 \tag{14.8}$$

$$h_c = 30\sum_{m=1}^{12}\sum_{t=1}^{24} h^c_{mt}, \quad \text{where} \quad h^c_{mt} = 1 \quad \text{if} \quad T^s_{mt} > T_c, \\ h^c_{mt} = 0 \quad \text{otherwise} \tag{14.9}$$

$$h_h = 30\sum_{m=1}^{12}\sum_{t=1}^{24} h^h_{mt}, \quad \text{where} \quad h^h_{mt} = 1 \quad \text{if} \quad T^s_{mt} < T_h, \\ h^h_{mt} = 0 \quad \text{otherwise} \tag{14.10}$$

$$T^s_{mt} = \frac{\left(T^s_{mmax} + T^s_{mmin}\right)}{2} + \frac{\left(T^s_{mmax} - T^s_{mmin}\right)}{2\left[\frac{0.96 \ \sin\pi}{12}(t-11) + \frac{0.14 \ \sin\pi}{6(t-11)}\right]}, \\ t = 1,2,\ldots,24; m = 1,2,\ldots, \tag{14.11}$$

$$T^s_{mmax} = \alpha_1 T^a_{mmax} + \alpha_2 r + \alpha_3 ER \tag{14.12}$$

$$T^s_{mmin} = \alpha_4 T^a_{mmin} + \alpha_5 r + \alpha_6 ER \tag{14.13}$$

where

C and E are the total energy cost (\$) and total energy use (kWh) for cooling and heating by the A/C, respectively;

p_c and p_h are the unit energy price for cooling and heating (\$/kWh), respectively;

m_c and m_h are the air mass for cooling and heating (kg), respectively;

c and ρ are the heat capacity (J/(kg °C)) and density of air (kg/m^3), respectively;

A and H are the building floor area (m^2) and average height (m), respectively;

μ_c and μ_h are the energy efficiency for cooling and heating (-), respectively;
CDH and HDH are the annual cooling degree-hour (°C h) and heating degree-hour (°C h) for the cooling and heating load of the A/C, respectively;
A_c and A_h are adjustment coefficients (-) for the energy use for cooling and heating, respectively, accounting for other related factors such as additional energy use for other thermal mass of the building itself;
T^s_{mt} is the (average) near-surface air temperature (°C) surrounding the building at time t in a day for month m;
T_c and T_h are thermostat setting point temperatures (°C) (comfort range) for cooling and heating, respectively;
h_c and h_h are total hours for cooling and heating (h), respectively;
h^c_{mt} and h^h_{mt} are dummy variables (-) for the hours that require cooling and heating, respectively;
r_c and r_h are air exchange rates (1/h) for cooling and heating, respectively;
T^s_{mmax} and T^s_{mmin} are the maximum and minimum near-surface air temperatures (°C) surrounding the building in a day for month m, respectively;
T^a_{mmax} and T^a_{mmin} are the maximum and minimum global air temperatures (°C) in the urban area climate region where the building is located in a day for month m, respectively;
r is the solar reflectivity (-), one of the most important thermal properties of pavement adjacent to the building; and
ER is the evaporation rate of water (mm/h) in the pavement adjacent to the building (in cm/h).

Determining the near-surface air temperature of the pavement is complicated in that it is determined from multiple factors including climate (i.e., ambient air temperature, solar radiation, wind speed, and humidity), pavement thermal properties (i.e., solar reflectivity, solar emissivity, heat capacity, and thermal conductivity), pavement area and distance from the pavement to the building, evaporation rate, shading over the pavement, etc. [20,66–71,73,74]. To simplify this calculation, the model considers only the solar reflectivity (r) and evaporative cooling rate (ER) variables for the cool pavements, with all other pavement factors constant [115].

A one-floor residential building with floor area $A = 185$ m^2 (2000 ft^2) and average height $H = 2.5$ m (8.2 ft) is assumed for the model calculations.

Table 14.1 Model parameter values suggested for future analysis (not validated)

Parameter	Value	Parameter	Value
A (m^2)	185	ρ (kg/m^3)	1.205
H (m)	2.5	c (J/(kg °C))	1006
p_c ($/kWh)	0.15	p_h ($/kWh)	0.15
T_c (°C)	26	T_h (°C)	18
μ_c	0.7	μ_h	0.7
A_c	1.2	A_h	1.2
r_c (/h)	0.125	r_h (/h)	0.0625
α_1	1.2	α_4	1.1
α_2	−15	α_5	−3
α_3	−25	α_6	−4

The cool (reflective pavement or evaporative pavement or both) pavement is installed on a pavement segment on the street near the building (for example, ~2 m from the building). Other parameters for the model that are based on values obtained in the literature [181,182,184], or are assumed based on experience and need to be validated using measured or other simulated data, are shown in Table 14.1.

14.3 THERMAL LOAD

From the model framework developed above, it is known that the thermal load (CDH and HDH) is the most important factor influence the building energy use. Since the model developed is not validated, only an example thermal load is calculated from the measured data of near-surface air (5 in (12.5 cm) above surface) temperatures on the nine test sections in Davis, California, built for this dissertation study (see Figure 4.3).

The base cooling and heat temperatures were selected as 26 and 18 °C (the estimated comfort range for Davis, California, see Table 13.2). The example measured near-surface air temperatures at 5 in (12.5 cm) above the surface on various pavements (paver, asphalt, and concrete) are presented in Figures 14.1–14.4 for March and July. The thermal loads (CDH and HDH and total = CDH + HDH) of each month in a year are calculated and presented in Table 14.2 and plotted in Figures 14.5 and 14.6. The CDH is high in summer and low in winter; the HDH is high in winter and low in summer, as expected. The total thermal load (CDH + HDH) is generally high in summer and slightly low in winter in Davis, California. Different pavement types have influence on the CDH and HDH, but not very

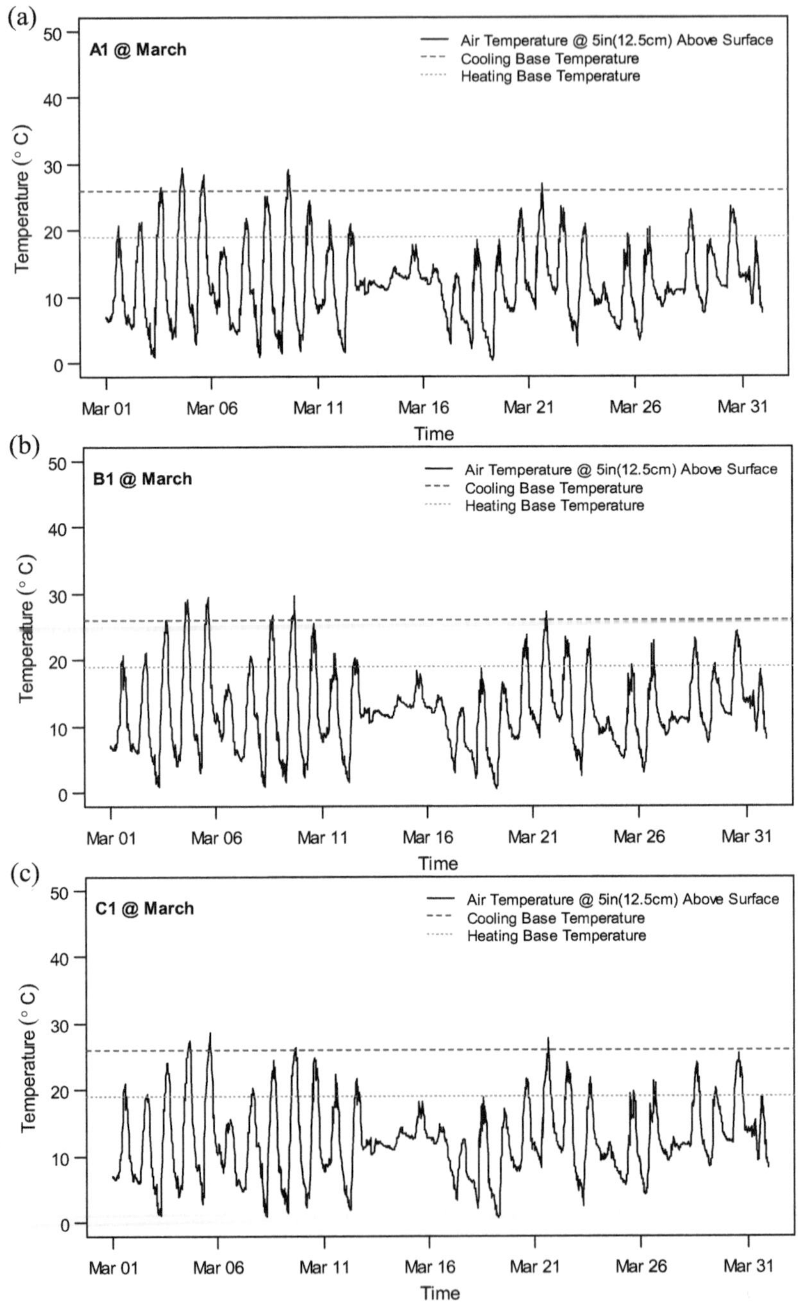

Figure 14.1 Example near-surface air temperatures at 5 in (12.5 cm) above the surface on various impermeable pavements (A1, B1, and C1) in March. (a) Paver. (b) Asphalt. (c) Concrete.

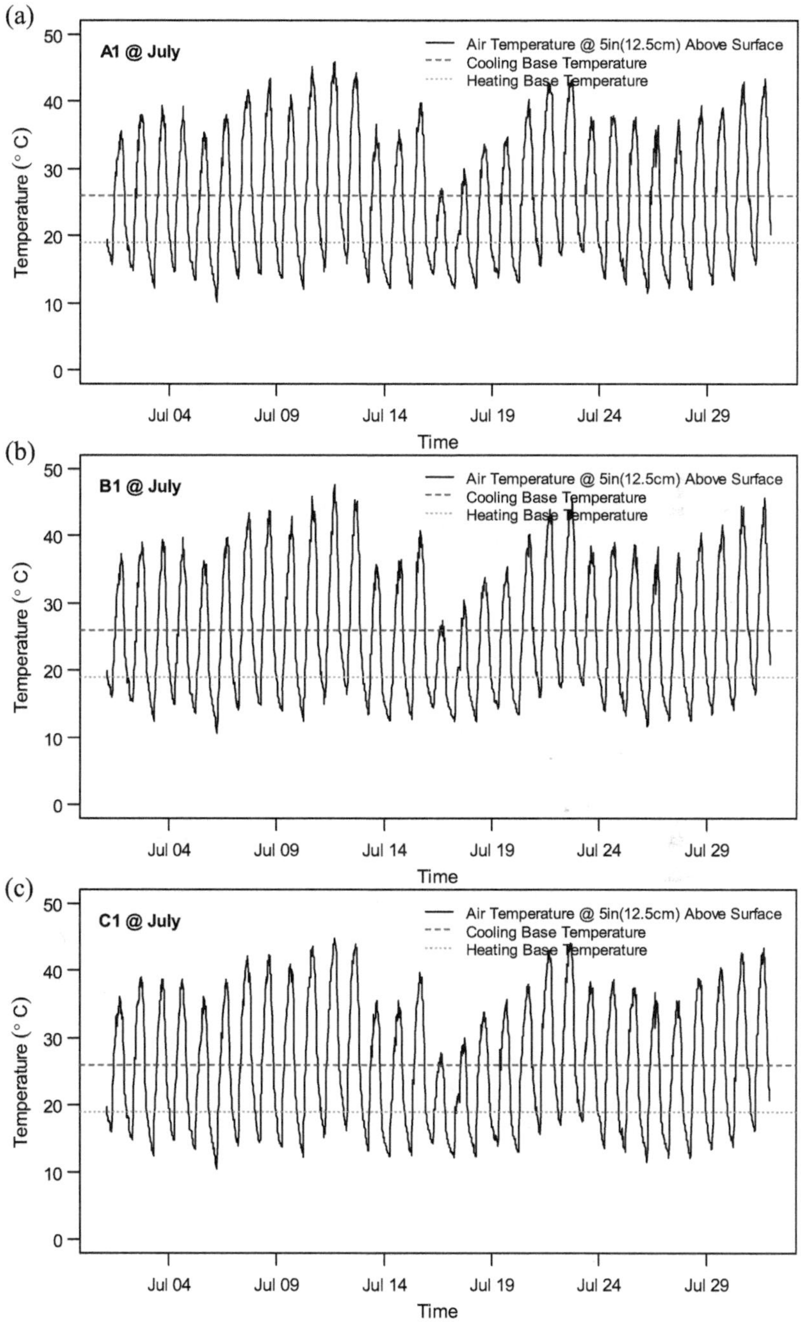

Figure 14.2 Example near-surface air temperatures at 5 in (12.5 cm) above the surface on various impermeable pavements (A1, B1, and C1) in July. (a) Paver. (b) Asphalt. (c) Concrete.

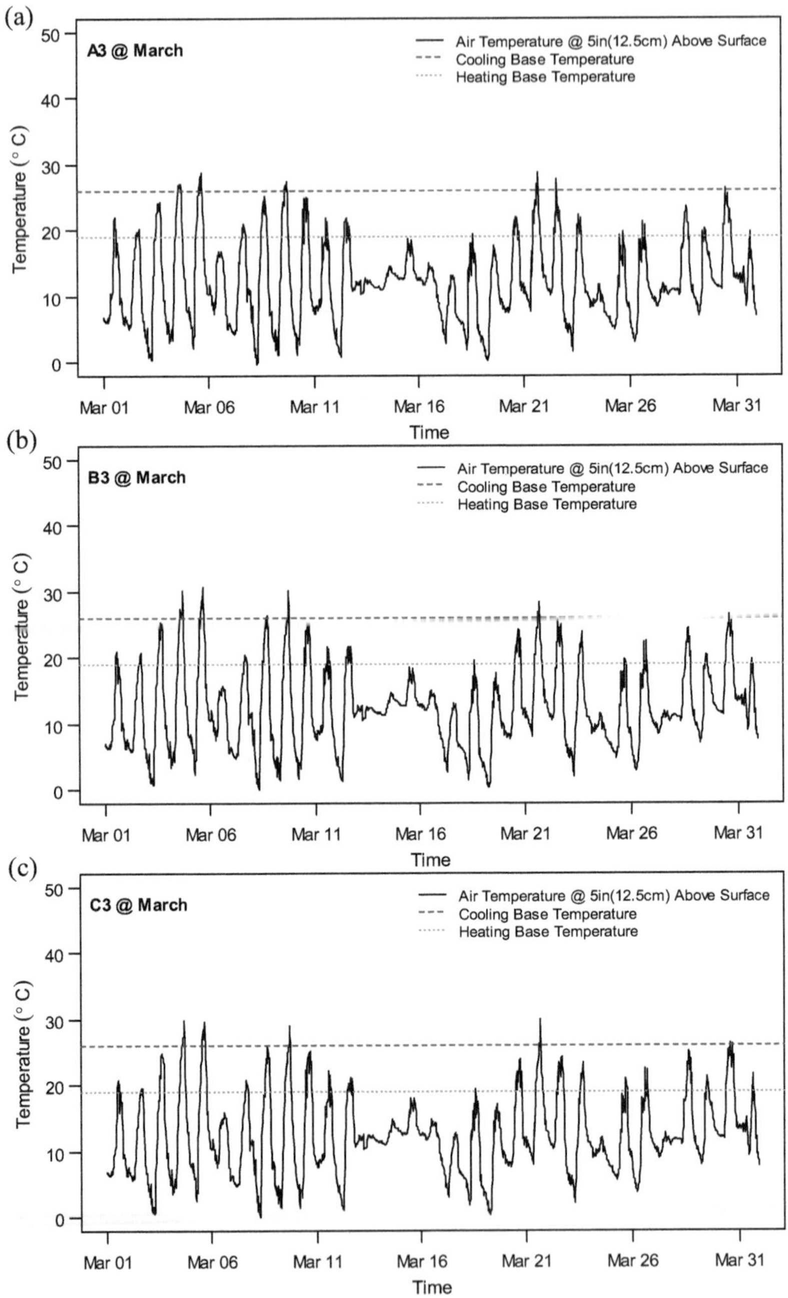

Figure 14.3 Example near-surface air temperatures at 5 in (12.5 cm) above the surface on various permeable pavements (A3, B3, and C3) in March. (a) Paver. (b) Asphalt. (c) Concrete.

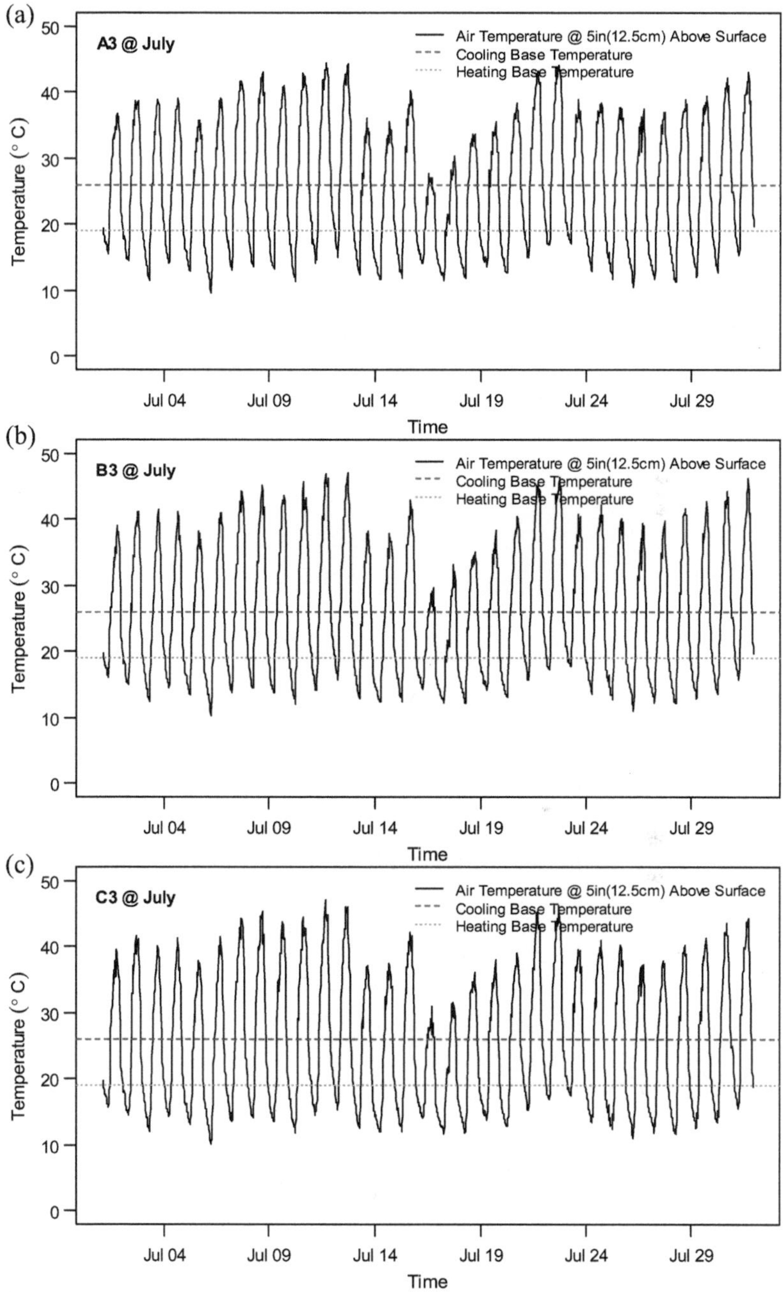

Figure 14.4 Example near-surface air temperatures at 5 in (12.5 cm) above the surface on various permeable pavements (A3, B3, and C3) in July. (a) Paver. (b) Asphalt. (c) Concrete.

Table 14.2 Thermal load of different pavement sections in different months for near-surface air at 5 in (12.5 cm) above surface

Section	Paver A1			Paver A2			Paver A3			Asphalt B1			Asphalt B2			Asphalt B3			Concrete C1			Concrete C2			Concrete C3		
Month	**CDH**	**HDH**	**Total**	**CDH**	**HDH**	**Total**	**CDH**	**HDH**	**Total**	**CDH**	**HDH**	**Total**	**CDH**	**HDH**	**Total**	**CDH**	**HDH**	**Total**	**CDH**	**HDH**	**Total**	**CDH**	**HDH**	**Total**	**CDH**	**HDH**	**Total**
January	0	183	183	0	180	180	0	183	183	8	181	189	0	176	176	0	181	181	0	177	177	0	182	182	0	181	181
February	27	132	159	12	133	145	18	134	152	24	130	154	11	128	139	15	131	146	22	125	147	17	131	148	11	130	141
March	25	131	156	20	134	154	16	135	151	24	130	154	21	131	152	25	133	158	17	128	145	20	132	152	24	132	156
April	73	106	180	66	108	173	67	112	179	65	101	167	70	104	175	70	106	176	65	100	165	70	103	173	69	104	173
May	90	77	166	88	81	169	90	82	172	86	73	159	93	77	170	103	78	181	85	74	159	95	79	174	97	80	178
June	108	60	168	110	65	175	113	65	178	110	58	168	115	61	177	121	61	182	107	60	167	118	63	181	124	64	187
July	131	52	183	135	58	193	132	57	189	138	49	187	141	53	194	147	51	198	130	51	181	138	56	194	146	56	201
August	146	58	205	149	63	212	146	64	210	162	52	214	153	57	210	163	59	221	146	54	200	154	60	214	159	62	220
September	126	65	191	125	71	196	120	72	192	139	60	199	126	66	193	142	67	209	120	61	181	128	69	197	131	70	201
October	73	54	127	70	59	129	69	59	128	95	81	176	79	87	166	94	89	183	81	82	163	82	88	170	82	89	171
November	10	143	153	9	142	151	11	143	154	21	140	161	10	139	148	13	141	154	5	135	140	13	138	151	10	138	148
December	0	183	183	0	184	184	0	186	186	0	182	182	0	181	181	0	184	184	0	180	180	0	182	182	0	186	186
Annual	**810**	**1244**	**2054**	**783**	**1277**	**2060**	**783**	**1291**	**2073**	**873**	**1237**	**2110**	**820**	**1261**	**2081**	**892**	**1280**	**2172**	**778**	**1227**	**2006**	**834**	**1284**	**2119**	**853**	**1290**	**2143**

Unit for thermal load (CDH, HDH, total): °C h.

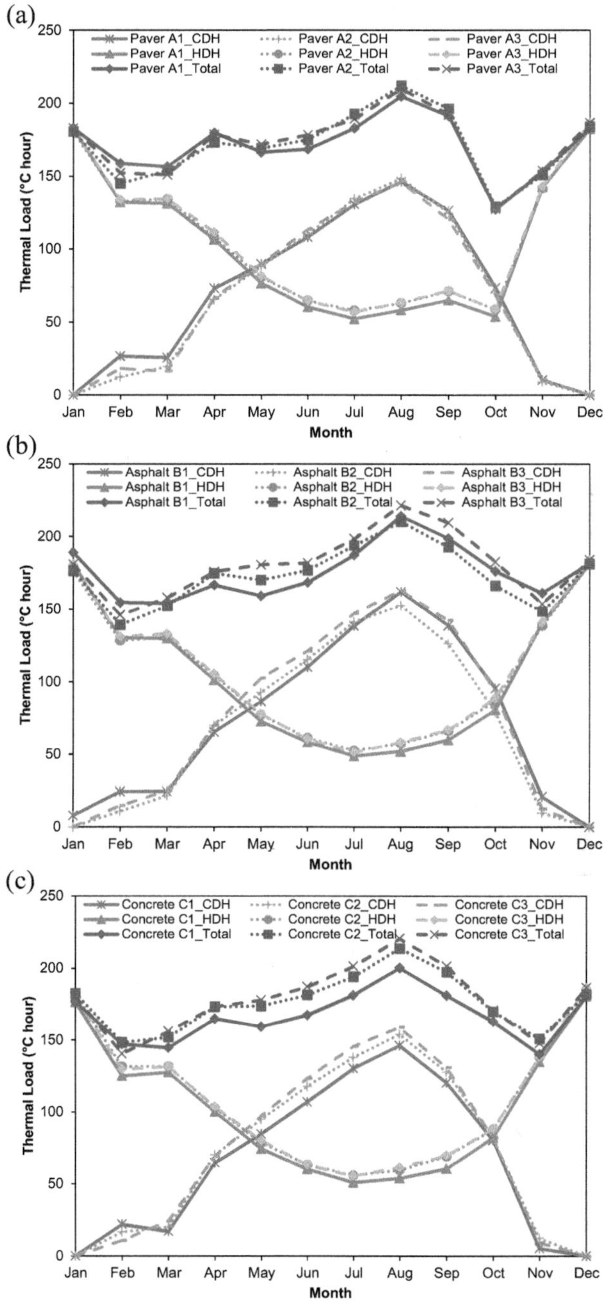

Figure 14.5 Thermal loads (CDH and HDH and total = CDH + HDH) for near-surface air at 5 in (12.5 cm) above the surface for each month in a year for each type of pavement. (a) Paver. (b) Asphalt. (c) Concrete.

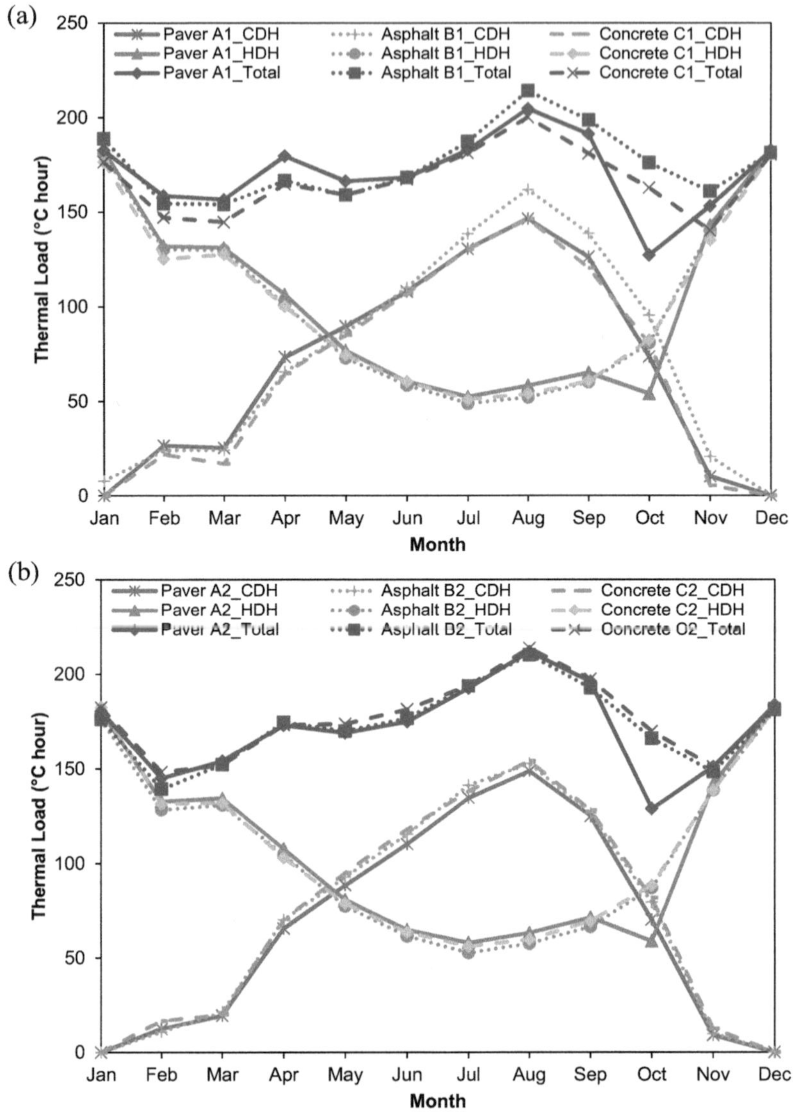

Figure 14.6 Thermal loads (CDH and HDH and total = CDH + HDH) for near-surface air at 5 in (12.5 cm) above the surface for each month in a year for comparison between types of pavement. (a) Impermeable pavements (paver A1, asphalt B1, and concrete C1). (b) Permeable pavements (paver A2, asphalt B2, and concrete C2).

Table 14.3 Annual thermal load for various pavement sections

Section	Albedo	CDH	HDH	Total
A1	0.28	810	1244	2054
A2	0.25	783	1277	2060
A3	0.28	783	1291	2073
B1	0.09	873	1237	2110
B2	0.08	820	1261	2081
B3	0.08	892	1280	2172
C1	0.29	778	1227	2006
C2	0.18	834	1284	2119
C3	0.26	853	1290	2143

Unit for thermal load (CDH, HDH, total): °C h.

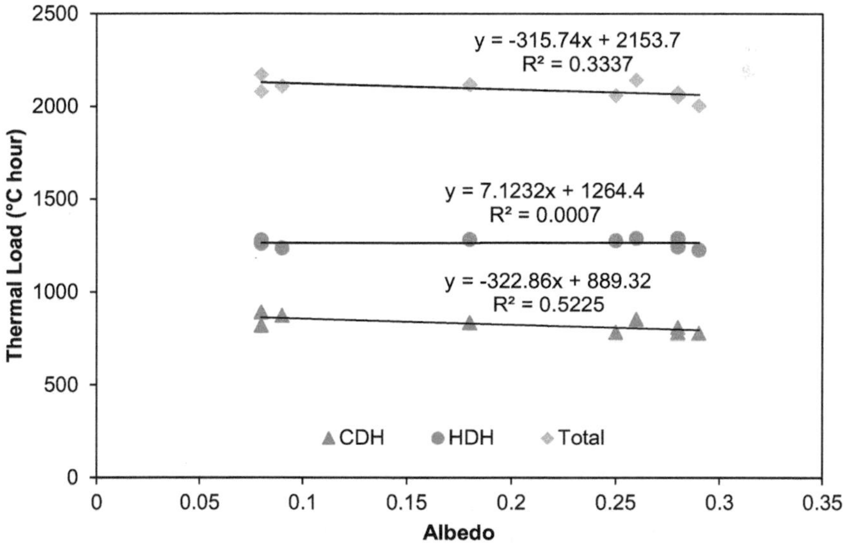

Figure 14.7 Annual thermal load versus albedo.

significantly. Moreover, the influence on the thermal load of pavement types is not constant (for example, positive on CDH and negative on HDH, positive in summer and negative in winter). The correlation of annual thermal load and albedo is listed in Table 14.3 and plotted in Figure 14.7. The increased albedo reduces the CDH but increases slightly the HDH and reduces the total thermal load on an annual basis. All these impacts are not very significant for the test site in the open area in Davis, California. These figures are examples of the types of input data that could be developed for future use to aid building designers in selecting pavement types and ground-level HVAC locations. They would need inclusion of the reflected solar radiation where applicable using data similar to those shown in Chapters 10 and 11and from research done by other groups (e.g., [154,198]).

14.4 LIMITATIONS

The model presented will provide only a first-level analysis of the potential benefit from energy savings for cooling and heating in buildings using very simple assumptions based on near-surface air temperatures, assuming that they are controlled by the pavement surrounding the building. Radiation from the pavement onto building wall surfaces and solar radiation onto roofs should also be considered, as their influence on building energy use might be more important than the influence of the near-surface air temperature. In addition, city-wide urban heat island air temperature changes contributed by individual pavement sections in the city should also be considered.

No impact is priceless, however; and no option is costless. Even though benefits from net energy savings can be gained from constructing cool pavements, the potential costs of removal of old pavement and installation of the new pavement and extra expenditures on the new materials and/or equipment for the new cool pavements should be taken into account in a comprehensive cost–benefit analysis over the whole lifetime of the pavement to ensure that a positive net benefit will be obtained [121,199]. The complete cost–benefit analysis should consider both the benefits to the building owners and occupants and the costs to the authorities or the same owners who are responsible for the pavements. In addition to the benefits from energy savings, other potential impacts associated with cool pavements should also be considered, including reducing cooling energy demand in the vehicles driving on the pavements, improving human thermal comfort and health (potentially increasing walking and bicycling), and improving air quality (reduced ground-level ozone). These potential impacts, although difficult to quantify, should be taken into account when performing a complete multidimensional cost–benefit analysis [48,49] for a given project. An even more complete analysis would consider additional environmental impacts such as greenhouse gas emissions and categorized pollutant emissions of both cool and baseline pavements, and the same environmental impacts from the source of the energy used for both heating and cooling, which will vary widely, particularly depending on the method used to generate electricity for A/C use (coal, gas, hydroelectric, nuclear, etc.) and the method used to heat the buildings (gas, oil, electric, etc.).

Additional research is needed to validate and apply the model, better define its limitations, develop guidance for users, and provide guidance for developing inputs for a more comprehensive analysis of the life cycle costs and environmental impacts considering the pavement as well as the buildings.

14.5 SUMMARY AND CONCLUSIONS

Cool pavements have been identified by the U.S. Environmental Protection Agency as a major strategy to mitigate heat island effects, with one of the main benefits being that they could lower air temperatures and consequently reduce cooling-related energy use and associated greenhouse gas emissions. Some applications have been implemented in a number of states in the United States with promotion from various levels of government. On the downside, cool pavements can also potentially increase energy use and greenhouse gas emissions if more heating is required during cold periods. A simple model has been developed for use to assess whether cool pavements reduce building energy use based only on near-surface air temperature and with no consideration of city-wide urban heat island effects on the ambient air temperature change. Models of this type, once better developed and validated, can be used to estimate the energy used in buildings for both cooling and heating on an annual basis for a given project.

An example thermal load was calculated from the measured data of near-surface air (5 in (12.5 cm) above the surface) temperatures on the nine test sections built for this dissertation study. The CDH is high in summer and low in winter; the HDH is higher in winter and lower in summer, as expected. The total thermal load (CDH + HDH) is generally higher in summer and slightly lower in winter in Davis, California. Various pavement types have influence on the thermal load (CDH and HDH), but not very significantly. Moreover, the influences on the thermal load of pavement types are not constant (for example, positive on CDH and negative on HDH, positive in summer and negative in winter). The correlation of annual thermal load and albedo was examined. The increased albedo reduces the CDH but increases slightly the HDH and reduces the total thermal load on an annual basis. All these impacts are not very significant for the test site in the open area in Davis, California. These figures are examples of the types of input data that could be developed for future use to aid building designers in selecting pavement types and ground-level HVAC locations. They would need inclusion of the reflected solar radiation where applicable using data similar to those shown in Chapters 10 and 11 and provided by some research by other groups (e.g., [154,198]).

Careful project investigation and location- and technology-specific designs are required before policy requirements are set and there is wide implementation, to provide better assurance that there will be a net benefit from cool pavements in terms of building energy use savings. A "do nothing/no change" alternative should always be included as one of the policy options for mitigating heat island effects using cool pavements until such time as the net benefits have been proven and all uncertainties have been reduced to an acceptable level.

CHAPTER 15

Summary, Conclusions, and Recommendations

Contents

15.1 SUMMARY AND CONCLUSIONS

15.1.1 Field Measurement of Albedo for Reflective Pavement

Nine 4-m by 4-m (13-ft by 13-ft) experimental sections were designed and constructed at Davis, California, including asphalt pavement, concrete pavement, and interlocking concrete paver, with a focus on permeable structures with impermeable pavements as controls. These specifically built test sections (with some other existing pavements) were used to measure the fundamental material properties, including albedo, permeability, thermal properties, and evaporation rate (for some of the materials used in the experimental sections) and to empirically examine the thermal behavior of various pavement types in various seasons and under various moisture conditions and their impact on near-surface air and building surfaces, the results of which were presented in Chapters 4–10 and are summarized here.

Pavement Materials for Heat Island Mitigation
ISBN 978-0-12-803476-7
http://dx.doi.org/10.1016/B978-0-12-803476-7.00015-5

Field measurements of albedo for the various pavement materials were performed on experimental test sections and other existing pavements (presented in Chapter 4). The albedos for the materials were compared, and the factors affecting the field measurement of albedo and the diurnal and seasonal changes in albedo were examined. The effect of albedo on pavement temperature was also examined using measured data of albedo and pavement surface temperatures. Three developments were achieved to enable the research on pavement albedo and thermal performance: (1) a new albedo measurement system, (2) new data documenting differences in albedo across pavement types and over pavement surfaces, and (3) a correlation between the cooling effects of increased albedo and solar radiation. The main conclusions drawn from the study include:

1. A new albedo measurement system using a dual pyranometer and automatic data collection system was developed. It can be used to conveniently measure albedo in the field and perform long-term monitoring of albedo when connected to a data acquisition system.
2. Albedo was measured for commonly used paving materials, including asphalt, concrete, and interlocking concrete paver surfacing materials, with various designs. These new data enhance the basic knowledge of albedo values for pavements as well as the other land-cover types measured (gravel, soil, and lawn), which can help reduce the uncertainty in understanding, evaluating, and modeling their thermal behavior and their consequences for human thermal comfort and building energy use.
3. This study found that the measured albedo of pavement materials is high in the early morning and in the late afternoon when the solar angle is low; it is low and constant over time at midday. This suggests that the albedo should be measured at midday on a clear day to obtain a stable and conservative value. No significant seasonal variation in albedo was found. Pavement albedo will change over time under weathering and trafficking, especially in the first month after construction. Cloud cover will negatively influence the measured albedo value. No impact of wind speed or air temperature on albedo was observed.
4. Albedo has great influence on the pavement surface high temperatures in the daytime (6 °C per 0.1 albedo change on a hot sunny day with solar radiation of 1000 W/m^2) and no significant impact on pavement low temperatures in the nighttime. An empirical relationship between the cooling effect of increased albedo on a pavement high temperature and solar radiation was developed. The cooling effect has a positive correlation with the peak solar radiation intensity in the daytime.

This simple correlation can help estimate the cooling effect of increased albedo or heating effect of reduced albedo on pavement for various climates and seasons with different solar radiation.

15.1.2 Field Measurement of Permeability for Permeable Pavement

The permeability of test sections was comparatively measured for porous asphalt, pervious concrete, and permeable interlocking concrete paver pavements using both the American Society for Testing and Materials (ASTM) C1701 and the National Center for Asphalt Technology (NCAT) methods (using constant-head and falling-head methods, respectively; see Chapter 5). The conclusions drawn from the study include:

1. For accurate permeability measurement, regardless of the method of measurement, water leakage must be prevented. It was found that silicone gel is superior for water sealing compared with the plumber's putty recommended by both methods.
2. Both the ASTM C1701 and the NCAT permeameter measurement methods can effectively be used to measure the permeability of all surface pavement types and their mix design will not significantly influence the measurement accuracy.
3. A weak correlation (with $R^2 = 0.52$) was observed between permeability measurements made by the NCAT permeameter and the ASTM C1701 method across all pavement surface types. The correlation was stronger ($R^2 = 0.72$ through 0.9) when the measured permeability values were for a single type of pavement surface material.
4. The permeability measured by the ASTM C1701 method was more conservative (i.e., lower than that with the NCAT permeameter) and on average about 25% of the values measured by the NCAT permeameter method.
5. The larger ring size used in the ASTM C1701 method or the double-ring method could reduce the variability of the permeability measurement.

15.1.3 Laboratory Measurement of Thermal Properties for Thermal Resistance Pavement

A multidimensional model for simulation of the transient temperature at any location on beam or cylinder specimens of various sizes subject to convective heat transfer was developed, and the thermal properties of specimens of various shapes and sizes were measured thorough back-calculation from the measured transient temperatures profile of the specimen (in Chapter 6).

1. The model and tool developed can accurately predict the transient temperature at any location on beam or cylinder specimens of various sizes subject to convective heat transfer. It can also be used, if the thermal properties are known, to simulate the transient temperature and predict the time it takes to reach a specified target testing temperature at any location for specimens of various shapes and sizes, when the specimen is preheated or precooled in a forced-convection oven or temperature chamber for mechanical or other temperature-related testing.
2. Based on the temperature simulation model developed, the developed and validated procedure for back-calculation can be used to easily obtain the thermal properties of a specimen of pavement material from its measured transient temperature profile, regardless of the shape and size of specimen. With this procedure, the thermal properties of novel pavement materials (various innovative materials such as porous concrete and high thermal resistance materials) can be easily measured and then used for evaluating and modeling the thermal performance of the built environment composed of these materials.
3. The developed and validated procedure was employed to measure the thermal properties of pavement materials using cylinder specimens, particularly some surface materials used for the experimental sections in this study. It was found that the dense-graded (i.e., nonporous) materials (concrete or asphalt) have a higher thermal conductivity and heat capacity than the open-graded (i.e., porous) materials. Concrete materials generally show a slightly higher thermal conductivity and heat capacity than asphalt materials. It should be noted that aggregate source will also influence the thermal properties, which was not examined in this study.
4. The lower thermal conductivity of open-graded porous materials increases the thermal resistance and thus the difficulty in conducting heat into in-depth pavement layers, which keeps the heat around the surface. Moreover, because of their lower heat capacity compared to dense-graded nonporous materials, the open-graded porous materials will be heated up to a higher temperature under the same amount of energy absorbed from solar radiation or the surroundings during daytime or hot periods. On the other side, just because of their lower heat capacity, the open-graded porous materials have less thermal energy (or heat) stored in the solid body around the surface to lose

into the cold ambient air during the nighttime or cold periods. Also, because of their lower thermal conductivity compared to the dense-graded nonporous materials, it is much more difficult for the open-graded materials to conduct heat to the surface from the underlying layers to supply more energy for loss. These two aspects will significantly increase the possibility that the open-graded porous materials will produce a lower surface temperature when the same amount of heat is lost into the ambient air during the nighttime or cold periods. This theoretically confirms that the permeable pavement composed of open-graded porous materials can be a potential strategy to counter the nighttime heat island effect because of its lower surface temperature and less heat released into the ambient air during the night. However, attention should be given to its potentially higher surface temperature during the daytime under dry conditions.

15.1.4 Outdoor Measurement of Evaporation Rate for Evaporative Cooling

The evaporation rate is an important factor that influences the effect of evaporative cooling of permeable pavements. It is determined by a complex system of factors, such as air temperature, relative humidity, water temperature, moisture content, air void content, size, and structure. To avoid the complex system, a simple experimental method was used to measure the average evaporation rates of various pavement materials under outdoor conditions at the test site in Davis, California (presented in Chapter 7). Based on the findings from this experimental study, the peak evaporation rate of bare water is about 2.0–2.5 mm/h during hot days in July at Davis California. During the first experiment day when more water was available near the surface of permeable materials, the evaporation rate ranged from 0.5 to 1.5 mm/h, which was much higher than that (0.1–0.3 mm/h) on the second and third day with less moisture available near the surface.

The findings imply that keeping the water near the surface by enhancing the capillary effect or sprinkling water on the surface or injecting water into the pavement to keep the water head near the surface will increase the evaporation rate and consequently produce a better evaporative cooling effect. The capillary effect depends on the air void content and structure of the surface materials and the size of the air voids. More experimental and theoretical studies are recommended to evaluate and optimally design the evaporative cooling effect of pavement materials.

15.1.5 Field Measurement of Thermal Performance of Various Pavements

Through the design, construction, and instrumentation of nine experimental sections of various pavement types at Davis, California, the seasonal thermal behavior and cooling effects of the pavement types were observed, with focus on the permeable pavements under both dry and wet conditions. Factors affecting the thermal behavior and cooling effect of permeable pavements were investigated in Chapter 8. Through measurements of the temperatures of pavement, surface air, and near-surface air, the thermal performance of the various pavements in different seasons (including asphalt, concrete, and paver; permeable and impermeable) was explored, and the thermal performance of the various permeable pavements under both dry and wet conditions in summer (compared with impermeable pavements) was comparatively examined.

The major conclusions drawn from the study presented in Chapter 8 include:

1. Concrete and paver pavements (albedos of 0.18–0.29) in this study showed lower surface peak temperatures than asphalt pavements (albedos of 0.08–0.09) by 10–25 °C during hot summer in Davis, California; asphalt pavement with a high albedo, through reflective coating or other treatments, could also produce a low surface temperature.
2. Under the dry condition, owing to the lower thermal conductivity and heat capacity, permeable pavements (including porous asphalt, pervious concrete, and pervious paver) were hotter in daytime but cooled faster and consequently got colder during the nights and helped mitigate the nighttime heat island effect, compared to impermeable pavements.
3. Permeable pavements (including porous asphalt, pervious concrete, and pervious paver) under the wet condition can have lower surface temperatures than impermeable pavements; the cooling effect depends highly on the availability of moisture near the surface layer and the evaporation rate.
4. The peak cooling effect of watering for the test sections was approximately 15–35 °C on the pavement surface temperature in the early afternoon during summer owing to the cool water and evaporation.
5. The overall average cooling effect of wetting alone (Wet = 1) over 1 week after irrigation was approximately 0.2–0.45 °C for the near-surface air on permeable pavements; for the surface it was approximately 1.2–1.6 °C and approximately 1.5–3.4 °C for the in-depth layers.

Based on the findings from this study, compared with impermeable pavements, permeable pavements (including pervious concrete pavement, permeable interlocking concrete paver, and porous asphalt pavement) have the potential of being cool pavements that produce lower temperatures and help to mitigate the local heat island effect. However, attention should be given to permeable pavements under dry conditions, which might produce a higher peak daytime temperature. Watering or irrigation and evaporation can help to reduce the daytime pavement surface temperature of permeable pavements and consequently mitigate the heat island effect and improve thermal comfort. The cooling effect depends on the availability of moisture around the pavement surface and will vanish over time as the water level decreases. As a pavement thermal management strategy, water collected from rain (where there is summertime rain, not in California) or irrigation can be injected into the permeable pavement during the late afternoons or evenings in summer. This is especially beneficial when the weather report forecasts a very hot day coming; this strategy can be conducted during the previous night to mitigate the heat wave coming the next day and improve thermal comfort.

15.1.6 Thermal Interactions between Pavement and Near-Surface Air

To investigate the thermal interactions between pavement and near-surface air for various pavement types, the temperature profiles of near-surface air were measured and the factors affecting the profiles were explored in Chapter 9.

It is noted that the temporal profiles of near-surface air have a pattern similar to that of ambient air temperature for both asphalt and concrete pavements. For the spatial profiles, the near-surface air temperatures gradually decrease as the distance from the pavement surface increases, with greatest slope (change rate) in the 10 in (25 cm) just above the pavement surface. In the range close to the pavement surface, the near-surface air temperatures are much higher than the ambient air temperature at ~2 m height, especially for the black asphalt pavement with higher surface temperature compared to higher reflectance concrete pavement. The high near-surface air temperatures would reduce the human thermal comfort, and they are more critical for babies and children, who are shorter and closer to the surface. In addition, the formation of ground-level ozone would be facilitated by these high near-surface air temperatures when smog (volatile organic compounds and nitrogen oxides) is emitted from vehicle exhaust pipes that are close to the pavement surface on roads or parking lots.

The wind speed will influence the shape of the temperature profiles of near-surface air. Lower wind speed will make the spatial profiles steeper owing to less heat dissipated by wind or airflow. The near-surface air temperatures with even higher distances from the surface will also be influenced by the pavement heat under low wind speed. This further implies that the heat effects of pavements will be more critical in environments with low wind speed, such as high-density urban areas, especially those with high-rise buildings that block prevailing winds.

It is revealed from the findings that during hot periods the near-surface air temperatures diminish gradually as the height above the surface increases, and the slope (change rate) also diminishes. The shapes of the spatial profiles are influenced by the surface temperature, ambient air temperature, and wind speed.

A simple model was developed to describe the spatial profiles of near-surface air temperatures, which characterizes the effects of surface temperature, ambient air temperature, and wind speed. The spatial profiles of near-surface air temperature could be predicted once the surface temperature, ambient air temperature, and wind speed are known.

15.1.7 Thermal Interactions between Pavement and Building Surfaces

The thermal interactions between pavement and building walls was preliminarily evaluated through both experiments and simulations, characterizing the impacts of various pavement types on building surface temperatures (presented in Chapter 10).

Four pavement sections were selected to conduct the experiments for measuring the temperature profiles of both wall and pavement surfaces. Experimental results from temperature sensors and thermal images indicated that there are some thermal interactions (through reflected and emitted radiation) between pavement and building walls. The thermal interactions increase the wall temperature to some extent and potentially will contribute increased energy use for cooling the building in hot periods. The nonuniform temperature of the pavement surface at various distances from the wall also provided some confirmation that there is thermal interaction between pavement and building walls.

Experimental results implied that increased pavement albedo will reduce the pavement surface temperature; however, the temperature of the building wall will be increased by the reflected energy from the pavement surface and consequently would potentially increase the energy use for cooling the building in hot periods. This risk might not be a big issue in open areas but would tend to be of greater significant for high-density urban areas where walls are next to the pavement. Therefore, special attention should be given to the application of high-reflectance pavement for mitigating the heat island effect, especially in high-density urban areas.

A simple numerical model was developed and used to simulate the heat transfer and thermal interaction between pavement and walls to obtain typical summer climate data in the dry and hot region of Sacramento, California.

Some simulation results were presented, including temperature profiles and view factor (i.e., the proportion of the radiation that leaves one surface and strikes another surface). More simulations can be performed with this and/or a modified and validated model to explore the effects of various parameters (materials, climates, etc.).

During the nighttime, the temperature on the wall surface is low and relatively uniform; however, the daytime temperature on the wall surface is high and nonuniform, and the lower middle part has a higher temperature owing to the thermal interactions between pavement and walls. This in part verified the results from the experiments presented previously.

The view factor is higher in the lower and middle parts of the wall and is lower for other parts. This helps to explain the temperature difference on the wall obtained from both experimental and simulation results.

Another more convincing and quantitative way of verifying the thermal impacts of various pavements on wall temperature would be to use a heat flux sensor (including both long- and short-wave radiation) to directly measure the heat flux on the wall surface at various heights and times and to directly compare the thermal impacts. However, such heat flux sensors are quite expensive (over $6000 per set of sensors). If more funding is available from any source, the direct measurement of heat flux (including both long- and short-wave radiation) on wall surfaces at various heights and times can be performed and would definitely produce quantitative and more convincing observations.

15.1.8 Numerical Simulation of Thermal Behavior of Cool Pavement Strategies

A framework for a general local microclimate model was developed using the finite element method and the finite difference method (in Chapters 11 and 12). This integrated local microclimate model includes both the pavement structure and the near-surface air and considers coupled processes of radiation, convection, conduction, shading, and evaporation. A simplified model was extracted from the general model to simulate the temporal and spatial distribution of temperatures of pavement and near-surface air. It can be used for various pavement structures under various climate conditions and time frames.

The model was numerically implemented and used to simulate the temporal and spatial distribution and variation of both pavement temperature and near-surface air temperature in the summer in a hot region (Sacramento, California) for validation against the field measurements from experimental test sections presented previously in this dissertation. The model was validated against field measurements for both asphalt and concrete pavements (with different materials and structures) and under various weather conditions. The simulated and measured results generally agree with one another for both asphalt and concrete pavements with different materials and structures and under various weather conditions. This indicates that the developed model can be used to simulate the temperature for the asphalt and concrete pavements for various weather conditions.

The validated model was then used to simulate temperature for a typical pavement structure under typical climate conditions in summer (July) in a hot region. Sensitivity analysis based on simulation was conducted on some variables. Findings about the specific effects of some factors showed complex correlations between these factors and the temperatures of pavement and near-surface air.

Findings obtained from the sensitivity analysis indicate that the temperatures of pavement and near-surface air are very sensitive to solar reflectivity (1 − absorptivity), solar radiation, and shading. The temperatures show relatively high sensitivity to wind speed, convection coefficient, evaporation, thermal conductivity, and thermal emissivity. Specific heat and density present low sensitivity. This implies that increasing solar reflectivity and reducing incident solar radiation (e.g., by shading, etc.) could be effective ways to reduce the pavement temperature. However, application of high-reflectivity pavement materials might increase the chance of causing a negative impact in terms of human thermal comfort and building energy use owing to reflected solar radiation.

According to the identified significant factors affecting the pavement thermal performance, potentially effective strategies include technologies such as high-reflectance pavement and thermal-resistance pavement (materials with low thermal conductivity conduct heat very poorly and leave the heat near the surface, e.g., porous materials for permeable pavements); management strategies such as shading, ventilation, and evaporation; and the combined use of these technologies and management strategies. The effectiveness of these strategies at improving the thermal environment at the street level when applied for a specific context can be evaluated using the model developed and validated in this study.

It should also be noted that while the technologies (e.g., modifying thermal variables, etc.) that cool down the pavement and near-surface air potentially provide a positive effect on pavement life, building energy use, and human thermal comfort during hot times (especially in the daytime during hot summers) and in hot climates, they might also lead to negative effects in terms of the same factors during cold seasons (e.g., nighttime in the winter) or in cold climates. Therefore, a comprehensive assessment of both positive and negative effects of cool pavements should be performed to ensure that the overall net benefits are positive.

15.1.9 Impacts of Cool Pavement Strategies on Human Thermal Comfort

The impacts of various pavement technologies and management strategies on human outdoor thermal comfort were investigated in Chapter 13. Some concepts related to the thermal comfort model and the human body energy balance model and thermal comfort index were introduced. The physiological equivalent temperature (PET) was then selected to evaluate and compare the effects of various pavement technologies and management strategies on the outdoor thermal environment for three climates (Sacramento and Los Angeles in California and Phoenix in Arizona). The pavement technologies and management strategies evaluated include high reflectance, evaporation, and both high reflectance and evaporation, as well as shading. The summer (July) and winter (January) climate data for those three locations were used for assessment.

Using the high-reflectance pavement or the high reflectance plus evaporation strategy will definitely reduce the pavement surface temperatures for all three locations. However, using the high-reflectance pavement would increase the mean radiant temperature owing to the increased reflected radiation hitting the human body. The increased mean radiant temperature would produce a higher PET compared to the baseline. This indicates that using high-reflectance pavement will reduce the pavement surface temperature and consequently might help decelerate the formation of ground-level ozone and help improve the air quality; however, increasing the pavement reflectance will increase the risk of reducing human thermal comfort during hot periods.

Enhancing the evaporation from pavement will help reduce both the pavement surface temperature and the mean radiant temperature and also help reduce the PET and improve the human thermal comfort in hot periods. This implies that enhancing the evaporation from pavement will be an effective way to reduce both the pavement surface temperature and the PET, helping improve both air quality and human thermal comfort in hot periods.

Shading alone will considerably reduce both the pavement surface temperature and the mean radiant temperature and thus significantly reduce the PET and improve human thermal comfort in hot periods. This indicates that enhancing the shading of pavement with trees or in other ways (e.g., canopy and solar panel) will be a very effective strategy to reduce both the pavement surface temperature and the PET, helping improve both air quality and human thermal comfort in hot periods.

Using cool pavement strategies might produce some benefits during hot summer periods, such as improving human thermal comfort and air quality. However, they also might have some negative impacts during cold winter times. The results obtained for winter did demonstrate this concern. For some areas such as Sacramento, California, it is hot in summer and cold in winter. Some cool pavement strategies used to improve the summer thermal environments might make the cold winter even colder. Therefore, a strategy that can help reduce the summer hot temperatures but not reduce the winter cold temperature is desirable, such as evaporation only in summer or tree shading only in summer, etc.

15.1.10 Impact of Cool Pavement Strategies on Building Energy Use

A simplified model framework has been developed for assessing whether cool pavements reduce energy use for a very simple building, considering only the use of a heater/air conditioner (presented in Chapter 14). Models of this type (but it is hoped more sophisticated than the model used in this dissertation) can be used to estimate the energy use in buildings for both cooling and heating on an annual basis.

With validation, the model can be used to evaluate energy use in various locations with various climates under one baseline and three alternative pavement scenarios (high reflectance, evaporation, and high reflectance plus evaporation).

An example thermal load was calculated from the measured data of near-surface air temperatures at 5 in (12.5 cm) above the surface on the nine test sections built for this dissertation study. The cooling degree hour (CDH) is high in summer and low in winter; the heating degree hour (HDH) is high in winter and low in summer, as expected. The total thermal load (CDH + HDH) is generally high in summer and slightly lower in winter in Davis, California. Various pavement types have influence on the thermal load (CDH and HDH), but they are not very significant. Moreover, the influences on the thermal load of pavement types are not constant (for example, positive on CDH and negative on HDH, positive in summer and negative in winter). This makes the annual thermal loads (CDH, HDH, and total) not significantly different for different pavements. The increased albedo reduces the CDH but increases slightly the HDH and reduces the total thermal load on an annual basis. All these impacts are not very significant for the test site in the open area in Davis, California. These results are exclusively for near-surface air temperature effects and ignored radiation from the surface to the building and city-wide urban heat island effects on air temperature. The relative environmental impact of pavement near buildings from near-surface effects depends on building technologies and energy sources used for cooling and heating.

15.2 RECOMMENDATIONS FOR THE APPLICATION OF COOL PAVEMENT STRATEGIES

Based on the findings from this dissertation study, some preliminary recommendations for the application of cool pavement strategies for mitigating near-surface heat island effects and improving the outdoor thermal environment are given as follows:

1. Pave less and plant more. For some areas such as parking lots and alleys, the sites could be partly paved, and more grass and/or trees could be planted on the sites to reduce the negative impacts, with examination of the potential negative impacts of the application of grass and/or trees.
2. Pave smart if it has to be paved. Permeable pavements (integrated with irrigation systems during hot dry seasons), including porous concrete pavement, porous asphalt pavement, and permeable interlocking concrete paver, as well reinforced grass paver, could be good alternatives for paving, if applicable, to both manage the storm-water runoff and potentially help mitigate the heat island effect and improve the thermal environment.

3. Care should be taken with the application of high-reflectance pavements. High-reflectance pavements can be used in open areas to help mitigate the heat island effects. However, special attention should be given when applied in high-density areas or areas with frequent walking and cycling human occupancy.
4. Consider evaporation and shading. Evaporation and shading (using tree, canopy and solar panel, etc.) could be very effective strategies to help improve the thermal environment in hot climates, if effective evaporation and shading can be achieved on pavements.

15.3 RECOMMENDATIONS FOR FUTURE STUDY

Recommendations for the future study on this topic include, but are not limited to:

1. Conduct a life cycle cost analysis and/or cost–benefit analysis, as well as an environmental life cycle assessment, on various cool pavement strategies. These can help answer the following important questions for implementation: (1) which one(s) has the lowest life cycle costs? (2) Which one(s) produces the lowest cost–benefit ratio? (3) Which one(s) has the lowest life cycle environmental impact (including materials production, construction, maintenance, use phase, and end-of-use)?
2. Do more in-depth quantitative analysis of potential impacts that are discussed in this dissertation, such as energy use (e.g., buildings and vehicles), air quality (ground-level ozone), water quality (e.g., pollution and temperature influence on some species), pavement life (e.g., rutting, fatigue, aging, etc.), and other potential unintended influences (e.g., rainfall), using experimental and modeling investigations, to achieve a relatively comprehensive evaluation on the impacts of cool pavement strategies.
3. Evaluate further technologies of cool pavements, including evaporation rate/water retention in permeable pavements (porous materials), various reflective coating/colorful binders (including cooling effect and durability and cost), and experimental testing on combined effects of different strategies, such as thermal resistance (low thermal conductivity), evaporative cooling and high reflectance, or shading.
4. Measure the thermal properties for more construction materials and investigate the factors affecting the thermal properties (including aggregate type and source, different fillers and admixture).

5. Conduct multiscale (including local street level, block, and city) simulation studies to evaluate the cooling effects of various cool pavement strategies on urban heat islands, integrating both pavements and buildings/vehicles as well as near-surface air.
6. Guidelines should be developed for the application of various cool pavement strategies under various contexts after completion of the first five recommendations listed above.

REFERENCES

[1] USGBC. LEED 2009 for new construction and major renovations rating system. 2009.

[2] Gartland L. Heat islands: understanding and mitigating heat in urban areas. London: Earthscan Press in UK and USA; 2008.

[3] EPA. Reducing urban heat islands: compendium of strategies – cool pavements. US EPA; 2008.

[4] Akbari H, Rose L, Taha H. Characterizing the fabric of the urban environment: a case study of Sacramento, California. Berkeley, CA: Lawrence Berkeley National Laboratory; 1999. p. 65.

[5] Akbari H, Pomerantz M, Taha H. Cool surfaces and shade trees to reduce energy use and improve air quality in urban areas. Sol Energy 2001;70(3):295–310.

[6] Hoehner CM, et al. Perceived and objective environmental measures and physical activity among urban adults. Am J Prev Med 2005;28(2):105–16.

[7] Handy SL, Cao XY, Mokhtarian PL. The causal influence of neighborhood design on physical activity within the neighborhood: evidence from Northern California. Am J Health Promot 2008;22(5):350–8.

[8] Xing Y, Handy SL, Mokhtarian PL. Factors associated with proportions and miles of bicycling for transportation and recreation in six small US cities. Transp Res Part D-Transp Environ 2010;15(2):73–81.

[9] Pucher J, Dill J, Handy S. Infrastructure, programs, and policies to increase bicycling: an international review. Prev Med 2010;50:S106–25.

[10] Saelens BE, Handy SL. Built environment correlates of walking: a review. Med Sci Sports Exerc 2008;40(7):S550–66.

[11] Handy S, Cao XY, Mokhtarian PL. Self-selection in the relationship between the built environment and walking – empirical evidence from northern California. J Am Plan Assoc 2006;72(1):55–74.

[12] Handy SL, et al. How the built environment affects physical activity – views from urban planning. Am J Prev Med 2002;23(2):64–73.

[13] Hisada Y, Matsunaga N, Ando S. Summer and winter structures of heat island in Fukuoka metropolitan area. Int J Ser B-Fluids Therm Eng 2006;49(1):65–71.

[14] Navigant Consulting Inc., B.T.P. Building technologies program, assessment of international urban heat island research – literature review of international studies on urban heat island countermeasures. Draft report. Washington, DC: Office of Energy Efficiency and Renewable Energy: USDOE; 2011.

[15] Sheridan S, et al. Heat vulnerability in California under future climate change (A spatial synoptic classification approach). 2011.

[16] Akbari H. Energy saving potentials and air quality benefits of urban heat island mitigation. Lawrence Berkeley National Laboratory; 2005. p. 19.

[17] Sailor DJ. Urban heat islands, opportunities and challenges for mitigation and adaptation. In: North American urban heat island summit. Toronto, Canada; 2002.

[18] Yamashita Y, et al. Ozone and temperature response of a chemistry climate model to the solar cycle and sea surface temperature. J Geophys Res Atmos 2010;115.

[19] Bloomer BJ, et al. Observed relationships of ozone air pollution with temperature and emissions. Geophys Res Lett 2009;36(9):L09803.

[20] Pomerantz M, Akbari H, Harvey J. Cooler reflective pavements give benefits beyond energy savings: durability and illumination. Berkeley, CA: Lawrence Berkeley National Laboratory; 2000.

[21] Chen M-Z, Wei W, Wu S-P. On cold materials of pavement and high-temperature performance of asphalt concrete. Mater Sci Forum 2009;620–622:379–82.
[22] Nishizawa T, et al. Finite element model analysis of thermal stresses of thick airport concrete pavement slabs. Transp Res Rec 2009;2095:3–12.
[23] Johnston DP, Surdahl RW. Influence of mixture design and environmental factors on continuously reinforced concrete pavement cracking. Transp Res Rec 2007;2020:83–8.
[24] Christen A, Vogt R. Energy and radiation balance of a central European city. Int J Climatol 2004;24(11):1395–421.
[25] Woolley J, Harrington C, Modera M. Swimming pools as heat sinks for air conditioners: model design and experimental validation for natural thermal behavior of the pool. Build Environ;46(1):187–195.
[26] Akbari H, Matthews HD. Global cooling updates: reflective roofs and pavements. Energy Build December 2012;55:2–6. http://dx.doi.org/10.1016/j.enbuild.2012.02.055.
[27] Akbari H, Matthews HD, Donny S. The long-term effect of increasing the albedo of urban areas. Environ Res Lett 2012;7(2):024004.
[28] Akbari H, Menon S, Rosenfeld A. Global cooling: increasing world-wide urban albedos to offset CO_2. Clim Change 2009;94(3):275–86.
[29] Pomerantz M, et al. Paving materials for heat island mitigation. Berkeley, CA: Lawrence Berkeley National Laboratory; 1997.
[30] Tran N, et al. Strategies for design and construction of high-reflectance asphalt pavements. Transp Res Rec 2009;2098:124–30.
[31] Boriboonsomsin K, Reza F. Mix design and benefit evaluation of high solar reflectance concrete for pavements. Transp Res Rec 2007;2011:11–20.
[32] Belshe M, et al. The urban heat island effect and impact of asphalt rubber friction course overlays on Portland cement concrete pavements in the Phoenix area. New Orleans, Louisiana: ASCE; 2008.
[33] Gui J, et al. Impact of pavement thermophysical properties on surface temperatures. J Mater Civ Eng 2007;19(8):683–90.
[34] Huang H, Ooka R, Kato S. Urban thermal environment measurements and numerical simulation for an actual complex urban area covering a large district heating and cooling system in summer. Atmos Environ 2005;39(34):6362–75.
[35] Annaratone D. Engineering heat transfer. Heidelberg: Springer; 2010.
[36] Asawa T, Hoyano A, Nakaohkubo K. Thermal design tool for outdoor spaces based on heat balance simulation using a 3D-CAD system. Build Environ 2008;43(12):2112–23.
[37] Bejan A. Convection heat transfer. Hoboken, NJ: Wiley; 2004.
[38] Jones D, et al. Laboratory testing and modeling for structural performance of fully permeable pavements under heavy traffic: final report. Davis, California: University of California Pavement Research Center; 2010.
[39] Grant RF, Izaurralde RC, Chanasyk DS. Soil-temperature under different surface managements – testing a simulation-model. Agric For Meteorol 1995;73(1–2):89–113.
[40] Avedisian CT, et al. Htd. Heat transfer in microgravity: presented at the 1993 ASME Winter Annual Meeting, New Orleans, Louisiana, November 28-December 3, 1993, New York, NY: American Society of Mechanical Engineers; 1993. vi, 181 p.
[41] Sauer T. Numerical analysis. Boston: Pearson; 2012.
[42] Epperson JF. An introduction to numerical methods and analysis. Hoboken, NJ: Wiley-Interscience; 2007.
[43] Arpaci VS, Kao S-H, Selamet A. Introduction to heat transfer. Upper Saddle River, NJ: Prentice Hall; 1999.
[44] Arpaci VS, Larsen PS. Convection heat transfer. Englewood Cliffs, NJ: Prentice-Hall; 1984. xiii, 512 p.

[45] Bayaz*to*glu Ylz, Arpaci VS, American Society of Mechanical Engineers. Heat transfer division, fundamentals of heat transfer in electromagnetic, electrostatic, and acoustic fields: presented at the 29th National Heat Transfer Conference, Atlanta, Georgia, August 8–11, 1993. Htd. 1993. New York, NY: American Society of Mechanical Engineers; v, 83 p.
[46] ASTM. ASTM C1549-09 standard test method for determination of solar reflectance near ambient temperature using a portable solar reflectometer. American Society for Testing and Materials; 2009.
[47] ASTM. ASTM E1918-06 standard test method for measuring solar reflectance of horizontal and low-sloped surfaces in the field. American Society for Testing and Materials; 2006.
[48] James D. Pricing nature: cost-benefit analysis and environmental policy. Australas J Environ Manag 2010;17(3):187–8.
[49] Almansa C, Martinez-Paz JM. What weight should be assigned to future environmental impacts? A probabilistic cost benefit analysis using recent advances on discounting. Sci Total Environ 2011;409(7):1305–14.
[50] Li H, Harvey J, Jones D. Material characterization for mechanistic design of fully permeable pavements: low-compaction subgrade and open-graded base materials. In: Transportation research board 91th annual meeting. Washington, DC; 2012.
[51] Chai L, et al. Hydraulic performance of fully permeable highway shoulder for storm water runoff management. J Environ Eng 2012;138(7):711–22.
[52] Mun S. Sound absorption characteristics of porous asphalt concrete pavements. Can J Civ Eng 2010;37(2):273–8.
[53] Cooley L, Brown E, Watson D. Evaluation of open-graded friction course mixtures containing cellulose fibers. Transp Res Rec J Transp Res Board 2000;1723(-1):19–25.
[54] Brattebo BO, Booth DB. Long-term stormwater quantity and quality performance of permeable pavement systems. Water Res 2003;37(18):4369–76.
[55] Wanphen S, Nagano K. Experimental study of the performance of porous materials to moderate the roof surface temperature by its evaporative cooling effect. Build Environ 2009;44(2):338–51.
[56] Arnold CL, Gibbons CJ. Impervious surface coverage – the emergence of a key environmental indicator. J Am Plan Assoc 1996;62(2):243–58.
[57] Yamagata H, et al. Water reuse in Japan. 2007.
[58] Yamagata H, et al. Heat island mitigation using water retentive pavement sprinkled with reclaimed wastewater. Water Sci Technol 2008;57:763–71.
[59] Ministry to promote water-retentive pavement for heat island control. Available from: http://www.japanfs.org/en/pages/025750.html; 2004 [cited 2010 September 9].
[60] Wang DC, et al. Benefit analysis of permeable pavement on sidewalks. Int J Pavement Res Technol 2010;3(4):207–15.
[61] Kevern JT, Schaefer VR, Wang KJ. Temperature behavior of pervious concrete systems. Transp Res Rec 2009;2098:94–101.
[62] Ongel A, Harvey JT. Analysis of 30 years of pavement temperatures using the enhanced integrated climate model (EICM). Pavement Research Center, Institute of Transportation Studies, University of California Berkeley, University of California Davis; 2004.
[63] Mallick RB, Chen B-L, Bhowmick S. Harvesting energy from asphalt pavements and reducing the heat island effect. Int J Sustain Eng 2009;2(3):214–28.
[64] Hasebe M, Kamikawa Y, Meiarashi S. Thermoelectric generators using solar thermal energy in heated road pavement. In: Thermoelectrics, 2006. ICT '06. 25th International conference on. 2006.

[65] Golden JS, et al. A comparative study of the thermal and radiative impacts of photovoltaic canopies on pavement surface temperatures. Sol Energy 2007;81(7): 872–83.
[66] Golden JS, Kaloush KE. Mesoscale and microscale evaluation of surface pavement impacts on the urban heat island effects. Int J Pavement Eng 2006;7(1):37–52.
[67] Pomerantz M, et al. The effect of pavements' temperatures on air temperatures in large cities. Berkeley, CA: Lawrence Berkeley National Laboratory; 2000.
[68] Ting M, Koomey J, Pomerantz M. Preliminary evaluation of the lifecycle costs and market barriers of reflective pavements. Berkeley, CA: Lawrence Berkeley National Laboratory; 2001. p. 61.
[69] Levinson R, Akbari H. Effects of composition and exposure on the solar reflectance of Portland cement concrete. Berkeley, CA: Lawrence Berkeley National Laboratory; 2001. p. 61.
[70] Pomerantz M, et al. Examples of cooler reflective streets for urban heat-island mitigation: Portland cement concrete and chip seals. Berkeley, CA: Lawrence Berkeley National Laboratory; 2003. p. 24.
[71] Kaloush KE, Carlson JD, Phelan PE. The thermal and radiative characteristics of concrete pavements in mitigating urban heat island effects. Tempe, Arizona: National Center of Excellence on Sustainable Material and Renewable Technology (SMART) Innovations, Arizona State University; 2008. p. 139.
[72] Kawakami A, Kubo K. Accelerated loading tests on the durability of cool pavement at PWRI. In: Third international conference on accelerated pavement testing. APT 2008. Impacts and benefits from APT programs. Madrid, Spain: Centro de Estudios y Experimentacion de Obras Publicas, Spain; 2008. p. 25.
[73] Takahashi K, Yabuta K. Road temperature mitigation effect of "Road cool," a water-retentive material using blast furnace slag. In: JFE technical report. Japan: JFE; 2009.
[74] Wong CW, et al. A study on the effectiveness of heat mitigating pavement coatings in Singapore. In: The second international conference on countermeasures to urban heat islands. California, U.S: Berkeley; 2009.
[75] Synnefa A, et al. Measurement of optical properties and thermal performance of coloured thin layer asphalt samples and evaluation of their impact on the urban environment. In: The second international conference on countermeasures to urban heat islands. California, US: Berkeley; 2009.
[76] Kevern JT, Haselbach L, Schaefer VR. Hot weather comparative heat balances in pervious concrete and impervious concrete pavement systems. In: The second international conference on countermeasures to urban heat islands. California, US: Berkeley; 2009.
[77] Kubo K, Kido H, Ito M. Study on pavement technologies to mitigate the heat island effect and their effectiveness. Tsukuba, Japan: Public Works Research Institute; 2006.
[78] Tan S-A, Fwa T-F. Influence of pavement materials on the thermal environment of outdoor spaces. Build Environ 1992;27(3):289–95.
[79] Lin T-P, Ho Y-F, Huang Y-S. Seasonal effect of pavement on outdoor thermal environments in subtropical Taiwan. Build Environ 2007;42(12):4124–31.
[80] Yilmaz H, et al. Determination of temperature differences between asphalt concrete, soil and grass surfaces of the city of Erzurum, Turkey. Atmosfera 2008;21(2):135–46.
[81] Kinoshita S, Yoshida A, Okuno N. Evaporation performance analysis for water retentive material based on outdoor heat budget and transport properties. In: The second international conference on countermeasures to urban heat islands. California, US: Berkeley; 2009.
[82] Nakayama T, Fujita T. Cooling effect of water-holding pavements made of new materials on water and heat budgets in urban areas. Landsc Urban Plan 2010;96(2): 57–67.

[83] Furumai H, et al. Recent application of rainwater storage and harvesting in Japan; 2003. p. 7.
[84] Mallick RB, Chen B-L, Bhowmick S. Reduction of urban heat island effect through harvest of heat energy from asphalt pavements. In: The second international conference on countermeasures to urban heat islands. California, US: Berkeley; 2009.
[85] Cao X, et al. Cooling principle analysis and performance evaluation of heat-reflective coating for asphalt pavement (11-1745). In: Transportation research board 90th annual meeting. Washington, DC; 2011.
[86] Sha A. Proceedings of international workshop on energy and environment in the development of sustainable asphalt pavements. In: International workshop on energy and environment in the development of sustainable asphalt pavements. China: Changan University in Xian; 2010.
[87] Zhang X. Research on heat reflection and thermal resistance technology of asphalt pavement cooling mechanism and its application. In: School of transportation science and engineering. China: Institute of Technology: Harbin; 2011. p. 127.
[88] Nikolopoulou M, Lykoudis S. Use of outdoor spaces and microclimate in a mediterranean urban area. Build Environ 2007;42(10):3691–707.
[89] Tian L, et al. Numerical study of indoor air quality and thermal comfort under stratum ventilation. Prog Comput Fluid Dyn 2008;8(7–8):541–8.
[90] Peeters L, et al. Thermal comfort in residential buildings: comfort values and scales for building energy simulation. Appl Energy 2009;86(5):772–80.
[91] Jun Y, et al. Thermal comfort properties of wearing caps from various textiles. Text Res J 2009;79(2):179–89.
[92] Hitchings R. Studying thermal comfort in context. Build Res Inf 2009;37(1):89–94.
[93] Gossauer E, Wagner A. Thermal comfort and satisfaction at workplaces – a field study in office buildings. Bauphysik 2008;30(6):445–52.
[94] Becker R, Paciuk M. Thermal comfort in residential buildings – failure to predict by standard model. Build Environ 2009;44(5):948–60.
[95] Streblow R, et al. Evaluation of thermal comfort at inhomogeneous environmental conditions. Bauphysik 2009;31(1):38–41.
[96] Ribeiro NL, et al. Assessment of thermal comfort indexes, physiological parameters and thermal gradient of native sheep. Eng Agric 2008;28(4):614–23.
[97] Lenzuni P, Freda D, Del Gaudio M. Classification of thermal environments for comfort assessment. Ann Occup Hyg 2009;53(4):325–32.
[98] Ho SH, Rosario L, Rahman MM. Three-dimensional analysis for hospital operating room thermal comfort and contaminant removal. Appl Therm Eng 2009;29(10):2080–92.
[99] Hens HSLC. Thermal comfort in office buildings: two case studies commented. Build Environ 2009;44(7):1399–408.
[100] Catalina T, Virgone J, Kuznik F. Evaluation of thermal comfort using combined CFD and experimentation study in a test room equipped with a cooling ceiling. Build Environ 2009;44(8):1740–50.
[101] Yau YH, Chew BT. Thermal comfort study of hospital workers in Malaysia. Indoor Air 2009;19(6):500–10.
[102] Xu XG, et al. Thermal comfort in an office with intermittent air-conditioning operation. Build Serv Eng Res Technol 2010;31(1):91–100.
[103] Tseliou A, et al. An evaluation of three biometeorological indices for human thermal comfort in urban outdoor areas under real climatic conditions. Build Environ 2010;45(5):1346–52.
[104] Indraganti M, Rao KD. Effect of age, gender, economic group and tenure on thermal comfort: a field study in residential buildings in hot and dry climate with seasonal variations. Energy Build 2010;42(3):273–81.

[105] Ahmed KS. Comfort in urban spaces: defining the boundaries of outdoor thermal comfort for the tropical urban environments. Energy Build 2003;35(1):103–10.

[106] He J, Hoyano A, Asawa T. A numerical simulation tool for predicting the impact of outdoor thermal environment on building energy performance. Appl Energy 2009;86(9):1596–605.

[107] Oliver-Solà J, et al. Environmental optimization of concrete sidewalks in urban areas. Int J Life Cycle Assess 2009;14(4):302–12.

[108] Li H. Evaluation of cool pavement strategies for heat island mitigation. In: Civil and environmental engineering. Davis, CA: University of California; 2012. p. 387.

[109] Markvart T, Castañer L. Practical handbook of photovoltaics: fundamentals and applications. New York: Elsevier Advanced Technology; 2003.

[110] Nellis G, Klein SA, Ebrary Inc. Heat transfer. Cambridge, New York: Cambridge University Press; 2009. p. xxxvii, 1107 p.

[111] Santamouris M. Cooling the cities – a review of reflective and green roof mitigation technologies to fight heat island and improve comfort in urban environments. Sol Energy May 2014;103:682–703. http://dx.doi.org/10.1016/j.solener.2012.07.003.

[112] Mrawra DM, Luca J. Thermal properties and transient temperature response of full-depth asphalt pavements. Des Rehabil Pavements 2002;1809:160–71.

[113] Schindler AK, et al. Concrete pavement temperature prediction FHWA HIPERPAV and case studies with the models. Cem Concr Compos 2004;26(5):463–71.

[114] Hermansson A. Mathematical model for calculation of pavement temperatures – comparison of calculated and measured temperatures. Assess Eval Pavements 2001;1764:180–8.

[115] Li H, Harvey J. Numerical simulation and sensitivity analysis of asphalt pavement temperature and near-surface air temperature using integrated local modeling (11-3125). In: Transportation research board 90th annual meeting. Washington, DC; 2011.

[116] Hermansson A. Simulation model for calculating pavement temperatures including maximum temperature. Pavement Manag Monit 2000;1699:134–41.

[117] Ali-Toudert F, Mayer H. Numerical study on the effects of aspect ratio and orientation of an urban street canyon on outdoor thermal comfort in hot and dry climate. Build Environ 2006;41(2):94–108.

[118] Huang JH. Prediction of air temperature for thermal comfort of people in outdoor environments. Int J Biometeorol 2007;51(5):375–82.

[119] Atmaca I, Kaynakli O, Yigit A. Effects of radiant temperature on thermal comfort. Build Environ 2007;42(9):3210–20.

[120] Nova Lynx Corporation. 240–8104 Albedometer. Available from: http://www.novalynx.com/240-8104.html; 2010 [cited 2011 July 10].

[121] Kula E, Evans D. Dual discounting in cost-benefit analysis for environmental impacts. Environ Impact Assess Rev 2011;31(3):180–6.

[122] Kayhanian M, et al. Permeability measurement and scan imaging to assess clogging of pervious concrete pavements in parking lots. J Environ Manag 2012;95(1):114–23.

[123] Prowell B, Dudley M. Evaluation of measurement techniques for asphalt pavement density and permeability. Transp Res Rec J Transp Res Board 2002;1789:36–45.

[124] Williams SG. Field permeability measurements of coarse-graded asphalt pavements. In: International society for asphalt pavements (ISAP) symposium on asphalt pavements and the environment. Zurich, Switzerland; 2008.

[125] Maupin G. Asphalt permeability testing: specimen reparation and testing variability. Transportation research record. J Transp Res Board 2001;1767:33–9.

[126] Maupin G. Asphalt permeability testing in Virginia. Transp Res Rec J Transp Res Board 2000;1723(-1):83–91.

[127] Cooley AJ. Permeability of superpave mixtures: evaluation of field permeameters. NCAT; 1999.
[128] Wong NH, et al. Thermal evaluation of vertical greenery systems for building walls. Build Environ 2010;45(3):663–72.
[129] Jo JH, et al. An integrated empirical and modeling methodology for analyzing solar reflective roof technologies on commercial buildings. Build Environ 2010;45(2):453–60.
[130] Asaeda T, Ca VT. Characteristics of permeable pavement during hot summer weather and impact on the thermal environment. Build Environ 2000;35(4):363–75.
[131] Xu T, et al. Quantifying the direct benefits of cool roofs in an urban setting: reduced cooling energy use and lowered greenhouse gas emissions. Build Environ 2012;48:1–6.
[132] Santamouris M, et al. Using cool paving materials to improve microclimate of urban areas – design realization and results of the flisvos project. Build Environ 2012;53:128–36.
[133] Tan Siew A, Fwa Tien F. Pavement evaluation for thermal/glare comfort during footdrills. Build Environ 1997;32(3):257–69.
[134] Mahmoud AHA. Analysis of the microclimatic and human comfort conditions in an urban park in hot and arid regions. Build Environ 2011;46(12):2641–56.
[135] Kültür S, Türkeri N. Assessment of long term solar reflectance performance of roof coverings measured in laboratory and in field. Build Environ 2012;48:164–72.
[136] Tapkın S. The effect of polypropylene fibers on asphalt performance. Build Environ 2008;43(6):1065–71.
[137] Ibrahim H, Wahhab A-A, Hasnain J. Laboratory study of asphalt concrete durability in Jeddah. Build Environ 1998;33(4):219–30.
[138] Abo-Qudais S, Al-Shweily H. Effect of antistripping additives on environmental damage of bituminous mixtures. Build Environ 2007;42(8):2929–38.
[139] Roy SK, Poh KB, Northwood Do. Durability of concrete – accelerated carbonation and weathering studies. Build Environ 1999;34(5):597–606.
[140] Ramadhan RH, Al-Abdul Wahhab HI. Temperature variation of flexible and rigid pavements in Eastern Saudi Arabia. Build Environ 1997;32(4):367–73.
[141] Carlson JD, et al. Determining thermal conductivity of paving materials using cylindrical sample geometry. J Mater Civ Eng 2010;22(2):186–95.
[142] Mrawira DM, Luca J. Effect of aggregate type, gradation, and compaction level on thermal properties of hot-mix asphalts. Can J Civ Eng 2006;33(11):1410–7.
[143] Luca J, Mrawira D. New measurement of thermal properties of superpave asphalt Concrete. J Mater Civ Eng 2005;17(1):72–9.
[144] Xu QW, Solaimanian M. Modeling temperature distribution and thermal property of asphalt concrete for laboratory testing applications. Constr Build Mater 2010;24(4): 487–97.
[145] Nguyen QT, Di Benedetto H, Sauzéat C. Determination of thermal properties of asphalt mixtures as another output from cyclic tension-compression test. Road Mater Pavement Des 2012;13(1):85–103.
[146] Wolfe RK, Heath GL, Colony DC. University of toledo time temperature model laboratory and field validation. Toledo, Ohio: Dept. of Industrial Engineering, Univ. of Toledo; 1980. Rep. No. FHWA/OH-80/006.
[147] Highter WH, Wall DJ. Thermal properties of some asphaltic concrete mixes. Transp Res Board 1984;968:38–45.
[148] Tan SA, et al. Determination of thermal properties of pavement materials and unbound aggregates by transient heat conduction. J Test Eval 1997;25(1):15–22.
[149] Solaimanian M, Bolzan P. Analysis of the integrated model of climatic effects on pavements. Washington, DC: Strategic Highway Research Program. U.S. Federal Highway Administration, Research Rep; 1993.

[150] Rathore MM, Kapuno RR. Engineering heat transfer. Sudbury, Mass: Jones & Bartlett Learning; 2011.
[151] Kyurkchiev NV. Initial approximations and root finding methods. Berlin, New York: Wiley-VCH; 1998.
[152] Kevern JT, Schaefer VR, Wang KJ. Temperature behavior of pervious concrete systems. Transp Res Rec 2009;2098:94–101.
[153] Stempihar J, et al. Porous asphalt pavement temperature effects for urban heat island analysis. Transp Res Rec J Transp Res Board 2012;2293(1):123–30.
[154] Golden JS, Kaloush KE. A hot night in the big city: how to mitigate the urban heat island. Public Works 2005;136(13).
[155] U.S. Green Building Council. LEED for new construction and major renovations rating system. 2009.
[156] Roseen R, et al. Preliminary results of the examination of thermal impacts from stormwater BMPs. In: World environmental and water resources congress; 2010. p. 3424–51.
[157] Roseen R, et al. Assessment of winter maintenance of porous asphalt and its function for chloride source control. In: World environmental and water resources congress; 2013. p. 99–116.
[158] Roseen R, et al. Pervious concrete and porous asphalt pavements performance for stormwater management in northern climates. In: Cold regions engineering; 2009. p. 311–27.
[159] Roseen R, et al. Water quality and hydrologic performance of a porous asphalt pavement as a storm-water treatment strategy in a cold climate. J Environ Eng 2012;138(1):81–9.
[160] Houle K, et al. Examinations of pervious concrete and porous asphalt pavements performance for stormwater management in northern climates. In: World environmental and water resources congress; 2009. p. 1–18.
[161] Li H, Harvey J, Jones D. Cooling effect of permeable asphalt pavement under both dry and wet conditions. Transp Res Rec J Transp Res Board 2013;3(2372): 97–107.
[162] Li H, et al. The use of reflective and permeable pavements as a potential practice for heat island mitigation and stormwater management. Environ Res Lett 2013; 8(1):14.
[163] Li H, Harvey J, Kendall A. Field measurement of albedo for different land cover materials and effects on thermal performance. Build Environ 2013;59:536–46.
[164] Li H, Kayhanian M, Harvey J. Comparative field permeability measurement of permeable pavements using ASTM C1701 and NCAT permeameter methods. J Environ Manag 2013;118:144–52.
[165] Li H, Harvey J, Kendall A. Measurement of pavement solar reflectivity and effect on thermal performance. In: Transportation research board 92nd annual meeting. Washington, DC; 2013.
[166] Gelegenis JJ. A simplified quadratic expression for the approximate estimation of heating degree-days to any base temperature. Appl Energy 2009;86(10):1986–94.
[167] Lee DS, Lemieux T. Regression discontinuity designs in economics. J Econ Lit 2010;48(2):281–355.
[168] Abaqus. Abaqus analysis user's manual. SIMULIA; 2009.
[169] Cheremisinoff NP. Handbook of heat and mass transfer. Houston: Gulf Pub. Co; 1986. v, 1–4.
[170] Eckert ERG, Irvine TF. Progress in heat and mass transfer. Heat transfer reviews, 1970–1975. Chicago: Rumford Pub. Co; 1977. vii, 321 p.

[171] Eckert ERG. Heat transfer reviews, 1976–1986. New York: Wiley; 1990. vii, 681 p.
[172] American Society of Mechanical Engineers. Winter meeting (1992: Anaheim Calif.), I. Catton, and American Society of Mechanical Engineers. Heat transfer division. Heat and mass transfer in porous media: presented at the winter annual meeting of the American Society of Mechanical Engineers, Anaheim, California, November 8–13, 1992. Htd. 1992. New York: The Society; v, 61 p.
[173] Incropera FP, DeWitt DP. Fundamentals of heat and mass transfer. 2nd ed. New York: Wiley; 1985. xxiii, 802 p.
[174] Bachelet D. MC1, a dynamic vegetation model for estimating the distribution of vegetation and associated ecosystem fluxes of carbon, nutrients, and water: technical documentation: version 1.0. General technical report PNW. Portland, OR: US. Dept. of Agriculture, Forest Service, Pacific Northwest Research Station; 2001. 95 p.
[175] Batchelor GK. An introduction to fluid dynamics. Cambridge: U.P; 1967. xviii, 615 p.
[176] Middleman S. An introduction to fluid dynamics: principles of analysis and design. New York: Wiley; 1998. xiv, 513 p.
[177] Mitchell AR, Griffiths DF. The finite difference method in partial differential equations. Chichester Eng, New York: Wiley; 1980. xii, 272 p.
[178] D'Acunto B. Computational partial differential equations for engineering science. Hauppauge, NY: Nova Science Publishers; 2010.
[179] Roache PJ. Computational fluid dynamics. Albuquerque, NM: Hermosa Publishers; Rev. print. ed. 1976. vii, 446 p.
[180] Chung TJ. Computational fluid dynamics. Cambridge, New York: Cambridge University Press; 2002. xxiii, 1012 p.
[181] Hermansson A. Mathematical model for paved surface summer and winter temperature: comparison of calculated and measured temperatures. Cold Reg Sci Technol 2004;40(1–2):1–17.
[182] Adkins DF, Merkley GP. Mathematical model of temperature changes in concrete pavements. J Transp Eng 1990;116(3):349–58.
[183] Herb WR, et al. Ground surface temperature simulation for different land covers. J Hydrol 2008;356(3–4):327–43.
[184] Barber ES. Calculation of maximum pavement temperatures from weather reports. Highw Res Board Bull 1957;168:8. Washington, DC.
[185] Li H. Research on relationship between rutting and temperature field of asphalt pavement. In: Civil engineering. China, Nanjing: Southeast University; 2007. p. 80.
[186] Bean EZ, Hunt WF, Bidelspach DA. Evaluation of four permeable pavement sites in eastern North Carolina for runoff reduction and water quality impacts. J Irrig Drain Eng-Asce 2007;133(6):583–92.
[187] Huang X, Li H. Simulation of rutting behavior of asphalt pavement based on real temperature field. In: Transportation research board 87th annual meeting. Washington, DC; 2008.
[188] NREL. National solar radiation database. National Renewable Energy Laboratory; 2010.
[189] Borges ARJ, Carvalho JL. Aerodynamic and thermal interaction of an element of a very rough surface with a turbulent boundary-layer flow. J Wind Eng Ind Aerod 1996;65(1–3):273–8.
[190] Suga K, Craft TJ, Iacovides H. An analytical wall-function for turbulent flows and heat transfer over rough walls. Int J Heat Fluid Flow 2006;27(5):852–66.
[191] Gagge AP, Fobelets AP, Berglund LG. A standard predictive index of human response to the thermal environment. ASHRAE Trans 1986;92(2B):709–31.

[192] Gagge AP, Stolwijk J, Nishi Y. A standard predictive index of human response to the thermal environment. ASHRAE Trans 1971;77(1):247–62.
[193] Brager GS, de Dear RJ. Thermal adaptation in the built environment: a literature review. Energy Build 1998;27(1):83–96.
[194] Matzarakis A, Mayer H. Heat stress in Greece. Int J Biometeorol 1997;41(1):34–9.
[195] Lin T-P, Matzarakis A, Hwang R-L. Shading effect on long-term outdoor thermal comfort. Build Environ 2010;45(1):213–21.
[196] Lin T-P, Matzarakis A. Tourism climate and thermal comfort in Sun Moon Lake, Taiwan. Int J Biometeorol 2008;52(4):281–90.
[197] Hwang R-L, Lin T-P. Thermal comfort requirements for occupants of semi-outdoor and outdoor environments in hot-humid regions. Archit Sci Rev 2007;50(4):357–64.
[198] Yaghoobian N, Kleissl J. Effect of reflective pavements on building energy use. Urban Clim December 2012;2:25–42.
[199] Klie A. Valuation of environmental goods via willingness to pay – why cost-benefit analysis fails. Gaia-Ecol Perspect Sci Soc 2010;19(2):103–9.
[200] Gogula A, Hossain M, Romanoschi S, Fager GA. Correlation between the laboratory and field permeability values for the superpave pavements. In: Proceedings of the 2003 Mid-Continent Transportation Research Symposium, Ames, Iowa; 2003.

APPENDIX

Table A.1 ξ for various biot numbers (first 10 terms)

(1) Infinite plate

Bi	ξ_1	ξ_2	ξ_3	ξ_4	ξ_5	ξ_6	ξ_7	ξ_8	ξ_9	ξ_{10}
0.01	0.0998	3.1448	6.2848	9.4258	12.5672	15.7086	18.8501	21.9916	25.1331	28.2747
0.02	0.1410	3.1479	6.2864	9.4269	12.5680	15.7092	18.8506	21.9921	25.1335	28.2750
0.04	0.1987	3.1543	6.2895	9.4290	12.5696	15.7105	18.8517	21.9930	25.1343	28.2757
0.06	0.2425	3.1606	6.2927	9.4311	12.5711	15.7118	18.8527	21.9939	25.1351	28.2765
0.08	0.2791	3.1668	6.2959	9.4333	12.5727	15.7131	18.8538	21.9948	25.1359	28.2772
0.1	0.3111	3.1731	6.2991	9.4354	12.5743	15.7143	18.8549	21.9957	25.1367	28.2779
0.2	0.4328	3.2039	6.3148	9.4459	12.5823	15.7207	18.8602	22.0002	25.1407	28.2814
0.3	0.5218	3.2341	6.3305	9.4565	12.5902	15.7270	18.8655	22.0048	25.1447	28.2849
0.4	0.5932	3.2636	6.3461	9.4670	12.5981	15.7334	18.8707	22.0093	25.1486	28.2885
0.5	0.6533	3.2923	6.3616	9.4775	12.6060	15.7397	18.8760	22.0139	25.1526	28.2920
0.6	0.7051	3.3204	6.3770	9.4879	12.6139	15.7460	18.8813	22.0184	25.1566	28.2955
0.7	0.7506	3.3477	6.3923	9.4983	12.6218	15.7524	18.8866	22.0229	25.1606	28.2991
0.8	0.7910	3.3744	6.4074	9.5087	12.6296	15.7587	18.8919	22.0275	25.1645	28.3026
0.9	0.8274	3.4003	6.4224	9.5190	12.6375	15.7650	18.8971	22.0320	25.1685	28.3061
1	0.8603	3.4256	6.4373	9.5293	12.6453	15.7713	18.9024	22.0365	25.1724	28.3096
2	1.0769	3.6436	6.5783	9.6296	12.7223	15.8336	18.9547	22.0815	25.2119	28.3448
3	1.1925	3.8088	6.7040	9.7240	12.7966	15.8945	19.0061	22.1259	25.2510	28.3797
4	1.2646	3.9352	6.8140	9.8119	12.8678	15.9536	19.0565	22.1697	25.2896	28.4142
5	1.3138	4.0336	6.9096	9.8928	12.9352	16.0107	19.1055	22.2126	25.3276	28.4483
6	1.3496	4.1116	6.9924	9.9667	12.9988	16.0654	19.1531	22.2545	25.3650	28.4820
7	1.3766	4.1746	7.0640	10.0339	13.0584	16.1177	19.1992	22.2954	25.4016	28.5151
8	1.3978	4.2264	7.1263	10.0949	13.1141	16.1675	19.2435	22.3351	25.4374	28.5476
9	1.4149	4.2694	7.1806	10.1502	13.1660	16.2147	19.2862	22.3736	25.4724	28.5794
10	1.4289	4.3058	7.2281	10.2003	13.2142	16.2594	19.3270	22.4108	25.5064	28.6106
20	1.4961	4.4915	7.4954	10.5117	13.5420	16.5864	19.6439	22.7131	25.7923	28.8800
30	1.5202	4.5615	7.6057	10.6543	13.7085	16.7691	19.8361	22.9098	25.9896	29.0754
40	1.5325	4.5979	7.6647	10.7334	13.8048	16.8794	19.9576	23.0394	26.1250	29.2143
50	1.5400	4.6202	7.7012	10.7832	13.8666	16.9519	20.0392	23.1287	26.2206	29.3148

(2) Infinite cylinder

Bi	ξ_1	ξ_2	ξ_3	ξ_4	ξ_5	ξ_6	ξ_7	ξ_8	ξ_9	ξ_{10}
0.01	0.1412	3.8343	7.0170	10.1745	13.3244	16.4712	19.6164	22.7605	25.9041	29.0472
0.02	0.1995	3.8369	7.0184	10.1754	13.3252	16.4718	19.6169	22.7610	25.9044	29.0475
0.04	0.2814	3.8421	7.0213	10.1774	13.3267	16.4731	19.6179	22.7618	25.9052	29.0482
0.06	0.3438	3.8473	7.0241	10.1794	13.3282	16.4743	19.6189	22.7627	25.9060	29.0489
0.08	0.3960	3.8525	7.0270	10.1813	13.3297	16.4755	19.6199	22.7636	25.9068	29.0496
0.1	0.4417	3.8577	7.0298	10.1833	13.3312	16.4767	19.6210	22.7645	25.9075	29.0503
0.2	0.6170	3.8835	7.0440	10.1931	13.3387	16.4828	19.6261	22.7689	25.9114	29.0537
0.3	0.7465	3.9091	7.0582	10.2029	13.3462	16.4888	19.6311	22.7733	25.9153	29.0572
0.4	0.8516	3.9344	7.0723	10.2127	13.3537	16.4949	19.6362	22.7777	25.9191	29.0606
0.5	0.9408	3.9594	7.0864	10.2225	13.3611	16.5009	19.6413	22.7820	25.9230	29.0640
0.6	1.0184	3.9841	7.1004	10.2322	13.3686	16.5070	19.6464	22.7864	25.9268	29.0675
0.7	1.0873	4.0085	7.1143	10.2419	13.3761	16.5130	19.6515	22.7908	25.9307	29.0709

Continued

Table A.1 ξ for various biot numbers (first 10 terms)—cont'd

(2) Infinite cylinder

Bi	ξ_1	ξ_2	ξ_3	ξ_4	ξ_5	ξ_6	ξ_7	ξ_8	ξ_9	ξ_{10}
0.8	1.1490	4.0325	7.1282	10.2516	13.3835	16.5191	19.6566	22.7952	25.9345	29.0744
0.9	1.2048	4.0562	7.1421	10.2613	13.3910	16.5251	19.6617	22.7996	25.9384	29.0778
1	1.2558	4.0795	7.1558	10.2710	13.3984	16.5312	19.6667	22.8040	25.9422	29.0812
2	1.5994	4.2910	7.2884	10.3658	13.4719	16.5910	19.7172	22.8476	25.9806	29.1155
3	1.7887	4.4634	7.4103	10.4566	13.5434	16.6499	19.7671	22.8908	26.0187	29.1496
4	1.9081	4.6018	7.5201	10.5423	13.6125	16.7073	19.8160	22.9334	26.0564	29.1834
5	1.9898	4.7131	7.6177	10.6223	13.6786	16.7630	19.8640	22.9754	26.0937	29.2168
6	2.0490	4.8033	7.7039	10.6964	13.7414	16.8168	19.9107	23.0165	26.1303	29.2498
7	2.0937	4.8772	7.7797	10.7646	13.8008	16.8684	19.9560	23.0567	26.1664	29.2824
8	2.1286	4.9384	7.8464	10.8271	13.8566	16.9179	19.9999	23.0959	26.2017	29.3144
9	2.1566	4.9897	7.9051	10.8842	13.9090	16.9650	20.0422	23.1340	26.2362	29.3459
10	2.1795	5.0332	7.9569	10.9363	13.9580	17.0099	20.0829	23.1710	26.2698	29.3767
20	2.2880	5.2568	8.2534	11.2677	14.2983	17.3442	20.4037	23.4749	26.5561	29.6457
30	2.3261	5.3410	8.3771	11.4221	14.4748	17.5348	20.6020	23.6762	26.7568	29.8435
40	2.3455	5.3846	8.4432	11.5081	14.5774	17.6508	20.7284	23.8100	26.8957	29.9853
50	2.3572	5.4112	8.4840	11.5621	14.6433	17.7272	20.8136	23.9026	26.9942	30.0885

INDEX

Note: Page numbers followed by "f" and "t" denote figures and tables, respectively.

Kolibri 360 GmbH
Ferdinand-Jühlke-Straße 7
99095 Erfurt, Deutschland
produktsicherheit@kolibri360.de